CASE–BASED PREDICTIONS

An Axiomatic Approach to
Prediction, Classification and
Statistical Learning

World Scientific Series in Economic Theory
(ISSN: 2251-2071)

Series Editor: Eric Maskin *(Harvard University, USA)*

Published

*The complete list of the published volumes in the series can be found at
http://www.worldscibooks.com/series/sikm_series.shtml.

World Scientific Series in Economic Theory – Vol. 3

CASE-BASED PREDICTIONS

An Axiomatic Approach to Prediction, Classification and Statistical Learning

Itzhak Gilboa

Tel-Aviv University, Israel & HEC, Paris, France

David Schmeidler

Tel-Aviv University, Israel & Ohio State University, USA

World Scientific

NEW JERSEY · LONDON · SINGAPORE · BEIJING · SHANGHAI · HONG KONG · TAIPEI · CHENNAI

Published by

World Scientific Publishing Co. Pte. Ltd.
5 Toh Tuck Link, Singapore 596224
USA office: 27 Warren Street, Suite 401-402, Hackensack, NJ 07601
UK office: 57 Shelton Street, Covent Garden, London WC2H 9HE

British Library Cataloguing-in-Publication Data
A catalogue record for this book is available from the British Library.

World Scientific Series in Economic Theory — Vol. 3
CASE-BASED PREDICTIONS
An Axiomatic Approach to Prediction, Classification and Statistical Learning

ISBN-13 978-981-4366-17-5
ISBN-10 981-4366-17-X

In-house Editor: Samantha Yong

Typeset by Stallion Press
Email: enquiries@stallionpress.com

Printed in Singapore.

Foreword

In an important advance for economic theory, Itzhak Gilboa and David Schmeidler have developed a new way of modeling agents' beliefs about uncertain events: the case-based approach. Besides having considerable intuitive appeal, their theory avoids some of the well-known drawbacks of the standard Bayesian approach. This volume collects some of Gilboa and Schmeidler's leading articles on the subject, together with an introduction and overview that they have specially prepared for the volume.

Eric Maskin
Editor, World Scientific Series in Economic Theory

Foreword

Acknowledgments

We are very grateful to our co-authors, and to the journals' publishers for granting their consent to re-print the papers included in this volume. We also thank Katrin Kish for her meticulous proofreading.

Contents

Case-Based Predictions: Introduction*

Itzhak Gilboa and David Schmeidler

December 2010

1. Background

There are several approaches to formal modeling of uncertainty, knowledge, and belief. They differ in the way they represent what is known and what is not known, the formal entities that capture beliefs, the manner in which these beliefs are updated in the face of new evidence, and so forth. For example, classical statistical inference describes knowledge by a family of distributions, where the reasoner is assumed to know that the process is governed by one of these distributions, but she does not know which one. Evidence is modeled as realizations of random variables, and beliefs are updated according to classical techniques such as maximum likelihood estimation, the construction of confidence sets, etc. By contrast, the Bayesian approach represents knowledge as a set of states of the world, such that the reasoner knows the set but does not know which particular state in it obtains. Beliefs are represented by a prior probability measure over the state space, while evidence is modeled as events, namely subsets of states, such that belief revision consists in Bayesian updating of the prior probability to a posterior. A rather different approach considers rules to be the primary objects of knowledge, while evidence is modeled as particular instances in which rules may or may not hold. The rules represent beliefs, and they are updated, in light of new information, according to belief revision methods (see Alchourron, Gardenfors, and

---

*Introduction to "Case-Based Prediction", a volume to be published by World Scientific Publishers; Economic Theory Series, edited by Eric Maskin.

Makinson, 1985; Gardenfors, 1992). Another approach assumes that the object of knowledge are particular cases, or observations, where beliefs are indirectly expressed by the similarity one finds between different cases (Schank, 1986; Riesbeck and Schank, 1989). And one may also describe knowledge and belief by neural nets, fuzzy sets, and other methods.

Given the state of art in the behavioral and social sciences on the one hand, and the limitation of the different methods on the other hand, it stands to reason that no single method would dominate all the others. Rather, one would typically expect that each method would have some applications to which it is best suited, and others where it may be inconvenient or awkward. Indeed, the literature in statistics, machine learning, artificial intelligence, and engineering is rather pluralistic. Even economists, when conducting research, use several methods. However, when modeling a rational agent, the modern nickname of *homo economicus*, the latter is restricted to be Bayesian. This is partly a result of the success of game theory. The concept of strategic (Nash) equilibrium replaced and extended that of price equilibrium, and greatly enhanced the ability of economists to analyze interactive situations. The equilibrium as defined by Nash requires mixed strategies, that is, beliefs that are quantified probabilistically. Moreover, the applicability of games to economics was further extended with Harsany's modeling of incomplete information in a Bayesian way (introducing the concept of Bayesian equilibrium). Modeling economic agents as non-Bayesians may cast doubts on the usefulness of game theory and the validity of its economic conclusions.

Another source of support for the assumption of the Bayesian agent is its simplicity and its impressive axiomatization. (For the latter, see Ramsey, 1931; de Finetti, 1931, 1937; von Neumann and Morgenstern, 1944; Savage, 1954.[1]) In the Bayesian approach the objects of knowledge and belief, namely events, also model the evidence one may obtain. Further, this approach has only one type of updating, that is,

---

[1]We discuss the meaning and goals of axiomatizations below. See also Gilboa (2009) for general methodological discussions as well as descriptions of these classical contributions and their critique.

Bayesian updating. In comparison to the variety of classical statistical inference techniques, or the various theories of belief revision in rule-based paradigms, Bayesian updating of a prior to a posterior shines through as a simple, almost inevitable updating procedure that suffices for all purposes. Moreover, the Bayesian approach is tightly linked to a decision-making procedure, i.e., expected utility maximization, and the two can be jointly derived from very elegant axioms.

However, despite its elegance and generality, its axiomatic foundations and breadth of applications, the Bayesian approach has been criticized on several grounds. First, it has long been claimed that there are types of uncertainty that cannot be quantified by probabilities. (Knight, 1921; Keynes, 1921; Ellsberg, 1961; Schmeidler, 1989.) Following Knight, the literature distinguishes between situations of "risk", with known probabilities, and "uncertainty", where probabilities are not known. The Bayesian approach holds that any uncertainty can be reduced to risk, employing subjective probabilities. Yet, there is ample evidence that people often behave under uncertainty differently than under risk, and many authors also justify such behavior as rational. Specifically, it has been argued that the Bayesian approach is well suited to describe knowledge, but that it is poor at describing ignorance.

Second, while the existence of subjective probabilities can be justified by seemingly compelling axioms on behavior, the Bayesian approach says little about the origins of such probabilities. The axiomatic derivations suggest that one should have such beliefs, but not what they should be or where they should be derived from. Indeed, when one can provide a good account of the emergence of probabilistic beliefs, these beliefs tend to be objective, because there are good reasons to adopt them and not others. It is precisely when one finds little to say about the origin of beliefs that one needs to resort to subjectivity.

Third, in the Bayesian approach beliefs are defined on states or events. By contrast, in economics data are usually collected and presented as lists of observations or cases. If economic agents derive beliefs from data, it may be more intuitive to use models that formally distinguish between data and beliefs, or between observations and theories.

We find that these weaknesses of the Bayesian approach are related. If we have a model of how beliefs are generated, we would know when beliefs would take the form of probabilities, and also when one might seek other models of beliefs. Thus, we find that it would be fruitful to find out which probabilities are chosen by an individual when beliefs are probabilistic, and also which other models of beliefs can be useful when probabilities are too restrictive.

2. Alternative Theories

2.1. *Uncertainty*

In the early 1980s, the second author developed a theory of decision making under uncertainty that could accommodate non-quantifiable uncertainty, that is "uncertainty" in the language of Knight (1921), "true uncertainty" as referred to by Keynes (1921), or "ambiguity" if one adopts the term suggested by Ellsberg (1961). The theory (Schmeidler, 1986, 1989) involved "probabilities" that were not necessarily additive, with respect to which one can compute expectation using a notion of integration due to Choquet (1953–4). This was the first axiomatically-based general-purpose theory of decision making under uncertainty that generalized the Bayesian approach, and that could smoothly span the entire spectrum between the Bayesian model and a model of complete ignorance.

We later developed the theory of maxmin expected utility (Gilboa and Schmeidler, 1989), holding that a decision maker's beliefs are given by a set of probability measures ("multiple priors"), and decisions are being made so as to maximize the worst-case expected utility (when probabilities are taken from the prescribed set). This theory is also axiomatically-based, flexible enough to model any decision problem that the Bayesian approach can model, and allows for a continuum of degrees of uncertainty. At the same time, Bewley (2002) developed a theory that also relies on a set of probabilities, modeling a partial order that is defined by unanimity: preference for one option over another only occurs where the former has a higher expected utility than the latter according to each and every probability in the set. In the years that followed, additional models have been suggested, among them are the "smooth" model (Klibanoff, Marinacci, and Mukerji, 2005; Nau, 2006; Seo, 2008) and the model of "variational preferences" (Maccheroni, Mukerji, and Rustichini, 2006). For a survey, see Gilboa and Marinacci (2010).

It should be stressed that this line of research has not tackled the question of belief formation. The models mentioned above suggest various generalizations of the Bayesian approach. They are axiomatically based in the sense that one has characterizations of the modes of decision making that are compatible with the formal model. Hence, one can in principle tell whether a particular pattern of choices is compatible with each of these models. But they remain silent on the question of the origin and generation of beliefs, whether probabilistic or not.

2.2. *CBDT*

In the 1990s, we developed a theory of case-based decisions (CBDT, Gilboa and Schmeidler, 1995, 2001). The motivation was to take a fresh look at decision making under uncertainty, and focus on intuitive cognitive processes. Specifically, we sought to develop a formal, axiomatically-based theory that relies on the assessment of past cases rather than of future events. In doing so, we took an extreme approach, and veered away from any notion of belief. Thus, the agents in CBDT do not explicitly have beliefs about future paths that may unfold should they take various actions. Instead, they are postulated to choose actions that did well in similar cases in the past.

CBDT has several versions. In particular, one can define the notion of "similarity" over decision problems alone, over problem-act pairs, or over entire cases (where a case consists of a problem, an act, as well as a result). It has two versions, one using a summation over cases, and the other — averages; and it can be augmented by a theory of the behavior of an aspiration level that naturally pops up in the analysis.

While the more advanced versions of CBDT are general enough to embed Bayesian expected utility, the fundamental nature of the exercise was not a generalization of the classical theory. Rather, the

CBDT was focusing on a particular mode of behavior, awaiting a more general theory that would be able to elegantly encompass both Bayesian, probability-based decision making, and analogical, case-based decisions.

2.3. *The present project*

The present collection includes papers that deal, for the most part, with case-based predictions. The basic motivation is to use the conceptual basis and mathematical techniques that were developed for CBDT and to apply them to the question of belief formation. We wish to study situations in which beliefs explicitly exist, and might even be given as probability distributions, but to focus on the cases that gave rise to these beliefs. This project is based on the premise that studying the relationship between observations and beliefs may simultaneously shed light on the two questions discussed above: when do beliefs take a particular form, most notably, probabilities, and, when they do, which beliefs emerge from a given database of observations.

Most of the papers collected here do not deal with decisions at all. Rather, they discuss predictions as the outcome of the model. At times, there is an implicit assumption that these predictions are used according to a certain decision theory; specifically, probabilities are assumed to be used for expected utility maximization. Yet, the formal models ignore decisions. In this sense, this project is closer to statistics than to standard decision theory. On the other hand, it is closer to decision theory in terms of method, in particular in its focus on axiomatic derivations. Before we describe the project in more detail, a few words on the axiomatic approach are in order.

3. The Axiomatic Approach

3.1. *Axiomatizations*

Axioms generally refer to propositions that are accepted without proof, and from which other propositions are derived. Typically, the axioms are supposed to be simpler and more intuitive than their implications: the rhetorical use of axioms starts with propositions that are accepted and proceeds to those propositions that the listener is supposed to be convinced of. In mathematics and related fields, an "axiomatic system" refers to a set of conditions that captures the essence of a particular structure. Thus, the axioms abstract away from details, generalize the structure and show what are its essential building blocks that are necessary to certain conclusions of interest.

The use of "axiomatizations" in economics refers to conditions on observable data that imply, or even perfectly characterize a certain theory. For example, von-Neumann and Morgenstern's (vNM) axioms on decision under risk are equivalent to the existence of a utility function whose expectation is maximized. This usage of the term "axioms" has much in common with the previous usages. First, vNM's axioms such as transitivity or independence are supposed to be simpler and more intuitive than the explicit representation of expected utility theory. Such an axiomatization is useful for rhetorical purposes, in line with the use of axioms in logic: someone who accepts the axioms is compelled to accept their conclusions. This is obviously useful for normative purposes, because the nature of a normative exercise is precisely this: to convince the listener that a certain mode of behavior (such as expected utility maximization) is to be preferred. Moreover, the rhetoric of axioms is also useful for descriptive purposes: a scientist who argues that people tend to be expected utility maximizers will be more convincing if she uses simple, acceptable axioms than if she were to use a more complicated theory, despite the fact that the two may be equivalent.

Second, an axiomatization of expected utility theory such as vNM's also serves the purpose of dissection: the axioms state precisely what is assumed by the theory. This simplifies the task of testing whether the theory is correct, and it paves the way to refining or generalizing it in case it isn't. Indeed, a violation of expected utility theory can be analyzed in light of the axioms. One can find which axiom fails, and perhaps indicate why and how it can be relaxed.

However, axiomatizations in economics are assumed to satisfy an additional condition, which was inspired by the thinking of the logical positivists in the philosophy of science (see Carnap, 1922): the axioms are supposed to be stated in terms of observable data. The fact that such axioms imply, or better still, characterize, a theory stated in terms of theoretical concepts renders the latter meaningful. Thus, vNM's axiomatization is viewed as endowing the term "utility" with scientific meaning, showing how it can relate to observations. Relatedly, a theory that does not satisfy axioms stated in observable terms should be suspected of being unrefutable, and therefore non-scientific according to Popper (1934). Axiomatizations therefore guarantee that the game we play has a scientific flavor and that, at least in principle, theories can be tested, and competing theories are guaranteed to have different observational implications.

Despite the fact that logical positivism and Popper's notion of refutation have been seriously challenged within the philosophy of science, we find that they still serve as useful guidelines for economics. It is generally a good idea to ask ourselves questions such as, "What does this term mean precisely?" "Under which conditions will we admit that our theory is false?" The logical positivist ideal, and axiomatizations in particular, suggest a healthy exercise regime that helps us guarantee that we have satisfactory answers to such questions.

3.2. *Axiomatizing predictions*

The axiomatizations presented in this volume differ from classical axiomatizations in decision theory in that the observable data they refer to are not choices or preferences but predictions or beliefs. These data are a step removed from economic activity, which involves decisions such as buying and selling, and this fact is often viewed as a disadvantage, at least as far as economics is concerned. On the other hand, these data are independent of the particular decision model one has in mind, and they are closer to the prediction choices made by statisticians, forecasters, or classifiers. Importantly, the axiomatizations relate theoretical concepts to some data that can be observed, and they allow one to ask which pairs of theories are different in content and which pairs may seem to be different while they are, in fact, equivalent.

The prediction problems we are interested in are close to statistical inference. In statistics as well as in machine learning, one deals with questions that are fundamentally very similar to ours: how should one learn from a database of observations? Indeed, we will mention some techniques that are well-known in these fields, such as maximum likelihood estimation or kernel classification. However, the statistical literature does not typically address questions of axiomatization, and focuses instead on asymptotic behavior. Thus, there is a very rich theory about the prediction models that would guarantee satisfactory long-run behavior, but relatively little about the behavior of such models in small databases. Since there are many important situations in which one is asked to make predictions (or to take decisions) despite the paucity of truly relevant past observations, we find this problem to be of interest. We do not believe that one can expect theoretical arguments to provide clear-cut predictions out of thin air. But we hope that the axiomatic approach can at least guarantee that the totality of predictions one offers are coherent in a well-defined sense.

3.3. *Statistics and psychology*

Our analysis can be interpreted both descriptively and normatively. That is, one may ask which patterns of learning from data are likely to be observed by real people who make predictions in economic contexts; and one may also ask which patterns are desirable, or sensible, and which can serve as ideals of rational reasoning. There are many applications in which the two interpretations can coexist. For example, after observing 100 tosses of a tack, of which 70 resulted in the pin pointing up, it makes sense to predict the same outcome in the next toss. Moreover, this is what most people would do. Thus, the normative recommendation and the descriptive theory coincide in this case. However, there are situations in which the normative differs from the descriptive. In the above example, the "Gambler's fallacy" phenomenon (Tversky and Kahneman, 1974) shows that people might predict that a sequence ("run") of several Heads will be followed by a Tail. In such situations, we typically choose axioms according to their normative interpretation. In other words, when a modeling choice is to be made, we tend to find ourselves closer to statistics than to psychology. One reason for this preference is that axioms are readily applicable to the normative question of choosing among learning methods; by contrast, it is less obvious that one may gain much by axiomatizing the way people actually make predictions, especially when these differ from the normatively acceptable ones.

The difference between statistics and psychology notwithstanding, during the course of this project we were several times surprised to see how close the two can be. Starting from a psychologically-motivated research on similarity, we axiomatized formulae that turned out to be identical to those used in kernel classification and kernel estimation of probabilities. We likewise found ourselves axiomatizing the preference for theories with higher likelihood without ever suspecting that this is where the axioms would lead us. Thus, several times we asked ourselves which conditions it is likely to assume that real people satisfy, and we found that these conditions characterized well-known statistical techniques. As far as statistics is concerned, this means that some of these techniques, which were devised by statisticians without explicitly thinking in terms of axioms, ended up satisfying reasonable conditions. From a psychological point of view, these coincidences suggested that the human mind is probably a rather successful inference engine, in that general principles that make sense for human reasoning are also corroborated by statistical analysis.

These coincidences should not be overstated. There are surely many circumstances in which people tend to make silly predictions (cf. the Gambler's fallacy). On the other hand, there are many statistical techniques that are far from anything that can be viewed as a model of natural human reasoning. Moreover, the coincidences we find certainly do not imply that statistics and psychology are the same discipline. In fact, the opposite is true: statistics is mainly interested in developing techniques that are not obvious, namely, that go beyond the intuitive. By contrast, as far as inductive inference is concerned, psychology is interested in reasoning processes that tell us something new about the human mind, and these tend to correlate with less reasonable inferences.

It is therefore possible that the majority of real-life inferences are made by people in a very reasonable way that also corresponds to simple statistical techniques. Psychology would tend to focus on the remaining predictions that are not necessarily rational, and perhaps also not yet well understood. Statistics would ask how these predictions should be made, in ways that are too difficult for most people to come up with on their own. Thus, a sampling of real-life problems may suggest that the normative (statistics) is close to the descriptive (psychology), but a sampling of recent research in either discipline would suggest the opposite. As the axiomatic approach we propose is rather rudimentary, it probably covers only the basic problems, where psychology and statistics may not be far apart. It is our hope that, while this domain covers a small portion of recent research, it does correspond to a non-negligible portion of everyday predictions.

4. The Combination Principle

4.1. *Basic logic and results*

Applying the axiomatic approach to the problem of inductive inference, we wish to identify reasonable patterns of inferences drawn from databases of observations. Thus, we do not focus on a single database and delve into the particular inferences that it entails. Rather, we consider an abstract method of inference, or a function that assigns sets of conclusions to databases, and ask which conditions should one impose on such a method or function.

In several studies we used axioms that are different manifestations of *the combination principle*: if a certain conclusion is reached given two disjoint databases, the same conclusion should be reached given their union. The precise meaning of "conclusion" and "database" should be specified. In fact, they are modeled in several ways in the papers presented here. A *database* can be a set of cases; or an ordered list of cases; or a counter vector, specifying how many times each type of case has been observed. In the first type of models, "union" is simply the union of two sets; in the second, "union" refers to concatenation of ordered lists of cases. Finally, if a database is no more than a counter vector, the union of two such databases corresponds to the addition of the two vectors, generating a new vector.

In general, the three formulations differ and they may give rise to different operations on databases. However, in each of the axiomatic works that follow, we assumed that cases were exchangeable in an appropriate sense. If a database is a set, we define a notion of case-equivalence, and assume that each particular case has infinitely many "replicas", that is, infinitely many cases that are equivalent to it. If a database is an ordered list of cases, the set of all cases already includes replicas because it includes lists in which the same case is repeated as many times as one wishes. In this context, we also assume that any list induces the same predictions as any permutation thereof. Finally, if a database is a vector of non-negative integers, counting how many times a case of each type has been encountered, specific cases do not explicitly appear, and exchangeability of cases is built into the model, as well as the assumption that each case can have as many replicas as we would like to consider.

Thus, in all three formulations, we basically have in mind the same structure: there exist *types of cases*, and of each type we have observed a number of occurrence that is a non-negative integer. The order in which these cases were observed is immaterial. Should one wish, for example, that more recent cases would matter more than less recent ones, one could include the time of occurrence as one of the features of the case. The formal model, however, needs to assume that only the numbers of observations of each type matter.

It is a crucial and non-trivial assumption that cases can have as many replicas as one may imagine. For example, if one case is the financial crisis of 1929 and another — the crisis of 2008, one need not worry about the order in which they are listed, as each case contains enough information to describe its recency, and presumably its relevance. However, one may question how meaningful is it to consider a database in which the crisis of 1929 never occurred and that of 2008 occurred, say, 14 times. It is important to highlight that our axiomatic derivations do rely on the predictions that the reasoner would generate given each of the possible databases. In the case of global events such as wars, financial crises, and the like, the very formulation of the set of possible databases may lead us to question the relevance of the axiomatic derivation. By contrast, if one has in mind an application that is closer to cross sectional data, where different observations are causally independent, it is not too demanding to assume that each case may appear any number of times, and to require that the reasoner should make a prediction given any number of occurrences of each case.

Next, we have to clarify what is meant by a *prediction*. In some papers, predictions are weak orders, ranking certain alternatives as at-least-as-likely-as others. The alternatives may be the possible values of the next observation, giving rise to models of frequency-based prediction, kernel classification, and kernel estimation. Alternatively, the alternatives may be general theories, or statistical models, which are ranked for plausibility given the data. This interpretation allows us to derive maximum likelihood-based selection of theories, as well as refinements thereof such as Akaike's information criterion, minimum-description-length criterion, etc. In other papers, the prediction is that a probability vector (or measure) lies on a certain line segment, or, equivalently, that a certain random variable has a pre-specified expectation.

The combination principle thus takes different shapes, depending on the model to which it applies. Its specific incarnations are referred to as "the combination axiom" or "the concatenation axiom", depending on the context. The axiomatic derivations make use of other axioms as well. These typically include an Archimedean condition and a richness condition. However, the conceptually important assumptions seem to be (i) that prediction is meaningfully defined for all databases; and that (ii) the combination principle holds.

Under these assumptions, our results are that there exists a function s over pairs of cases, which we tend to interpret as a similarity function, such that prediction can be represented by (the maximization of) s-weighted summation or by s-weighted averaging. More explicitly, assume that I is a counter vector, so that $I(c) \in \mathbb{Z}_+$ is the number of times cases (observations) of type c have been encountered. The set of case-types may be infinite, but it is assumed that $\sum_c I(c) < \infty$ for all databases I. Assume that, given I, $\succsim_I$ is a weak order on a set of alternatives. Gilboa and Schmeidler (2003a) employs the combination principle, coupled with the other axioms, to derive the following similarity-weighted sum representation of $\succsim_I$: for each database I and any two alternatives a, b,

$$a \succsim_I b \Leftrightarrow \sum_c I(c)s(a, c) \geq \sum_c I(c)s(b, c). \tag{1}$$

Gilboa and Schmeidler (2010) modifies the combination principle to allow for a-priori biases for certain alternatives. Its leading interpretation is that elements such as a, b are theories, and these may differ in terms of their complexity. A preference for simpler theories may lead to a representation of the type

$$a \succsim_I b \Leftrightarrow w(a) + \sum_c I(c)s(a, c) \geq w(b) + \sum_c I(c)s(b, c) \tag{2}$$

which is axiomatized in this paper.

If predictions are more quantitative and take the form of probability vectors, Billot, Gilboa, Samet, and Schmeidler (2005) show that, as long as the domain of the prediction function is not limited to a single segment, the combination principle is equivalent to the following similarity-weighted averaging: for each case-type c there exists a number $s_c > 0$ and a probability vector p^c such that, for each I, the probability the reasoner chooses is

$$p(I) = \frac{\sum_c I(c)s_c p^c}{\sum_c I(c)s_c}.$$

An important special case of this rule is the following: one of finitely many states Ω will occur. Each past case c describes certain circumstances x_c and a realization of a state $\omega_c \in \Omega$. Given a new problem x_p, the probability of state ω is the similarity-weighted empirical frequency of ω in the past:

$$p(\omega|I) = \frac{\sum_c I(c)s(x_c, x_p)\mathbf{1}_{\{\omega_c = \omega\}}}{\sum_c I(c)s(x_c, x_p)}. \tag{3}$$

Clearly, this result does not apply if $|\Omega| = 2$, because in this case the range of the function I is included in a line segment. Gilboa, Lieberman, and Schmeidler (2006) provides an axiomatization of this formula in the two-outcome case.

4.2. *Examples*

There are several well-known statistical techniques that are special cases of the general representations above. Consider the simple similarity-weighted sum in (1), and suppose that the possible predictions (a, b) and past cases (c) belong to the same set, as in the case of a repeated roll of a die. Assume that the similarity function is the indicator function,

$$s(a, c) = \mathbf{1}_{\{a=c\}}$$

This means that only past occurrences of the very same prediction a may lend non-zero support to this prediction, and that all past cases are deemed equally relevant. In this case, the ranking according to (1) coincides with the ranking of possible predictions by their frequency in the past.

Next assume that the set to which past cases c and possible predictions a belong is infinite, such as $\mathbb{R}^k$. In this case, it makes sense to allow the function $s(a, c)$ to be positive also when a and c are close, though not necessarily identical. Then, the expressions on the right hand side of (1) are those used for kernel estimation of a density function (see Akaike, 1954; Rosenblatt, 1956; Parzen, 1962; Silverman, 1986; Scott, 1992, for a survey).

Along similar lines, assume that the prediction problem is a classification problem: each observation $c = (x_c, a_c)$ consists of data x_c (say, a point in $\mathbb{R}^k$) and a class a_c to which the point is known to belong. Given a new point with parameters x, the reasoner is asked to guess to which class a it belongs. Then, specifying

$$s(a, (x_c, a_c)) = k(x_c, x)\mathbf{1}_{\{a=a_c\}}$$

the formula (1) boils down to kernel classification with a kernel function k.

More interesting examples involve applications where the set of observations and the possible predictions have no common structure. For example, if cases c are past observations, and the predictions a are theories, they do not typically belong to the same set. However, the axiomatization suggests that, if the reasoner satisfies the axioms, one can find a function s that would describe the reasoner's predictions via (1) and thus, indirectly, reflect the reasoner's perception of the relationship between theories and observations. Specifically, assume that the function s is negative.[2] Define, for a theory a and an observation c,

$$p(c|a) = \exp(s(a,c))$$

so that

$$\log(p(c|a)) = s(a,c)$$

and (1) becomes equivalent to ranking of theories by the (log-) likelihood function.

The introduction of a-priori biases to the ranking of theories, such as the preference for simplicity, suggests that a constant should be added to the log-likelihood function as in (2). Clearly, this formulation includes as special cases the Akaike Information Criterion (AIC, Akaike, 1974) and Minimum Description Length criterion for model selection (MDL, see Wallace and Boulton, 1968; Wallace and Dowe, 1999; and Wallace, 2005 for a more recent survey). These do not fall under the category of (1), where there is no room for the function w. Indeed, ranking of theories by AIC or by MDL does not satisfy the combination principle as stated. Gilboa and Schmeidler (2010) invoke a weaker version of this principle to derive such ranking rules.

4.3. *Limitations*

The combination principle appears to be rather intuitive, and it is perhaps not too surprising that this principle is satisfied by a variety of statistical techniques. Yet, there are also many statistical techniques, as well as natural reasoning procedures, that do not satisfy it. These violations can be classified into three types: first, there are situations in which the principle is inappropriately applied. Given the generality of the principle, stated for general "conclusions" drawn from databases, there are situations where the principle may formally apply, yet it may not be very sensible to adhere to it. For example, Simpson's paradox

---

[2]The analysis in Gilboa and Schmeidler (2003) shows that this assumption involves no loss of generality as long as the number of theories is finite.

(see Simpson, 1951, or, for example, de Groot, 1975) is a well-known example in which a certain conclusion can be drawn from each of two databases but not from their union. As we argue in Gilboa and Schmeidler (2010), this is a mis-application of the principle, because in this example the theories concerned are not directly about the single data, but about certain patterns in the data. More specifically, the completeness axiom (which, in one shape or another, appears in all axiomatizations mentioned here) expects one to rank theories given each and every database, even if the database contains only one observation. This does not seem to be the case in Simpson's paradox, and thus we argue that this violation of the combination principle is due to a mis-application of the model. Put differently, the completeness axiom implicitly restricts the type of observation-prediction pairs to which the theory should be applied.

A second class of violations of the combination principle are those in which one considers a statistical technique and concludes that it is indeed a theoretical flaw that it fails to satisfy such a reasonable principle. According to our personal taste, this is the case with k-nearest neighbor techniques (Fix and Hodges, 1951, 1952; Cover and Hart, 1967), which violate the combination axiom (as in Gilboa and Schmeidler, 2003a) because the weight assigned to an observation does not depend solely on its inherent relevance, but also on its relative relevance, as compared to other observations. While this is ultimately a subjective judgment, we find that this violation is not among the merits of k-nearest neighbor techniques.

Finally, there are violations of the third type, in which one finds that the principle is too restrictive. Two main such categories are situations in which one learns the similarity function from the data, and when one engages in combination of induction and deduction. We view each of these as pointing to important directions for future research, and discuss them separately.

5. Learning the Similarity

Case-based reasoning relies on the similarity that one finds between cases. Where does this similarity function come from? Taking a descriptive interpretation, this question brings us to the domain of psychology, and it would suggest that the similarity function is not a fixed, immutable reasoning tool that one is born with. Rather, it is learnt from experience. For example, a physician may learn, through her experience, that for a particular diagnosis, weight and blood pressure are important features of similarity, whereas blood type is not. Thus, past cases do not only suggest what will be the outcome of future cases using the similarity function; they also indicate which similarity functions are better suited to perform this type of case-to-case induction.

Taking a normative viewpoint, closer to the statistical mindset, it stands to reason that one may update the similarity function based on data. Indeed, kernel methods typically change the kernel function as data accumulate, so that in larger databases a tighter kernel is used, allowing the prediction to be based mostly on the more relevant cases when there are sufficiently many of these. But beyond the sheer number of past cases, their content can also serve as a guide regarding the choice of the similarity/kernel function.

In Gilboa, Lieberman, and Schmeidler (2006) we formally introduce the notion of *empirical similarity*. This is defined as a similarity function that, within a pre-specified class of functions, minimizes the sum of squared errors one would have obtained, were one to use that similarity function in the past, predicting each outcome based on the rest of the database (the "leave-one-out" criterion). We develop the statistical theory for estimation of the similarity function, assuming that the process is indeed governed by a similarity-weighted-average of other (or past) observations.

The empirical similarity idea can also be used to analyze databases that are not necessarily believed to have been generated by a similarity-based process. Rather, one can suggest the idea as a statistical prediction technique that mimics the informal learning of similarity that human beings naturally engage in. Further, one can follow this line of reasoning and use the empirical similarity to define objective probabilities: probabilities are defined by similarity-weighted frequencies in past cases, where the similarity function is learnt from the same database. Thus, one can shed the subjective baggage of the psychological notion of similarity, and replace it by a notion of empirical similarity, which has a claim to objectivity similar to those of other statistical constructs. Gilboa, Lieberman, and Schmeidler (2009) is devoted to this definition of objective probabilities, and it also discusses the tension between the proposal to learn the similarity function and the combination principle that is violated by such learning.

6. Rule-Based Reasoning

We originally formulated the combination principle with case-to-case induction in mind. Somewhat to our surprise, we found that it can be reinterpreted for case-to-rule (or observation-to-theory) induction, and that it then basically coincides with maximum likelihood selection of models. Further, as mentioned above, the principle can be adapted to introduce considerations other than the likelihood function, such as simplicity or prior probability, into model selection.

However, when theories are selected based on past observations, and they are then used to forecast future observations, the combination principle does not seem appropriate. For example, if one uses observations of variables x, y in order to estimate a regression model $y = \alpha + \beta x + \varepsilon$, the selection of a model (or "theory" or "rule") boils down to the selection of the parameters α, β. Under the standard assumptions, least square estimators are maximum likelihood estimators, and the ranking of parameter values by the likelihood function will satisfy the combination principle. However, if the selected parameters are then used, via the regression equation, to predict the value of y for a new observation x, the combination principle will be violated. Moreover, this should be expected to be the case whenever one engages in combined inductive and deductive reasoning: inductive reasoning to find a model based on the data, and deductive reasoning to predict the data based on the selected model.

Thus, we find that the theory developed here has little to say about rule-based prediction. Moreover, Aragones, Gilboa, Postlewaite, and Schmeidler (2005) shows that the theory selection problem is a computational hard one. Specifically, when one introduces goodness of fit as well as simplicity as model selection criteria, very reasonable formulations of the problem render the selection of the "best" theory an NP-Hard problem. This implies that even the first step, of case-to-rule induction, may be too complicated to be performed, either by humans or by computers.

7. Summary

We believe that the axiomatic approach to inductive inference is important and useful. It helps us understand what theories actually assume; it highlights equivalences between different formulations of the same theory; it guarantees that theories have a clear empirical content; and it may ensure that the method one chooses for inductive reasoning is coherent and sensible even if the database is small and asymptotic results are of limited relevance. Ideally, one would like to have axiomatic derivations of all theories one uses, and use the axioms to help select a theory for specific classes of inductive inference problems.

Unfortunately, the results we report here indicate only partial success. All the axiomatic results rely very heavily on the combination principle. We find this principle a reasonable starting point, but it certainly cannot be considered a universal condition on inductive inference. Future research might find more flexible axiomatic approaches that would be able to generalize the theories presented here to include other types of reasoning.

The book is organized as follows. As background, we start with two axiomatic derivations of case-based decision theory. The first, Gilboa and Schmeidler (1995), is the original paper, highlighting the basic ideas. The second, Gilboa and Schmeidler (1997), extends the theory to incorporate act-similarity considerations, and introduces the mathematical structure of the combination principle. Several of the subsequent papers use this paper as their mathematical backbone.

Next, we consider inductive inference as modeled by an "at least as likely as" relation. The basic tools are given in Gilboa and Schmeidler (2003a). If the objects to be ranked are events in a given state space, one may hope to say more, as there is a measure-theoretic structure one may use. On the other hand, some auxiliary assumptions are inappropriate in this context. Gilboa and Schmeidler (2002) deals with this case, and with the combinatorial issues that arise if one wishes to obtain a probability function over events based on likelihood rankings that satisfy the combination principle.

The application of Gilboa and Schmeidler (2003a) to theory selection is limited to the maximum likelihood principle. As mentioned above, this is extended to an additive trade-off between simplicity and likelihood in Gilboa and Schmeidler (2010). This paper is therefore the next in the volume.

When the observable data are numerical probabilities, Billot, Gilboa, Samet, and Schmeidler (2005) provide an axiomatization of the similarity-weighted-frequency formula. It is followed by Billot, Gilboa, and Schmeidler (2008), which characterizes the exponential similarity function in the context of this formula.

However, one may only go so far when using theoretical considerations for the selection of a similarity function. In the final analysis, the choice of the function remains an empirical issue, which is what Gilboa, Lieberman, and Schmeidler (2006) is about. This paper also completes the axiomatization of the similarity-weighted frequency formula for the single-dimensional case. It is followed by Gilboa, Lieberman, and Schmeidler (2009), which offers the empirical similarity, coupled with the similarity-weighted formula, as a definition of objective probabilities.

While the empirical similarity papers suggest a new statistical technique and a new definition of objective probabilities, they also highlight the limitation of the axiomatizations provided here, as they focus on violating the combination principle (which lies at the heart of these axiomatizations). We then move to discuss the complexity of theory selection, in Aragones, Gilboa, Postlewaite, and Schmeidler (2005), which indicates another important direction in which the axiomatic theory of inductive inference may be enriched.

Finally, we conclude with two applications of the mathematical techniques developed in Gilboa and Schmeidler (1997, 2003) to other problems. Gilboa and Schmeidler (2003b) applies the method to the derivation of a utility function, coupled with the expected utility principle, in the context of a game, that is, without referring to lotteries other than the game offers. Gilboa and Schmeidler (2004) offers a definition of subjective probabilities limited to the distributions of given random variables, without reference to the (much larger) underlying state space. None of these two papers is related to the main project in terms of content. However, both contain results that may be useful for certain extensions, such as modelling memory in a continuous way.

References

Akaike, H. (1954), "An Approximation to the Density Function", *Annals of the Institute of Statistical Mathematics*, **6**: 127–132.

———— (1974), "A new look at the statistical model identification". *IEEE Transactions on Automatic Control*, **19** (6): 716–723.

Alchourron, C. E., P. Gardenfors and D. Makinson (1985), "On the Logic of Theory Change: Partial Meet Functions for Contraction and Revision", *Journal of Symbolic Logic*, **50**: 510–530.

Aragones, E., I. Gilboa, A. Postlewaite and D. Schmeidler (2005), "Fact-Free Learning", *American Economic Review*, **95**: 1355–1368.

Bewley, T. (2002), "Knightian Decision Theory: Part I", *Decisions in Economics and Finance*, **25**: 79–110.

Billot, A., I. Gilboa, D. Samet and D. Schmeidler (2005), "Probabilities as Similarity-Weighted Frequencies", *Econometrica*, **73**: 1125–1136.

Billot, A., I. Gilboa and D. Schmeidler (2008), "Axiomatization of an Exponential Similarity Function", *Mathematical Social Sciences*, **55**: 107–115.

Carnap, R. (1923), "Uber die Aufgabe der Physik und die Andwednung des Grundsatze der Einfachstheit", *Kant-Studien*, **28**: 90–107.

Choquet, G. (1953), "Theory of Capacities", *Annales de l'Institut Fourier*, **5**: 131–295.

Cover, T. and P. Hart (1967), "Nearest Neighbor Pattern Classification", IEEE *Transactions on Information Theory*, **13**: 21–27.

de Finetti, B. (1931), Sul Significato Soggettivo della Probabilità, *Fundamenta Mathematicae*, **17**: 298–329.

———— (1937), "La Prevision: ses Lois Logiques, ses Sources Subjectives", *Annales de l'Institut Henri Poincare*, **7**: 1–68 (translated in *Studies in Subjective Probability*, edited by H. E. Kyburg and H. E. Smokler, Wiley, 1963).

de Groot, M. H. (1975), *Probability and Statistics*, Reading, MA: Addison-Wesley Publishing Co.

Ellsberg, D. (1961), "Risk, Ambiguity and the Savage Axioms", *Quarterly Journal of Economics*, **75**: 643–669.

Fix, E. and J. Hodges (1951), "Discriminatory Analysis. Nonparametric Discrimination: Consistency Properties". Technical Report 4, Project Number 21-49-004, USAF School of Aviation Medicine, Randolph Field, TX.

—— (1952), "Discriminatory Analysis: Small Sample Performance". Technical Report 21-49-004, USAF School of Aviation Medicine, Randolph Field, TX.

Gardenfors, P. (1992), *Belief Revision: An Introduction*. Cambridge: Cambridge University Press.

Gilboa, I. and M. Marinacci (2010), "Ambiguity and the Bayesian Paradigm", mimeo.

Gilboa, I. and D. Schmeidler (1989), "Maxmin Expected Utility with a Non-Unique Prior", *Journal of Mathematical Economics*, **18**: 141–153.

—— (1995), "Case-Based Decision Theory", *Quarterly Journal of Economics*, **110**: 605–639.

—— (1997), "Act Similarity in Case-Based Decision Theory", *Economic Theory*, **9**: 47–61.

—— (2001), *A Theory of Case-Based Decisions*. Cambrdige: Cambridge University Press.

—— (2002), "A Cognitive Foundation of Probability", *Mathematics of Operations Research*, **27**: 68–81.

—— (2003a), "Inductive Inference: An Axiomatic Approach", *Econometrica*, **71**: 1–26.

—— (2003b), "Expected Utility in the Context of a Game", *Games and Economic Behavior*, **44**: 184–194.

—— (2004), "Subjective Distributions", *Theory and Decision*, **56**: 345–357.

—— (2010), "Likelihood and Simplicity: An Axiomatic Approach", *Journal of Economic Theory*, **145**: 1757–1775.

Gilboa, I., O. Lieberman and D. Schmeidler (2006), "Empirical Similarity", *Review of Economics and Statistics*, **88**: 433–444.

—— (2009), "On the Definition of Objective Probabilities by Empirical Similarity", *Synthese*.

Hume, D. (1748), *Enquiry into the Human Understanding*. Oxford, Clarendon Press.

Keynes, J. M. (1921), *A Treatise on Probability*. London: MacMillan and Co.

Klibanoff, P., M. Marinacci, and S. Mukerji (2005), "A Smooth Model of Decision Making under Ambiguity," *Econometrica*, **73**: 1849–1892.

Knight, F. H. (1921), *Risk, Uncertainty, and Profit*. Boston, New York: Houghton Mifflin.

Maccheroni, F., M. Marinacci and A. Rustichini (2006), "Ambiguity Aversion, Robustness, and the Variational Representation of Preferences," *Econometrica*, **74**: 1447–1498.

Nau, R. F. (2006), "Uncertainty Aversion with Second-Order Utilities and Probabilities", *Management Science* **52**: 136–145.

von Neumann, J. and O. Morgenstern (1944), *Theory of Games and Economic Behavior.* Princeton, NJ: Princeton University Press.

Parzen, E. (1962), "On the Estimation of a Probability Density Function and the Mode", *Annals of Mathematical Statistics*, **33**: 1065–1076.

Popper, K.R. (1934), *Logik der Forschung*; English edition (1958), *The Logic of Scientific Discovery.* London: Hutchinson and Co. Reprinted (1961), New York: Science Editions.

Ramsey, F. P. (1931), "Truth and Probability", *The Foundation of Mathematics and Other Logical Essays.* New York: Harcourt, Brace and Co.

Riesbeck, C. K. and R. C. Schank (1989), *Inside Case-Based Reasoning.* Hillsdale, NJ, Lawrence Erlbaum Associates, Inc.

Rosenblatt, M. (1956), "Remarks on Some Nonparametric Estimates of a Density Function", *Annals of Mathematical Statistics*, **27**: 832–837.

Royall, R. (1966), *A Class of Nonparametric Estimators of a Smooth Regression Function.* Ph.D. Thesis, Stanford University, Stanford, CA.

Savage, L. J. (1954), *The Foundations of Statistics.* New York: John Wiley and Sons.

Schank, R. C. (1986), *Explanation Patterns: Understanding Mechanically and Creatively.* Hillsdale, NJ: Lawrence Erlbaum Associates.

Schmeidler, D. (1986), "Integral Representation without Additivity." *Proceedings of the American Mathematical Society*, **97**: 255–261.

————— (1989), "Subjective Probability and Expected Utility without Additivity", *Econometrica*, **57**: 571–587.

Seo, K. (2009), "Ambiguity and Second-Order Belief", *Econometrica*, **77**: 1575–1605.

Scott, D. W. (1992), *Multivariate Density Estimation: Theory, Practice, and Visualization.* New York: John Wiley and Sons.

Silverman, B. W. (1986), *Density Estimation for Statistics and Data Analysis.* London and New York: Chapman and Hall.

Simpson, E. H. (1951). "The Interpretation of Interaction in Contingency Tables". *Journal of the Royal Statistical Society, Ser. B* **13**: 238–241.

Tversky, A. and D. Kahneman (1974), "Judgment under Uncertainty: Heuristics and Biases", *Science*, **185**: 1124–1131.

Wallace, C. S. (2005), *Statistical and Inductive Inference by Minimum Message Length* Series: Information Science and Statistics, Springer.

Wallace, C. S. and D. M. Boulton (1968), "An Information Measure for Classification", *The Computer Journal*, **13**, 185–194.

Wallace, C. S. and D. L. Dowe (1999), "Minimum Message Length and Kolmogorov Complexity", *The Computer Journal*, **42**, 270–283.

Chapter 1

Case-Based Decision Theory*

Itzhak Gilboa and David Schmeidler

Reprinted from *The Quarterly Journal of Economics*, *110* (1995):
605–639.

This paper suggests that decision-making under uncertainty is, at least partly, case-based. We propose a model in which cases are primitive, and provide a simple axiomatization of a decision rule that chooses a "best" act based on its past performance in similar cases. Each act is evaluated by the sum of the utility levels that resulted from using this act in past cases, each weighted by the similarity of that past case to the problem at hand. The formal model of case-based decision theory naturally gives rise to the notions of satisficing decisions and aspiration levels.

In reality, all arguments from experience are founded on the similarity which we discover among natural objects, and by which we are induced to expect effects similar to those which we have found to follow from such objects.... From causes which appear *similar* we expect similar effects. This is the sum of all our experimental conclusions (Hume, 1748).

*We are grateful to many people for conversations and discussions that influenced this work. In particular, we benefited from insightful conversations with Eva Gilboa, who also exposed us to case-based reasoning, and Akihiko Matsui and Kimberly Katz, who also referred us to Hume. A special thank-you is due to Benjamin Polak who served as a referee. His two reports on earlier versions of the paper are longer than the final product. His suggestions reshaped the paper and made it much more accessible to a reader who is not one of the authors. We are also grateful to the faculty and students of the Institute for the Learning Sciences at Northwestern University, faculty and guests at the Santa Fe Institute, as well as to Max Bazerman, Avraham Beja, Edward Green, Ehud Kalai, Morton Kamien, Edi Karni, Simon Kasif, James Peck, Stanley Reiter, Ariel Rubinstein, Michael Sang, Karl Schlag, Andrei Shleifer, Costis Skiadas, Steven Tadelis, and Amos Tversky for comments and references. Partial financial support from NSF Grants Nos. SES-9113108 and SES-9111873, the Alfred Sloan Foundation, and the Suntory Foundation are gratefully acknowledged.

1

1.1. Introduction

Expected utility theory enjoys the status of an almost unrivaled paradigm for decision-making in the face of uncertainty. Relying on such sound foundations as the classical works of Ramsey (1931), de Finetti (1937), von Neumann and Morgenstern (1944), and Savage (1954), the theory has formidable power and elegance, whether interpreted as positive or normative, for situations of given probabilities ("risk") or unknown ones ("uncertainty") alike.

While evidence has been accumulating that the theory is too restrictive (at least from a descriptive viewpoint), its various generalizations only attest to the strength and appeal of the expected utility paradigm. With few exceptions, all suggested alternatives retain the framework of the model, relaxing some of the more "demanding" axioms while adhering to the more "basic" ones. (See Machina (1987), Harless and Camerer (1994), and Camerer and Weber (1992) for extensive surveys.)

Yet it seems that in many situations of choice under uncertainty, the very language of expected utility models is inappropriate. For instance, in many decision problems under uncertainty, states of the world are neither naturally given, nor can they be simply formulated. Furthermore, often even a comprehensive list of all possible outcomes is not readily available or easily imagined. The following examples illustrate.

Example 1:   As a benchmark, we first consider Savage's famous omelet problem (Savage, 1954, pp. 13–15): Savage is making an omelet using six eggs. Five of them are already opened and poured into a bowl. He is holding the sixth and has to decide whether to pour it directly into the bowl, or to pour it into a separate, clean dish to examine its freshness. This is a decision problem under uncertainty, because Savage does not know whether the egg is fresh or not. Moreover, uncertainty matters: if the egg is fresh, he will be better off pouring it directly into the bowl, saving the need to wash another dish. On the other hand, a rotten egg would result in losing the five eggs already in the bowl; thus, if the egg is not fresh, he would prefer to pour it into the clean dish.

In this example, uncertainty may be fully described by two states of the world: "the egg is fresh" and "the egg isn't fresh." Each of these states "resolves all uncertainty" as prescribed by Savage. Not only are there relatively few relevant states of the world in this example, they are also "naturally" given in the description of the problem. In particular, they can be defined independently of the acts available to the decision-maker. Furthermore, the possible outcomes can be easily defined. Thus, this example falls neatly into "decision-making under uncertainty" in Savage's model.

Example 2: A couple has to hire a nanny for their child. The available acts are the various candidates for the job. The agents do not know how each candidate would perform if hired. For instance, each candidate may turn out to be negligent or dishonest. Coming to think about it, they realize that other problems may also occur. Some nannies are treating children well, but cannot be trusted with keeping the house in order. Others appear to be just perfect on the job, but are not very loyal and may quit the job on short notice.

The couple is facing uncertainty regarding the candidates' performance on several measures. However, there are a few difficulties in fitting this problem into the framework of expected utility theory (EUT). First, imagining all possible outcomes is not a trivial task. Second, the "states of the world" do not naturally suggest themselves in this problem. Furthermore, if the agents should try to construct them analytically, their number and complexity would be daunting: every state of the world should specify the exact performance of each candidate on each measure.

Example 3: President Clinton has to decide on military intervention in Bosnia-Herzegovina. (A problem that he is facing while this paper is being written, revised, and re-revised.) The alternative acts are relatively clear: one may do nothing; impose economic sanctions; use limited military force (say, only air strikes), or opt for a full-blown military intervention. Of course, the main problem is to decide what are the likely short-run and long-run outcomes of each act. For instance, it is not exactly clear how strong are the military forces of the warring factions in Bosnia; it is hard to judge how many casualties each military option would involve, and what would be the public opinion response; there is some uncertainty about the reaction of Russia, especially if it goes through a military coup.

In short, the problem is definitely one of decision under uncertainty. But, again, neither all possible eventualities, nor all possible scenarios are readily available. Any list of outcomes or of states is bound to be incomplete. Furthermore, each state of the world should specify the result of each act at each point of time. Thus, an exhaustive set of the states of the world certainly does not naturally pop up.

In Example 1, expected utility theory seems a reasonable description of how people think about the decision problem. By contrast, we argue that in examples such as 2 and 3, EUT does not describe a plausible cognitive process. Should the agent attempt to "think" in the language of EUT, she would have to imagine all possible outcomes and all relevant states. Often the definition of a state of the world would involve conditional statements, attaching outcomes to acts. Not only would the number of states be huge, the states themselves would not be defined in an intuitive way.

Moreover, even if the agent managed to imagine all outcomes and states, her task would by no means be done. Next she would have to assess the utility of each outcome, and to form a prior over the state space. It is not clear how the utility and the prior are to be defined, especially since past experience appears to be of limited help in these examples. For instance, what is the probability that a particular candidate for the job in Example 2 will end up being negligent? Or being both negligent and dishonest? Or, considering Example 3, what are the chances that a military intervention will develop into a full-blown war, while air strikes will not? What is the probability that a scenario that no expert predicted will eventually materialize?

It seems unlikely that decision-makers can answer these questions. Expected utility theory does not describe the way people "really" think about such problems. Correspondingly, it is doubtful that EUT is the most useful tool for predicting behavior in applications of this nature. A theory that will provide a more faithful description of how people think would have a better chance of predicting what they will do. How do people think about such decision problems, then? We resort to Hume (1748), who argued that "From causes which appear *similar* we expect similar effects. This is the sum of all our experimental conclusions." That is, the main reasoning technique that people use is drawing analogies between past cases and the one at hand.[1]

Applying this idea to decision-making, we suggest that people choose acts based on their performance in similar problems in the past. For instance, in Example 2 a common, and indeed very reasonable, thing to do is to ask each candidate for references. Every recommendation letter provided by a candidate attests to his/her performance (as a nanny) in a different problem. In this example, the agents do not rely on their own memory; rather, they draw on the experience of other employers. Each past "case" would be judged for its similarity; for instance, serving as a nanny to a month-odd toddler is somewhat different from the same job when a two-year-old child is concerned. Similarly, the house, the neighborhood, and other factors may affect the relevance of past cases to the problem at hand. Thus, we expect the agents to put more weight on the experience of people whose decision problem was "more similar" to theirs. Furthermore, they may rely more heavily on the experience of people they happen to know, or judge to have tastes similar to their own.

Next consider Example 3. While military and political experts certainly do try to write down possible "scenarios" and to assign likelihood to them, this is by no means the only reasoning technique used. (Nor is it necessarily the most compelling a priori or the most successful a posteriori.) Very often the reasoning used is by analogies to past cases. For instance, proponents of military intervention tend to cite the Gulf War as a "successful" case. They stress the similarity of the two problems, say, as local conflicts in post-cold-war world. Opponents

---

[1] We were first exposed to this idea as an explicit theory in the form of case-based reasoning (Schank, 1986; Riesbeck and Schank, 1989), to which we owe the epithet "case-based." Needless to say, our thinking about the problem was partly inspired by case-based reasoning. At this early stage, however, there does not seem to be much in common — beyond Hume's basic idea — between our theory and case-based reasoning. It should be mentioned that similar ideas were also expressed in the economics literature by Keynes (1921), Selten (1978), and Cross (1983).

adduce the Vietnam War as a case in which military intervention is generally considered to have been a mistake. They also point to the similarity of the cases, for instance to the "peace-keeping mission" mentioned in both.

Specifically, we suggest the following theory, which we dub "case-based decision theory" (CBDT). Assume that a set of "problems" is given as primitive, and that there is some measure of similarity on it. The problems are to be thought of as descriptions of choice situations, as "stories" involving decision problems. Generally, an agent would remember some of the problems that she and other agents encountered in the past. When faced with a new problem, the similarity of the situation brings this memory to mind, and with it the recollection of the choice made and the outcome that resulted. We refer to the combination of these three — the problem, the act, and the result — as a case. Thus, "similar" cases are recalled, and based on them each possible decision is evaluated. The specific model we propose and axiomatize here evaluates each act by the sum, over all cases in which it was chosen, of the product of the similarity of the problem to the one at hand and the resulting utility. (Utility will be assumed scaled such that zero is a default value.)

Formally, a *case* is a triple (q, a, r), where q is a problem, a is an act, and r a result.[2] Let M, the memory, be a set of such cases. A decision-making agent is characterized by a *utility* function u, which assigns a numerical value to results, and a *similarity* function s, which assigns nonnegative values to pairs of problems. When faced with a new problem p, our agent would choose an act a which maximizes

$$(*) \qquad U(a) = U_{p,M}(a) = \sum_{(q,a,r) \in M} s(p, q) u(r),$$

where the summation over the empty set is taken to yield zero.

In CBDT, as in EUT, acts are ranked by weighted sums of utilities. Indeed, this formula so resembles that of expected utility theory that

---

[2]We implicitly assume that the description of a problem includes the specification of available acts. In particular, we do not address here the problem of identifying which acts are, indeed, available in a given problem, or identifying a decision problem in the first place.

one may suspect CBDT to be no more than EUT in a different guise. However, despite appearances, the two theories have little in common. First, note some mathematical differences between the formulae. In CBDT there is no reason for the coefficients $s(p, \cdot)$ to add up to 1 or to any other constant. More importantly, while in EUT every act is evaluated at *every* state, in CBDT each act is evaluated over a different set of cases. To be precise, if $a \neq b$, the set of elements of M summed over in $U(a)$ is disjoint from that corresponding to $U(b)$. In particular, this set may well be empty for some a's.

On a more conceptual level, in expected utility theory the set of states is assumed to be an exhaustive list of all possible scenarios. Each state "resolves all uncertainty," and, in particular, attaches a result to each available act. By contrast, in case-based decision theory the memory contains only those cases that actually happened. Each case provides information only about the act that was chosen in it, and the evaluation of this act is based on the actual outcome that resulted in this case. Hence, to apply EUT, one needs to engage in hypothetical reasoning, namely to consider all possible states and the outcome that would result from each act in each state. To apply CBDT, no hypothetical reasoning is required.

As opposed to expected utility theory, CBDT does not distinguish between "certain" and "uncertain" acts. In hindsight, an agent may observe that a particular act always resulted in the same outcome (i.e., that it seems to "involve no uncertainty"), or that it is uncertain in the sense that it resulted in different outcomes in similar problems. But the agent is not assumed to "know" a priori which acts involve uncertainty and which do not. Indeed, she is not assumed to know anything about the outside world, apart from past cases.

CBDT and EUT also differ in the way they treat new information and evolve over time. In EUT new information is modeled as an event, i.e., a subset of states, which has obtained. The model is restricted to this subset, and the probability is updated according to Bayes' rule. By contrast, in CBDT new information is modeled primarily by adding cases to memory. In the basic model, the similarity function calls for no update in the face of new information. Thus, EUT implicitly assumes that the agent was born with knowledge of and beliefs over all possible scenarios, and her learning consists of ruling out scenarios which are no longer possible. On the other hand, according to CBDT, the agent was born completely ignorant, and she learns by expanding her memory. (In the sequel we will also briefly discuss learning that is reflected in changes of the similarity judgments.) Roughly, an EUT agent learns by observing what cannot happen, whereas a CBDT agent learns by observing what can.

The framework of CBDT provides a natural way to formalize both the idea of frequentist belief formation (insofar as it is reflected in behavior) and the idea of satisficing. Although beliefs and probabilities do not explicitly exist in this model, in some cases they may be implicitly inferred from the number of summands in (∗). That is, if the decision-maker happens to choose the same act in many similar cases, the evaluation function (∗) may be interpreted as gathering statistical data, or as forming a frequentist prior. However, CBDT does not presuppose any a priori beliefs. Actual cases generate statistics, but no beliefs are assumed in the absence of data.

If an agent faces similar problems repeatedly, it is natural to evaluate an act by its average past performance, rather than by a mere summation as in (∗). Both decision criteria can be thought of as performing "implicit" induction: they are ways to learn from past cases which decision should be made in a new problem. A case-based decision-maker does not explicitly formulate "rules." She could never arrive at any "knowledge" regarding the future. (Indeed, this is also in line with Hume's teachings.) But she may come to behave as if she realized, or at least believed in certain regularities.

Case-based decisions may result in conservative or uncertainty-averse behavior. For example, if each act $a \in A$ only ever results in a particular outcome r_a, then the agent will only try new acts until she finds one that yields $u(r_a) > 0$. Thereafter, she will choose this act over and over again. She will be satisfied with the "reasonable" act a (so defined by $U(a) > 0$), and will not attempt to maximize her utility function u. Thus, CBDT has some common features with the notion of "satisficing" decisions of Simon (1957) and March and Simon (1958), and may be viewed as formalizing this idea. Specifically, the number zero on the utility scale may be interpreted as the agent's "aspiration level": so long as it is not reached, she keeps experimenting; once this level is obtained, she is satisfied.

Further discussion may prove more useful after a formal presentation of our model, axioms, and results. We devote Section 1.2 to this purpose. In Section 1.3 we discuss the model and its axiomatization. Further discussion, focusing on the comparison of CBDT to EUT, is relegated to Section 1.4. Section 1.5 presents some economic applications. In Section 1.6 we suggest some variations on the basic theme, and discuss avenues for further research.

1.2. The Model

Let P and A be finite and nonempty sets, of *problems* and of *acts*, respectively. To simplify notation, we will assume that all the acts A are available at all problems $p \in P$. It is straightforward to extend the model to deal with the more general case in which for each $p \in P$ there is a subset $A_P \subseteq A$ of available acts. Let R be a set of *outcomes* or *results*. For convenience, we include in R an outcome r_0 to be interpreted as "this act was not chosen." The set of *cases* is $C \equiv P \times A \times R$.

Given a subset of cases $M \subseteq C$, denote its projection on P by H. That is,

$$H = H(M) = \{q \in P | \exists a \in A, \ r \in R, \ \text{such that } (q, a, r) \in M\}.$$

H will be referred to as the *history* of problems.

A *memory* is a subset $M \subseteq C$ such that

(i) for every $q \in H(M)$ and $a \in A$, there exists a unique $r = r_M(q, a)$ such that $(q, a, r) \in M$;

(ii) for every $q \in H(M)$ there is a unique $a \in A$ for which $r_M(q, a) \neq r_0$.

A memory M may be viewed as a function, assigning results to pairs of the form (problem, act). For every memory M, and every $q \in H = H(M)$, there is one act that was actually chosen at q — with an outcome $r \neq r_0$ defined by the past case — and the other acts will be assigned r_0.

The definition of memory makes two implicit simplifying assumptions, which entail no loss of generality: first, we assume that no problem $p \in P$ may be encountered more than once. However, the fact that two formally distinct problems may be "practically identical" (as far as the agent is concerned) can be reflected in the similarity function. Second, we define memory to be a *set*, implying that the *order* in which cases appear in memory is immaterial. Yet, if the description of a problem is informative enough, for instance, if it includes a time parameter, a set is as informative as a sequence.

To simplify exposition, we will henceforth assume (explicitly) that $R = \mathscr{R}$ (the reals) and (implicitly) that it is already measured in "utiles." That is, our axioms should be interpreted as if R were scaled so that the "utility" function be the identity. Furthermore, we will assume that $r_0 = 0$. (See Section 1.3 for a discussion of these assumptions.) We do not distinguish between the actual outcome 0 and r_0. In particular, it is possible that for some $q \in H(M), r_M(q, a) = 0 = r_0$ for all $a \in A$.

Though by no means necessary, it may be helpful to visualize a memory, which is a function from $A \times H$ to $\mathscr{R}$, as a matrix. That is, choosing arbitrary orderings of A and of $H = H(M)$, a memory M can also be thought of as a $(k \times n)$-real-valued matrix, in which the $k \equiv |A|$ rows correspond to acts, and the $n \equiv |H|$ columns — to problems in H. In such a matrix every column contains at most one nonzero entry. Conversely, every $(k \times n)$-matrix which satisfies this condition corresponds to some memory M' with $H(M') = H$. Thus, every such matrix may be viewed as a *conceivable* memory, which may differ from the actual one in terms of the acts chosen at the various problems, as well as the results they yielded.

We assume that, when the agent has memory M and is confronted with problem p, she chooses an act in accordance with a preference relation $\geq_{p,M} \subseteq A \times A$. We further assume that the evaluation of an act is based only on the outcomes which resulted from the act. This assumption has two implications. First, for a given memory, each act may be identified with its "act profile," that is, with a vector in $\mathscr{R}^H$, specifying the results it yielded in past problems. Thus, a memory matrix M induces a preference order over k vectors in $\mathscr{R}^H$, namely, its rows.

Second, we require that the preference between two real-valued vectors not depend on the memory which contains them. Formally, for $x, y \in \mathcal{R}^H$, assume that M and M' are such that $H(M) = H(M') = H$, and that each of x and y corresponds to a row in the matrix M and to a row in the matrix M'. Then we require that $x \geq_{p,M} y$ iff $x \geq_{p,M'} y$.

Under these assumptions we can simply postulate a preference order $\geq_{p,H}$ on $\mathcal{R}^H$, which depends only on p and the observed problems $H (p \notin H)$. One interpretation of this preference order is that the agent can not only rank acts given their actual profile, but also provide preferences among hypothetical act profiles. (See a discussion of this point in the following section.)

However, we will not assume that $\geq_{p,H}$ is a complete order on $\mathcal{R}^H$. Consider two distinct act profiles $x, y \in \mathcal{R}^H$, assigning $x(q) \neq 0$ and $y(q) \neq 0$, respectively, to some $q \in H$. Naturally, these cannot be compared even hypothetically: for any memory M, at most one act may be chosen in problem q, and therefore at most one act may have a value different from 0 in its act profile for any given q. In other words, there is no memory matrix in which both x and y appear as rows. We therefore restrict the partial order $\geq_{p,H}$ to compare act profiles which are *compatible* in the sense that they could appear in the same memory matrix. Formally, given $x, y \in X$, let $x * y \in \mathcal{R}^H$ be defined as a coordinatewise product; i.e., $(x * y)(q) = x(q)y(q)$ for $q \in H$. Using this notation, two act profiles x, y are compatible if $x * y = 0$ or $x = y$.

Our first axiom states that compatibility is necessary and sufficient for comparability. Since compatibility is not a transitive relation, this axiom implies that neither is $\geq_{p,H}$.

A1. COMPARABILITY OF COMPATIBLE PROFILES. For every $p \in P$ and every history $H = H(M)$, for every $x, y \in \mathcal{R}^H$, x and y are compatible iff $x \geq_{p,H} y$ or $y \geq_{p,H} x$.

The following three axioms will guarantee the additively separable representation of $\geq_{p,H}$ on $\mathcal{R}^H$.

A2. MONOTONICITY. For every $p, H, x \geq y$ and $x * y = 0$ implies that $x \geq_{p,H} y$.

A3. CONTINUITY. For every p, H, and $x \in \mathcal{R}^H$, the sets $\{y \in \mathcal{R}^H | y \geq_{p,H} x\}$ and $\{y \in \mathcal{R}^H | x \geq_{p,H} y\}$ are closed (in the standard topology on $\mathcal{R}^H$).

A4. SEPARABILITY. For every p, H and $x, y, z, w \in \mathscr{R}^H$, if $(x + z) *$
$(y + w) = 0, x \geq_{p,H} y$, and $z \geq_{p,H} w$, then $(x + z) \geq_{p,H} (y + w)$.

A2 is a standard monotonicity axiom. It will turn out to imply that the similarity function is nonnegative. Without it one may obtain a numerical representation as in $(*)$ where the similarity function is not constrained in sign. A3 is a continuity axiom. It guarantees that, if $x_k \geq_{p,H} y$ and $x_k \rightarrow x$, then $x \geq_{p,H} y$ also holds (and similarly $x_k \leq_{p,H} y$ implies that $x \leq_{p,H} y$).

From a conceptual viewpoint, the separability axiom A4 is our main assumption. It states that preferences can be "added up." That is, if two act profiles, x and z, are (weakly) preferred to two others, y and w, respectively, then the sum of the former is (weakly) preferred to the sum of the latter, provided that such preferences are well defined. It is powerful enough to preclude cyclical strict preferences. Moreover, A4 will play a crucial role in showing that the numerical representation is additive across cases, as well as that the effect of each past case may be represented by the product of the utility of the result and the similarity of the problem.

We do not attempt to defend A4 as "universally reasonable." On the contrary, we readily agree that it may be too restrictive for some purposes.[3] For instance, one may certainly consider an additive functional with a case-dependent utility, as in theories of state-dependent expected utility theory, or a nonseparable functional. Alternatively, one may allow the similarity function to be modified according to the results that the agent has experienced. For the time being we merely offer an axiomatization of *a* case-based decision theory, which may be viewed as a "first approximation." The main role of the axioms above is not to convince the reader that our theory is reasonable. Rather, our main goal is to show that the theoretical concept of "similarity," combined with U-maximization, is in principle derivable from observed preferences.

The first result can finally be presented.

---

[3] Note, however, that A4 may appear very restrictive partly because of our simplifying assumption that results are represented by utiles.

Theorem 1:   *The following two statements are equivalent*:

(i) *A1–A4 hold*;
(ii) *For every $p \in P$ and every H there exists a function*

$$s_{p,H} : H \to \mathcal{R}_+$$

$$x \geq_{p,H} y \quad \text{iff} \quad \sum_{q \in H} s_{p,H}(q) x(q) \geq \sum_{q \in H} s_{p,H}(q) y(q)$$

for all compatible $x, y \in \mathcal{R}^H$.

Furthermore, in this case, for every p, H, the function $s_{p,H}$ is unique up to multiplication by a positive scalar.

Setting $s(p, q) = s_{p,H}(q)$, Theorem 1 gives rise to U-maximization for a given set of problems H. That is, considering the actual memory M the agent possesses at the time of decision p, she would choose an act that maximizes the formula $(*)$ with $s(p, q) = s_{p,H}(q)$ and $H = H(M)$. However, this similarity function may depend on the set of problems H. The next axiom ensures that the similarity measure is independent of memory. Specifically, A5 compares the relative importance of two problems, q_1 and q_2, in two histories, H^1 and H^2. It requires that the similarity weights assigned to these problems in the two histories be proportional.

A5. SIMILARITY INVARIANCE. For every $p, q_1, q_2 \in P$ and every two memories M^1, M^2 with $q_1, q_2 \in H^i \equiv H(M^i)(i = 1, 2)$ and $p \notin H^i (i = 1, 2)$, let v_j^i stand for the unit vector in $\mathcal{R}^{H^i} (i = 1, 2)$ corresponding to $q_j (j = 1, 2)$. (That is, v_j^i is a vector whose q_jth component is 1 and its other components are 0.) Then, denoting the symmetric part of $\geq_{p,H}$ by $\approx_{p,H}$,

$$x, y \in \mathcal{R}^{H^1}, \quad z, w \in \mathcal{R}^{H^2}, \quad x \approx_{p,H^1} y, \quad z \approx_{p,H^2} w$$

and

$$x + \alpha v_1^1 \approx_{p,H^1} y + \beta v_2^1$$

imply that

$$z + \alpha v_1^2 \approx_{p,H^2} w + \beta v_2^2$$

whenever the compared profiles are compatible.

Equipped with A5, one may define a single similarity function that represents preferences given any history.

Theorem 2:   *The following two statements are equivalent*:

(i) *A1–A5 hold.*
(ii) *There exists a function $s\colon P^2 \to [0,1]$ such that for all $p \in P$, every memory M with $p \notin H = H(M)$ and every compatible $x, y \in \mathscr{R}^H$,*

$$x \geq_{p,H} y \quad iff \quad \sum_{q \in H} s(p,q)x(q) \geq \sum_{q \in H} s(p,q)y(q).$$

Furthermore, in this case, for every p, the function $s(p, \cdot)$ is unique up to multiplication by a positive scalar.

Note that the decision rule axiomatized here is U-maximization as discussed in the introduction. The proofs of both theorems are given in Appendix 1.

1.3. Discussion

1.3.1. *The model*

Subjective Similarity. The similarity function in our model is derived from preferences, and is thus "subjective." That is, different individuals will typically have different preferences, which may give rise to different similarity functions, just as preferences give rise to subjective probability in the works of de Finetti (1937) and Savage (1954). Yet, for some applications one may wish to have a notion of "objective similarity," comparable to "objective probability."

Anscombe and Aumann (1963) define objective probability as a nickname for a subjective probability measure, which happens to be shared by all individuals involved. By a similar token, if the subjective similarity functions of all relevant agents happen to coincide, we might dub this common function objective similarity. Alternatively, one may argue that objectivity of a certain cognitive construct — such as probability or similarity — entails more than a mere (and perhaps coincidental) identity of its subjective counterpart across individuals. Indeed, some feel that objectivity requires some justification. Be that as it may, objective similarity is in particular also the subjective similarity of those individuals who accept it.

For purposes of objective similarity judgments, as well as for normative applications, our similarity function may be too permissive. For instance, we have not required it to be symmetric. One may wonder under what conditions can the similarity function $s(p, \cdot)$ of Theorem 2 be rescaled (separately for each p) so that $s(p, q) = s(q, p)$ for all $p, q \in P$. It turns out that a necessary and sufficient condition is that for all $p, q, r \in P$,

$$s(p, q)s(q, r)s(r, p) = s(p, r)s(r, q)s(q, p).$$

(Note that this condition does not depend on the choice of $s(p, \cdot)$, $s(q, \cdot)$, and $s(r, \cdot)$.)[4] However, in view of psychological evidence (Tversky, 1977), this can be unduly restrictive for a descriptive theory of subjective similarity.

Other Interpretations. In the development of CBDT we advance a certain cognitive interpretation of the functions u and s. However, the theory can also accommodate alternative, behaviorally equivalent interpretations. First, consider the function u. We assumed that it represents fixed preferences, and that memory may affect choices only by providing information about the u-value that certain acts yielded in the past. Alternatively, one may suggest that memory has a direct effect on preferences. According to this interpretation, the utility function is the aggregate U, while the function u describes the way in which U changes with experience. For instance, if the agent has a high aspiration level — corresponding to negative u values — she will like an option

---

[4]This condition can be translated to original data, namely, to observed preferences. Such a formulation will be more cumbersome without offering any theoretical advantage. Since the similarity functions are derived from preferences in an essentially unique manner, we may use them in the formulation of additional axioms without compromising the behavioral content of the latter. Indeed, A5 could also be more elegantly formulated in terms of the derived similarity functions.

less, the more she used it in the past, and will exhibit change-seeking behavior. On the other hand, a low aspiration level — positive u values — would make her "happier" with an option, the more she is familar with it, and would result in habit formation. In Gilboa and Schmeidler (1993) we develop a model of consumer choices based on this interpretation.

The function s can also have more than one cognitive interpretation. Specifically, when the agent is faced with a decision problem, she may not recall all relevant cases. The probability that a case be recalled may depend on its salience, the time that elapsed since it was encountered, and so forth. Thus, our function s should probably be viewed as reflecting both probability of recall and "intrinsic" similarity judgments.[5]

When "behavior" is understood to mean "revealed preference" (as opposed to, say, speech), one probably cannot hope to disentangle various cognitive interpretations based on behavioral data. Whereas specific applications may favor one interpretation over another, predictions of behavior would not depend on the cognitive interpretation chosen.

Hypothetical Cases. Consider the following example. An agent has to drive to the airport in one of two ways. When she gets there safely, she learns that the other road was closed for construction. A week later she is faced with the same problem. Regardless of her aspiration level, it seems obvious that she will choose the same road again. (Road constructions, at least in psychologically plausible models, never end.)

Thus, relevant cases may also be hypothetical, or counterfactual. ("If I had taken the other way, I would never have made it.") Hypothetical cases may endow a case-based decision-maker with reasoning abilities she would otherwise lack. It seems that any knowledge the agent possesses and any conclusions she deduces from it can, inasmuch

---

[5] Some readers expressed preference for the terms "relevance" or "weight" over "similarity." Others insisted that we should use "payoff" rather than "utility." We find these alternative terms completely acceptable.

as they are relevant to the decision at hand, be reflected by hypothetical cases.

Average Performance. The functional U gathers data in an additive way. For instance, assuming that all problems are equally similar, an act that was tried ten times with a u-value of 1 will be ranked higher than an act that was tried only once and resulted in a u-value of 5. One may therefore be interested in a decision rule that maximizes the following functional:

$$(**) \qquad V(a) = \sum_{(q,a,r)\in M} s'(p,q)u(r),$$

where

$$s'(p,q) = \begin{cases} \dfrac{s(p,q)}{\sum_{(q',a,r)\in M} s(p,q')} & \text{if well defined} \\ 0 & \text{otherwise} \end{cases}$$

and $s(p,q)$ is the similarity function of Section 1.2. According to this formula, for every act a the similarity coefficients $s'(p,q)$ add up to one (or to zero). Note that this similarity function depends not only on the problem encountered in the past, but also on the acts chosen at different problems.

Observe that V is discontinuous in the similarity values at zero. For example, if an act a was chosen in a single problem q and resulted in a very desirable outcome, it will have a high V-value as long as $s(p,q) > 0$ but will be considered a "new act," with zero V-value, if $s(p,q) = 0$. In the Bosnia example, for instance, V maximization may lead to different decisions depending on whether the Gulf War is considered to be "remotely relevant" or "completely irrelevant."[6] By contrast, the functional U is continuous in the similarity values.

Of special interest is the case where $s(p,q) = 1$ for all $p, q \in P$. In this case, V is simply the average utility of each act. The condition

---

[6]Observe, however, that discontinuity can only occur if all past cases are at most remotely relevant.

$s(\cdot,\cdot) = 1$ means that, at least as far as the agent's preferences reveal, all problems are basically identical. In this case, this variant of case-based decision theory is equivalent to "frequentist expected utility theory": the agent chooses an act with maximal "expected" utility, where the outcome distribution for each act is assumed to be given by the observed frequencies. (Note also that in this particular model the discontinuity at $s(\cdot,\cdot) \equiv 0$ does not pose a problem, since $s(\cdot,\cdot) \equiv 1$.) In Appendix 2 we provide an axiomatization of V-maximization.

The Definition of Acts. Case-based decision-makers may appear to be extremely conservative and boring creatures: once an act achieves their aspiration level, they stick to it. Our agent, it would seem, is an animal that always eats the same food at the same place, chooses the same form of entertainment (if at all), and so forth.

Although this is true at some level of description, it does not have to be literally true. For instance, the act that is chosen over and over again need not be "Have lunch at X"; it may also be "Have lunch at a place I did not visit this week." Repetition at this level of description will obviously generate an extremely diverse lunch pattern.

1.3.2. *The axiomatization*

Observability of Preferences and Hypothetical Questions. Whenever we encounter our agent, she has a certain memory M and can only exhibit preferences complying with $\geq_{p,M}$. It is therefore natural to ask, in what sense is the relation $\geq_{p,H}$ observable?

An experimenter may try to access the agent's preferences for different memories by confronting her with (i) counterfactual choices among acts, or (ii) actual choices among "strategies." In the first case, the agent may be asked to rank acts not only based on their *actual* act profiles, but also based on act profiles they *may have had*. Thus, she may be asked, "Assume act a yielded r in problem q. Would you still prefer it to act b?" In the second case, the agent may only be given the set of problems H, and then be asked to choose a strategy, that is, to make her choice contingent upon the act profiles which were not revealed to her.

In both procedures, one may distinguish between two levels of hypothetical (or conditional) questions. Suppose that the agent prefers act a to b, and consider the following types of questions.

I. Remember the case $c = (p, a, r)$, where you chose a and got r? Well, assume that the outcome were t instead of r. Would you still prefer a to b?

II. Remember the case $c = (p, a, r)$ where you chose a and got r? Well, now imagine you actually chose another act a' and received t. Would you still prefer a to b? How about a' to b?

In Section 1.2 we implicitly assumed that questions of both types can be meaningfully answered. Yet one may argue that questions of type II are too hypothetical to serve as foundations of any behavioral decision theory. While the agent has no control over the outcome r, she may insist that in problem p she would never have tried act a' and that the preference question is meaningless.

Appendix 2 presents a model in which answers to questions only of the first type are assumed. We provide an axiomatic derivation of the linear evaluation functional with a similarity function which, unlike that in Theorem 1, depends not only on the problems encountered, H, but also on the actions that were chosen in each. This more general functional form allows us to axiomatize V-maximization as a special case.

Derivation of Utility. The axiomatization provided here presupposes that the set of results is $\mathscr{R}$, and that results are measured in utiles, namely, that the utility function is linear. Thus, our axiomatic derivation of the notion of similarity and the CBDT functional relies on a supposedly given notion of utility, in a manner that parallels de Finetti's (1937) axiomatization of subjective probability together with expected utility maximization. Needless to say, the concept of a utility function is also a theoretical construct that calls for an axiomatic derivation from observable data. Ideally, one would like to start out with a model that presupposes neither similarity nor utility, and to derive them simultaneously, in conjunction with the CBDT decision rule. Such a derivation would also highlight the fact that the utility function, like the similarity function in Theorem 1, may, in general, depend on p, H. However, to keep the axiomatization simple, we do not follow this track here.

1.4. CBDT and EUT

Complementary Theories. We do not consider case-based decision theory "better" than or as a substitute for expected utility theory. Rather, we view them as complementary theories. The classical derivation of EUT, as well as the derivation of CBDT in this paper, are behavioral in that the theoretical constructs in these models are induced by observable (in principle) choices. Yet the scope of applicability of these theories may be more accurately delineated if we attempt to judge the psychological plausibility of the various constructs. Two related criteria for classification of decision problems may be relevant. One is the problem's description; the second is its relative novelty.

If a problem is formulated in terms of probabilities, for instance, EUT is certainly a natural choice for analysis and prediction. Similarly, when states of the world are naturally defined, it is likely that they would be used in the decision-maker's reasoning process, even if a (single, additive) prior cannot be easily formed. However, when neither probabilities nor states of the world are salient (or easily accessible) features of the problem, CBDT may be more plausible than EUT.

We may thus refine Knight's (1921) distinction between risk and uncertainty by introducing a third category of "ignorance": risk refers to situations where probabilities are given; uncertainty to situations in which states are naturally defined, or can be simply constructed, but probabilities are not. Finally, decision under ignorance refers to decision problems for which states are neither (i) "naturally given" in the problem nor (ii) can they be easily constructed by the decision-maker. EUT is appropriate for decision-making under risk. In the face of uncertainty (and in the absence of a subjective prior) one may still use those generalizations of EUT that were developed to deal with this problem specifically, such as nonadditive probabilities (Schmeidler, 1989) and multiple-priors (Bewley, 1986; Gilboa and Schmeidler,

1989). However, in cases of ignorance, CBDT is a viable alternative to the EUT paradigm.

Classifying problems based on their novelty, one may consider three categories. We suggest that CBDT is useful at the extremes of the novelty scale, and EUT in the middle. When a problem is repeated frequently enough, such as whether to stop at a red traffic light, the decision becomes almost automated and "rule-based." Such decisions may be viewed as a special type of case-based decisions. Indeed, a rule can be thought of as a summary of many cases, from which it was probably derived in the first place.[7] When deliberation is required, but the problem is familiar, such as whether to buy insurance, it can be analyzed "in isolation": its own history suffices for the formulation of states of the world and perhaps even a prior, and EUT (or some generalization thereof) may be cognitively plausible. Finally, if the problem is unfamiliar, such as whether to get married or to invest in a politically unstable country, it needs to be analyzed in a context- or memory-dependent fashion, and CBDT is again a more accurate description of the way decisions are made.

Reduction of Theories. While CBDT may be a more natural framework in which to model satisficing behavior, EUT can be used to explain this behavior as well. For instance, the Bayesian-optimal solution to the famous "multi-armed bandit" problem (Gittins, 1979) may not ever attempt to choose certain options. In fact, it is probably possible to provide an EUT account of any application in which CBDT can be used, by using a rich enough state space and an elaborate enough prior on it. Conversely, one may also "simulate" an expected utility maximizer by a case-based decision-maker whose memory contains a sufficiently rich set of hypothetical cases: given a set of states of the world Ω and a set of consequences R, let the set of acts be $A = R^\Omega = \{a: \Omega \to R\}$. Assume that the agent has a utility function $u: R \to \mathscr{R}$ and a probability measure μ on Ω. (Where Ω is a measurable space. For simplicity, it may be assumed finite.) The corresponding case-based decision-maker would have a hypothetical case for each pair of a state

---

[7] See the discussion of "ossified cases" in Riesbeck and Schank (1989) and of induction and rules in Gilboa and Schmeidler (1993b).

of the world ω and an act a:

$$M = \{((\omega, a), a, a(\omega)) | \omega \in \Omega, \ a \in A\}.$$

Letting the similarity of the problem at hand to the problem (ω, a) be $\mu(\omega)$, U-maximization reduces to expected utility maximization. (Naturally, if Ω or R are infinite, one would have to extend CBDT to deal with an infinite memory.) Furthermore, Bayes' update of the probability measure may also be reflected in the similarity function: a problem whose description indicates that an event $B \subseteq \Omega$ has occurred should be set similar to degree zero to any hypothetical problem (ω, a), where $\omega \notin B$.

Since one can mathematically embed CBDT in EUT and vice versa, it is probably impossible to choose between the two on the basis of predicted observable behavior.[8] Each is a refutable theory *given a description of a decision problem*, where its axioms set the conditions for refutation. But in most applications there is enough freedom in the definition of states or cases, probability or similarity, for each theory to account for the data. Moreover, a problem that is formulated in terms of states has many potential translations to the language of cases and vice versa. It is therefore hard to imagine a clear-cut test that will select the "correct" theory.

To a large extent, EUT and CBDT are not competing theories; they are different *languages*, in which specific theories are formulated. Rather than asking which one of them is more accurate, we should ask which one is more convenient. The two languages are equally powerful in terms of the range of phenomena they can describe. But for each phenomenon, they will not necessarily be equally intuitive. Furthermore, the specific theories we develop in these languages need not provide the same predictions given the same observations. Hence we believe that there is room for both languages.

Asymptotic Behavior. One may wonder whether, when the same problem is repeated over and over again, CBDT would converge to the choice prescribed by EUT for the one-shot problem with known

---

[8] Matsui (1993) formally proves an equivalence result between EUT and CBDT. His construction does not resort to hypothetical cases.

probabilities. This does not appear to be the case if we take CBDT to mean either U- or V-maximization.

Consider the following setup: $A = \{a, b\}, s(\cdot, \cdot) \equiv 1$. Assume that nature chooses the outcomes for each act by given distributions in an independent fashion. That is, there are two random variables R_a, R_b such that whenever the agent chooses $a(b)$, the outcome is chosen according to a realization of $R_a(R_b)$, independently of past choices and realizations. Further, assume the following distributions:

$$R_a = \begin{cases} 1 & 0.6 \\ -1 & 0.4 \end{cases}; \quad R_b = \begin{cases} 100 & 0.7 \\ -2 & 0.3 \end{cases}.$$

First consider a U-maximizer agent. At the beginning, both a and b have identical (empty) histories, and the decision is arbitrary. Suppose that the agent chooses a with probability .5, and that R_a results in $+1$. From then on she will choose a as long as the random walk generated by these choices is positive. Hence there is a positive probability that she will always choose a. Next consider a V-maximizer agent. Suppose that she first chose b, and that it resulted in the outcome -2. From that stage on this agent will always choose a.

Thus, for both decision rules we find that there is a positive probability that the agent will not maximize the "real" expected utility even in cases where objective probabilities are defined. Arguably, it is in these cases that EUT is most appealing. However, in Gilboa and Schmeidler (forthcoming) we show that, if the aspiration level is adjusted in an appropriate manner over time, U-maximization will converge to expected u maximization in the long run.[9]

Hypothetical Reasoning. Judging the cognitive plausibility of EUT and CBDT, one notes a crucial difference between them: CBDT, as opposed to EUT, does not require the decision-maker to think in hypothetical or counterfactual terms. In EUT, whether explicitly or implicitly, the decision-maker considers states of the world and reasons in propositions of the form, "If the state of the world were ω

---

[9] It was pointed out to us by Avraham Beja that, should one adapt our model to derive the utility function u axiomatically, the latter may or may not coincide with von-Neumann-Morgenstern utility function derived from choices among lotteries.

and I chose a, then r would result." In CBDT no such hypothetical reasoning is assumed.

Similarly, there is a difference between EUT and CBDT in terms of the informational requirements they entail regarding the utility function: to "implement" EUT, one needs to know the utility function u, i.e., its values for any consequence that may result from any act. For CBDT, on the other hand, it suffices to know the u-values of those outcomes that were actually experienced.

The reader will recall, however, that our axiomatic derivation of CBDT involved preferences among hypothetical act profiles. It might appear therefore that CBDT is no less dependent on hypothetical reasoning than EUT. But this conclusion would be misleading. First, one has to distinguish between elicitation of parameters by an outside observer, and application of the theory by the agent herself. While the elicitation of parameters such as the agent's similarity function may involve hypothetical questions, a decision-maker who knows her own tastes and similarity judgments need not engage in any hypothetical reasoning in order to apply CBDT. By contrast, hypothetical questions are intrinsic to the application of EUT.

Second, when states of the world are not naturally given, the elicitation of beliefs for EUT also involves inherently hypothetical questions. Classical EUT maintains that no loss of generality is involved in assuming that the states of the world are known, since one may always define the states of the world to be all the functions from available acts to conceivable outcomes. This view is theoretically very appealing, but it undermines the supposedly behavioral foundations of Savage's model. In such a construction, the set of "conceivable acts" one obtains is much larger than the set of acts from which the agent can actually choose. Specifically, let there be given a set of acts A and a set of outcomes X. The states of the world are X^A, i.e., the functions from acts to outcomes. The set of conceivable acts will be $\bar{A} = X^{(X^A)}$, that is, all functions from states of the world to outcomes. Hence the cardinality of the set of conceivable acts $\bar{A}$ is by two orders of magnitude larger than that of the actual ones A. Yet, using a model such as Savage's, one needs to assume a (complete) preference order on $\bar{A}$, while *in principle* preferences can be observed only between elements of A.

Differently put, such a "canonical construction" of the states of the world gives rise to preferences that are intrinsically hypothetical, and is a far cry from the behavioral foundations of Savage's original model.

In summary, in these problems both EUT and CBDT rely on hypothetical questions or on "contingency plans" for elicitation of parameters. The Savage questionnaire to elicit EUT parameters will typically involve a much larger and less intuitive set of acts than the corresponding one for CBDT. Furthermore, when it comes to application of the theory, CBDT clearly requires less hypothetical reasoning than EUT.

Cognitive and Behavioral Validity. CBDT may reflect the way people think about certain decision problems better than EUT. But many economists would argue that we should not care how agents think, as long as we know how they behave. Moreover, they would say, Savage's behavioral axioms are very reasonable; thus, it is very reasonable that people would behave *as if* they were expected utility maximizers. However, we claim that behavioral axioms which appear plausible assuming the EUT models are not as convincing when this very model is unnatural. For instance, Savage's "sure-thing principle" (his axiom P2) is very compelling when acts are given as functions from states to outcomes. But in examples such as 2 and 3 in the Introduction, outcomes and states are not given, and it is not clear what all the implications of the sure-thing principle are. It may even be hard to come up with an example of acts that are actually available in these examples, and such that the sure-thing principle constrains preferences among them. It is therefore not at all obvious that actual behavior would follow this seemingly very compelling principle. More generally, the predictive validity of behavioral axioms is not divorced from the cognitive plausibility of the language in which they are formulated.

1.5. Applications

This section is devoted to economic applications of case-based decision theory. All we could hope to provide here are some sketchy illustrations, which certainly fall short of complete models. Our goal is merely to suggest that CBDT may be able to explain some phenomena in a simpler and more intuitive way than EUT.

1.5.1. *To buy or not to buy*

Consider the following example. A firm is about to introduce two new products $\{1, 2\}$ into a market. When product i is introduced, the consumers face a decision problem p_i, with two possible acts $\{a, b\}$, where b stands for buying the product and a for abstaining from purchase. A consumer's decision to buy product i, say, a cereal or a soup, implies consumption on a regular basis in given quantities. The consumers are familiar with product 0 of the same firm. Product 1 is similar to both products 0 and 2, but the latter are not similar to each other. Finally, each consumer will derive a positive utility level from each product consumed.

   In this case, the order in which the products are introduced may make a difference. If the firm introduces product 1 and then product 2, both will be purchased. However, if product 2 is introduced first, a consumer's memory contains nothing that resembles it at the time of decision. Thus, her choice between a and b will be arbitrary, and she may decide not to buy the product. As a result, we expect a lower aggregate demand for product 2 if it is introduced first than if it is introduced after product 1.

   While EUT-based models could also provide such behavioral predictions, we find it more plausible that consumer decisions are directly affected by perceived similarities. Indeed, advertising techniques often seem to exploit and even manipulate the consumer's similarity judgments.

1.5.2. *Reputation*

Case-based consumer decisions give rise to aspects of reputation quite naturally. Consider a model with two products and two firms. Assume that product 1 is produced only by firm A. Product 2 is new. It is produced by both firms A and B. Other things being equal, firm A will have an edge in market 2 if it satisfies consumers' expectations in market 1 (i.e., if $U(A) > 0$). Thus, one would expect successful firms to enter new markets even if the technology needed in them is completely different from that used in the traditional ones.

An EUT explanation of the role of reputation would typically involve consumers' beliefs about the firms' rationality, as well as beliefs about the firms' beliefs about consumers. CBDT makes much weaker rationality assumptions in explaining this phenomenon.

1.5.3. *Introductory offers*

Another phenomenon that is close in nature is the introduction of new products at discounted rates. Again, one may explain the optimality of such marketing policies with "fully rational" expected utility consumers. For instance, in the presence of experimentation cost or risk aversion, a fully rational consumer may tend to buy the product at the regular price after having bought it at the introductory (lower) price. Yet if consumers are case-based decision-makers, the formation of habits is a natural feature of the model.

1.6. Variations and Further Research

Memory-Dependent Similarity. In reality, similarity judgments may depend on the results obtained in past cases. For instance, the agent may realize that certain attributes of a problem are more or less important than she previously believed. In Gilboa and Schmeidler (1994b) we dub this phenomenon "second-order induction," and discuss the relationship between CBDT and the process of induction in more detail.

When similarity is memory-dependent, two assumptions of our model may be violated. First, the separability axiom A4 may fail to hold. Second, the assumption that acts are ranked based on their act profiles may also be too restrictive. Specifically, the choice among acts need not satisfy independence of irrelevant alternatives, and it therefore cannot be represented by a binary relation over act profiles. Thus, CBDT in its present form does not describe how agents learn the similarity function.

Similar Acts. In certain situations, an agent may have some information regarding an act without having tried it in the past. For instance, the agent may consider buying a house in a neighborhood where she has owned a house before. The experience she had with a different, but similar, act is likely to color her evaluation of the one now available to her. Furthermore, some acts may involve a numerical parameter, such as "Offer to sell at a price p." One would expect the evaluation of such acts to depend on past performance of similar acts with a slightly different value of the parameter.

These examples suggest the following generalization of CBDT: consider a similarity function over (problem, act) pairs; given a certain memory and a decision problem, every act is compared — in conjunction with the current problem — to *all* (problem, act) pairs in memory, and a similarity weighted utility value is computed for it. Maximization of such a function is axiomatized in Gilboa and Schmeidler (1994a).

Act Generation. It is often the case that the set of available acts is not naturally given and has to be constructed by the agent. CBDT as presented here is not designed to deal with these problems, and it may certainly benefit from insights into the process of "act generation." In particular, the vast literature on planning in artificial intelligence may prove relevant to modeling of decision-making under uncertainty.

Changing Utility. The framework used in Section 1.2, in which outcomes are identified with utility levels, is rather convenient to convey the main idea, but it may also be misleading: it entails the implicit assumption that the utility function does not depend on the memory M, on time (which may be implicit in M), and so forth.

There may be some interest in a more general model, where the utility is allowed to vary with memory. In particular, the utility scale may "shift" depending on one's experience. Recall that the utility is normalized so as to set $u(r_0) = 0$. As mentioned above, one may refer to this value as the aspiration level of the decision-maker. A shift of the utility function is therefore equivalent to a change in the agent's aspiration level.

Adopting this cognitive interpretation, it is indeed natural that the aspiration level be adjusted according to past achievements. In Gilboa and Schmeidler (1994a) we axiomatize a family of decision rules that allow the aspiration level to be a linear function of the outcomes experienced in the past. However, some applications may resort to nonlinear adjustment rules as well. (See, for instance, Gilboa and Schmeidler (forthcoming).) The axiomatic foundations of aspiration level adjustments therefore call for further research.

Normative Interpretation. Our focus in this paper is on CBDT as a descriptive theory, which, for certain applications, may be more successful than EUT. Yet, in some cases CBDT may also be a more useful normative theory. While we share the view that it is desirable and "more rational" to think about all possible scenarios and reason about them in a consistent way, we also hold that a normative theory should be practical. For instance, if the state space is huge, and the agent does not entertain probabilistic beliefs over it, telling her that she *ought* to have a prior may be of little help.

If we believe that, in a given problem, applying EUT is not a viable option, we might at least attempt to improve case-based decisions. For instance, one may try to change one's similarity function so that it be symmetric, ignore primacy and recency effects (i.e., resist the tendency to assign disproportionate similarity weights to the first and the most recent cases), and so forth. It might even be argued that it is more useful to train professionals (doctors, managers, etc.) to make efficient and probably less biased case-based decisions rather than to teach them expected utility theory. However, such claims and the research that is needed to support them are beyond the scope of this paper.

Welfare Implications. A cognitive interpretation of CBDT raises some welfare questions. Is a satisficed individual "happier" than an unsatisficed one? Should the former be treated as richer simply because she has a lower aspiration level? Should we strive to increase people's aspiration levels, thereby prodding them to perform better? Or should we lower expectations so that they are content? We do not dwell on these questions here, partly because we have no answers to offer.

Strategic Aspects. In a more general model, one may try to capture manipulations of the similarity function. In phenomena as diverse as advertising and legal procedures, people try to influence other peoples' perceived similarity of cases. Moreover, an agent may wish to expose other agents to information selectively, in a way that will bring about certain modes of behavior on their part.

Procedures of Recall. CBDT may greatly benefit from additional psychological insights into the structure of memory and from empirical findings regarding the recollection process. For instance, one may hypothesize that the satisficing nature of decision-making is revealed not only in a dynamic context, but also within each decision: rather than computing the U-value of *all* possible acts, the agent may stop at the first act which obtains a positive U-value. There are, however, several ways in which "first" could be defined. For example, the agent may ask herself, "When did I choose this act?" and only after the evaluation of a given act will the next one be considered. Alternatively, she may focus on the problem and ask, "When was I in a similar situation?" and as the cases are retrieved from memory one by one, the function U is updated for all acts — until one act exceeds the aspiration level. These two models induce different decision rules.

Similarly, insight may be gained from analyzing the structure of a "decision problem" and the corresponding structure of the similarity function in specific contexts. Some psychological studies relating to this problem are Gick and Holyoak (1980, 1983) and Falkenhainer, Forbus, and Gentner (1989).

Other Directions. The model we present here should be taken merely as a first approximation. Just as EUT encountered the "paradoxes" of Allais (1953) and of Ellsberg (1961), the linear functional we propose here is likely to be found too restrictive in similar examples. Correspondingly, almost every generalization of EUT may have a reasonable counterpart for CBDT.

The main goal of this paper was to explore the possibility of a formal, axiomatically based decision theory, using a less idealized and at times more realistic paradigm than EUT. We believe that case-based decision theory is such an alternative.

Appendix 1: Proof of Theorems

Regarding both theorems, the fact that the axioms are necessary for the desired representation is straightforward. Similarly, the uniqueness of the similarity functions is simple to verify. We therefore provide here only proofs of sufficiency, that is, that our axioms imply the numerical representations.

Proof of Theorem 1:
Fix p, H, and denote $\geq \cdot = \geq_{p,H}$. We also use the notation $X = \mathscr{R}^H$ and identify it with $\mathscr{R}^n$ for $n = |H|$. W.l.o.g. assume that $H \neq \varnothing$. First, note the following.

Observation If $\geq \cdot$ satisfies A1 and A4, then

(i) for all $x, y \in X$ with $x * y = 0, x \geq \cdot - y \Leftrightarrow -y \geq \cdot - x$;
(ii) for all $x, y, z, w \in X$ with $x * y = 0, (x + z) * (y + w) = 0$ and $z \approx \cdot w, x \geq \cdot y \Leftrightarrow (x + z) \geq \cdot (y + w)$.

Proof:

(i) Assume that $x \geq \cdot y$. Consider $z = w = -(x + y)$, and use A4 (where $z \geq \cdot w$ follows from A1).
(ii) Under the provisions of the claim, $z \geq \cdot w$, and A4 implies that $x \geq \cdot y \Rightarrow (x + z) \geq \cdot (y + w)$.
   As for the converse, define $z' = -z$ and $w' = -w$. By (i), $-z \geq \cdot - w$. Thus, A4 can be used again to conclude $x \geq \cdot y$.

We now turn to the proof of Theorem 1. Define $\geq' \subseteq X \times X$ by

$$x \geq' y \Leftrightarrow (x - y) \geq \cdot 0 \quad \text{for all } x, y \in \mathscr{R}^n = X.$$

We need several lemmata whose proofs are rather simple. For brevity's sake we merely indicate which axioms and lemmata are used in each, omitting the details:

Lemma 1: *For $x, y \in X$ with $x * y = 0, x \geq' y$ iff $x \geq \cdot y$* (*the observation above*).

Lemma 2: $\geq'$ *is complete, i.e., for all $x, y \in X, x \geq' y$, or $y \geq' x$* (*the definition of $\geq'$ and A1*).

Lemma 3: $\geq'$ *is transitive (the definition of $\geq'$ and A4).*

Lemma 4: $\geq'$ *is monotone, i.e., for all $x, y \in X, x \geq y$ implies that $x \geq' y$ (the definition of $\geq'$ and A2).*

Lemma 5: $\geq'$ *is continuous, i.e., for all $x \in X$ the sets $\{y \in X | y \geq' x\}$, $\{y \in X | x \geq' y\}$ are closed in $\mathcal{R}^n$ (the definition of $\geq'$ and A3). (In view of Lemma 2, this is equivalent to the sets $\{y \in X | y >' x\}, \{y \in X | x \geq' y\}$ being open.)*

Lemma 6: $\geq'$ *satisfies the following separability condition: for all $x, y, z \in X, x \geq' y$ if and only if $(x + z) \geq' (y + z)$ (the definition of $\geq'$).*

Lemma 7: $\geq'$ *satisfies the following condition: for all $x, y, z, w \in X$, if $x \geq' y$ and $z \geq' w$, then $(x + z) \geq' (y + w)$ (the definition of $\geq'$ and A4).*

Lemma 8:   *If $x \geq' y$, then $x \geq' (x + y)/2 \geq' y$ (Lemmata 2, 3, and 6).*

Lemma 9:   *If $x \geq' y$ and $\alpha \in (0, 1)$, then $x \geq' \alpha x + (1 - \alpha)y \geq' y$ (successive application of Lemma 8, in conjunction with Lemmata 3 and 5).*

Lemma 10:   *For every $x \in X$, the sets $\{y \in X | y \geq' x\}, \{y \in X | x \geq' y\}$, $\{y \in X | y >' x\}$, and $\{y \in X | x >' y\}$ are convex. (Lemmata 3 and 9).*

Lemma 11:   *Define $A = \{x \in X | x \geq' 0\}$ and $B = \{x \in X | 0 >' x\}$ (where, as above, 0 denotes the zero vector in $X = \mathcal{R}^n$). Then A is nonempty, closed and convex; B is open and convex; $A \cap B = \varnothing$; and $A \cup B = \mathcal{R}^n$ (Lemmata 2, 5, and 10).*

Lemma 12:   *If $B = \varnothing$, the function $s(\cdot) \equiv 0$ satisfies the representation condition. If $B \neq \varnothing$, there exist a nonzero linear functional $S : \mathcal{R}^n \to \mathcal{R}$ and a number $c \in \mathcal{R}$ such that*

$$S(x) \geq c \quad \text{for all } x \in A$$

$$S(x) < c \quad \text{for all } x \in B$$

(in view of Lemma 11, a standard separating-plane argument).

Lemma 13:   *In the case $B \neq \varnothing$, the constant c in Lemma 12 is zero, hence $S(x) \geq 0$ if and only if $x \in A$ ($c \leq 0$ follows from Lemma 2; $c \geq 0$ is a result of Lemma 7).*

By Lemmata 1 and 7, for every compatible $x, y \in X, x \geq \cdot y$ iff $(x - y) \geq' 0$, i.e., iff $(x - y) \in A$. If $B = \varnothing$, Lemma 12 concludes the proof. If $B \neq \varnothing$, the function $s : H \to \mathscr{R}$ defined by S satisfies the desired representation by Lemma 13. Furthermore, it is nonegative by Lemma 4.

Remark: Note that we have also proved that A1, A3, and A4 are necessary and sufficient for a numerical representation as in $(*)$, where the similarity function s is not restricted to be nonnegative.

Proof of Theorem 2:
Theorem 1 guarantees that for every $p \in P$ and $H \subseteq P$ with $p \notin H$ there exists a function $s_H(P, \cdot) \equiv s_{p,H}(\cdot) : H \to \mathscr{R}_+$ such that

$$x \geq_{p,H} y \quad \text{iff} \quad \sum_{q \in H} s_H(p, q) x(q) \geq \sum_{q \in H} s_H(p, q) y(q)$$

for every compatible $x, y \in \mathscr{R}^H$. We wish to show that for every $p \in P$ there is a *single* function $s(p, \cdot)$ satisfying the condition above for every history $H \subseteq P \backslash \{p\}$.

Theorem 1 also states that each of the functions $s_H(p, \cdot)$ is unique, but only up to a positive multiplicative scalar. Thus, it suffices to show that for every $p \in P$ there exists a function $s(p, \cdot)$, such that for every $H \subseteq P \backslash \{p\}$ there exists a coefficient $\lambda_{p,H} > 0$ such that

$$s(p, q) = \lambda_{p,H} s_H(p, q) \quad \text{for all } q \in H.$$

Fix a problem $p \in P$. We first define the function $s(p, \cdot)$, and will then show that there are coefficients $\lambda_{p,H} > 0$ as required. Set $H^0 = P \backslash \{p\}$. Since P is finite, so is H^0. We set

$$s(p, q) = s_{H^0}(p, q) \quad \text{for all } q \neq p.$$

(One may generalize our results to an infinite set of problems, out of which only finitely many may appear in any history. In this case one may not use a maximal finite $H \subseteq P \backslash \{p\}$. Yet the proof proceeds in a similar manner. The only major difference is that the resulting similarity function may not be bounded.)

We will now show that for every H, the function $s_H(p, \cdot)$ provided by Theorem 1 is proportional to $s(p, \cdot)$ on the intersection of their domains, namely H. Let there be given any nonempty $H \subseteq P\backslash\{p\}$. (The case $H = \varnothing$ is trivial.) It will be helpful to explicitly state two lemmata:

Lemma 1:   *For every $q \in H, s_H(p, q) = 0$ iff $s_{H^0}(p, q) = 0$.*

Proof: Use axiom A5 with $H^1 = H^0, H^2 = H, q_1 = q_2 = q, x = y = 0$ (in $\mathscr{R}^{H^0}$), $z = w = 0$ (in $\mathscr{R}^H$) and $\alpha \neq \beta = 0$.

Lemma 2:   *For every $q_1, q_2 \in H$, with $s_H(p, q_1) > 0$,*

$$\frac{s_H(p, q_2)}{s_H(p, q_1)} = \frac{s_{H^0}(p, q_2)}{s_{H^0}(p, q_1)}.$$

Proof: Use axiom A5 with $H^1 = H^0, H^2 = H, x = y = 0$ (in $\mathscr{R}^{H^0}$), $z = w = 0$ (in $\mathscr{R}^H$), $\alpha = s_{H^0}(p, q_2)$ and $\beta = s_{H^0}(p, q_1)$.

We now turn to define the coefficients $\lambda_{p,H}$. Distinguish between two cases.

Case 1: $s_H(p, \cdot) \equiv 0$.
In this case, by Lemma 1, $s(p, q) = S_{H^0}(p, q) = 0$ for every $q \in H$. Hence any $\lambda_{p,H} > 0$ will do.

Case 2: $s_H(p, q) > 0$ for some $q \in H$.
In this case, choose such a q, and define $\lambda_{p,H} = [s_{H^0}(p, q)/s_H(p, q)] > 0$. By Lemmata 1 and 2, $\lambda_{p,H}$ is well defined.

Furthermore, it satisfies

$$\lambda_{p,H} s_H(p, q) = s_{H^0}(p, q) = s(p, q) \quad \text{for all } q \in H.$$

This completes the proof of Theorem 2.

Appendix 2: An Alternative Model

In this appendix we outline the axiomatic derivation of case-based decision theory where the similarity function depends not only on the problems encounted in the past, but also (potentially) on the acts chosen in these problems. We also show that V-maximization can be axiomatically derived.

We assume that the sets P, A, R, C, and M are defined and interpreted as in Section 1.2. Given a problem $p \in P$ and memory $M \subseteq C$, we define the sets of (problem, act) pairs encountered, the set of problems encountered, and the set of acts chosen, respectively, to be

$$E = E(M) = \{(q, a) | \exists\, r \in R,\ (q, a, r) \in M\}$$
$$H = H(M) = \{q \in P | \exists\, a \in A,\ (q, a) \in E\}$$

and

$$B = B(M) = \{a \in A | \exists\, q \in P,\ (q, a) \in E\}.$$

For each $a \in B$, let H_a denote the set of problems in which a was chosen; i.e., $H_a = \{q \in H | (q, a) \in E\}$. Let F_a be the set of *hypothetical acts*, i.e., all the act profiles an actual act a could have had: $F_a = \{x | x \colon H_a \to \mathscr{R}\}$. (Again, we identify the set of outcomes R with the real line and implicitly assume that it is measured in utiles.) We assume that $|B| \geq 2$ and define $F = \cup_{a \in B} F_a$.

For every p, E (with $|B| \geq 2$) we assume that $\geq_{p,E}\ \subseteq F \times F$ is a binary relation satisfying the following axioms. For simplicity of notation, $\geq_{p,E}$ will also be denoted by $\geq \cdot$ whenever possible.

A1′. ORDER. $\geq \cdot$ is reflexive and transitive, and for every $a, b \in B, a \neq b, x \in F_a$, and $y \in F_b, x \geq \cdot y$ or $y \geq \cdot x$.

A2′. CONTINUITY AND COMPARABILITY. For every $a, b \in B, a \neq b$ and every $x \in F_a$, the sets $\{y \in F_b | y \geq \cdot x\}$ and $\{y \in F_b | x > \cdot y\}$ are nonempty and open (in F_b endowed with the standard topology).

A2′ entails a "continuity" requirement by stipulating that these sets be open; the fact that they are also assumed nonempty is an

Archimedian condition which guarantees that the similarity function will not vanish on H_a for any $a \in B$.

A3′.   MONOTONICITY. For every $a, b \in B, a \neq b, x, z \in F_a$, and $y \in F_b$, if $x \geq z$ then $z \geq \cdot y$ implies that $x \geq \cdot y$, and $y \geq \cdot x$ implies that $y \geq \cdot z$.

A4′.   SEPARABILITY. For every $a, b \in B, a \neq b, x, z \in F_a$, and $y, w \in F_b$, if $z \approx \cdot w$, then $x \geq \cdot y \Leftrightarrow (x + z) \geq \cdot (y + w)$.

Theorem A2.1:   *The following two statements are equivalent:*

(i) *for every p and $E, \geq \cdot = \geq_{p,E}$ satisfies A1′–A4′;*

(ii) *for every p and E there exists a function $s = s_{p,E} : H \to \mathcal{R}_+$ such that*

— *for all $a \in B, \Sigma_{q \in H_a} s(q) \geq 0$;*
   and
— *for all $a \neq b, x \in F_a, y \in F_b$*

$$x \geq_{p,E} y \Leftrightarrow \sum_{q \in H_a} s(q) x(q) \geq \sum_{q \in H_b} s(q) y(q).$$

Furthermore, in this case, for every p and E, the function $s = s_{p,E}$ is unique up to multiplication by a positive scalar.

Note that A1′ requires that $\geq \cdot$ be transitive, which implies that acts belonging to the same space F_a be comparable. However, if $|B| \geq 3$, one may start out by assuming that transitivity holds only if all pairs compared belong to different spaces, and then consider the transitive closure of the original relation.

Proof (Outline). Fix $a \in B$, and consider the restriction of $\geq \cdot$ to F_a. It is easy to see that on $F_a, \geq \cdot$ is complete (hence a weak order), continuous and monotone (in the weak sense, i.e., $x \geq z$ implies that $x \geq \cdot z$).

Finally, if $x, y, z \in F_a$, we get

$$x \geq \cdot y \quad \text{iff} \quad (x + z) \geq \cdot (y + z).$$

We therefore conclude that for every $a \in B$ there is a function $s_a: H_a \to \mathscr{R}_+$ such that for all $x, y \in F_a$,

$$x \geq \cdot y \Leftrightarrow \sum_{q \in H_a} s_a(q)x(q) \geq \sum_{q \in H_a} s_a(q)y(q).$$

Furthermore, by A2', $\geq \cdot$ is nontrivial on each F_a, hence for some $q \in H_a, s_a(q) > 0$.

Thus, we have a numerical, additively separable representation of $\geq \cdot$ on each F_a separately. To obtain a global representation, comparing act profiles of different acts, we need to "calibrate" the various similarity functions $\{s_a\}_{a \in A}$, each of which is unique up to a positive multiplicative scalar.

One natural way to perform this calibration is to compare "constant" act profiles. That is, let 1_a denote the element of F_a consisting of 1's only ($1_a(q) = 1$ for all $q \in H_a$). Fix $a \in B$, and for each $b \in B$ let δ_b satisfy

$$1_a \approx \cdot \delta_b 1_b.$$

Define $s: H \to \mathscr{R}_+$ by

$$s(q) = \frac{s_b(q)}{\delta_b \sum_{q' \in H_b} s_b(q')}$$

for all $q \in H_b$. Observe that s is proportional to s_b on H_b for each $b \in B$. Thus, the s-weighted utility represents $\geq \cdot$ on F_b. Furthermore, the calibration above guarantees that, if $x \in F_a$ and $y \in F_b$ are two constant act-profiles,

$$x \geq \cdot y \Leftrightarrow \sum_{q \in H_a} s_a(q)x(q) \geq \sum_{q \in H_b} s_b(q)y(q).$$

To see that this representation holds in general, one may find for each act profile $x \in F_a$ a constant act profile $\bar{x} \in F_a$ such that $x \approx \cdot \bar{x}$ and complete the proof using transitivity of $\geq \cdot$.

Finally, it is straightforward to verify that the axioms are also necessary, and that the similarity function is unique up to a positive multiplicative scalar.

To obtain the representation by the functional V in $(**)$, consider the following axiom.

A6. EXPERIENCE INVARIANCE. For all $a, b \in B, 1_a \approx \cdot 1_b$.

Without judging its reasonableness, we note that A6 means that the "quality" of the experience is all that matters, rather than its "quantity." Specifically, imposing A6 on top of A1'–A4' guarantees that in the construction of s above, $\delta_b = 1$ for all $b \in B$. Thus, if two acts always yielded the same result, they would be equivalent, regardless of the number of times each was chosen. Preferences satisfying A6 focus on the "average performance" of each act, disregarding any accumulated measures of performance. In other words, A1'–A4' and A6 are necessary and sufficient conditions on $\geq \cdot$ to be representable by a functional V as in $(**)$ for given p, E. (To obtain the V representation using a single similarity function $s(p, q)$ for all sets E, one needs to impose an additional axiom corresponding to A5.) Formally,

Corollary A2.2:   *The following two statements are equivalent*:

(i) *for every p and $E, \geq \cdot = \geq_{p,E}$ satisfies A1'–A4' and A6*;
(ii) *for every E there exists a function $s_E : H^c \times H \to \mathcal{R}_+$ such that for every $p \notin H$*:
  — *for all $a \in B, \Sigma_{q \in H_a} s_E(p, q) = 1$;*
   and
  — *for all $a \neq b, x \in F_a$, and $y \in F_b$,*

$$x \geq_{p,E} y \Leftrightarrow \sum_{q \in H_a} s_E(p, q) x(q) \geq \sum_{q \in H_b} s_E(p, q) y(q).$$

Furthermore, in this case, for every E, the function s_E is unique.

Note that the average performance as measured by V may still be a weighted average. One may further demand that $\geq \cdot$ satisfy the following axiom:

A7. CONSTANT SIMILARITY. For every $a, b \in B, a \neq b, q, q' \in H_a$, $x \in F_b$,

$$v_q \geq \cdot x \quad \text{iff} \quad v_{q'} \geq \cdot x,$$

where $v_q, v_{q'}$ stand for the corresponding unit vectors in F_a.

It is rather straightforward to show that A1′–A4′, A6 and A7 are necessary and sufficient conditions for $\geq$ · to be representable by a simple average. Specifically, if a preference order which is representable by V also satisfies A7, the intrinsic similarity function s in (∗∗) (before normalization) is constant, and the functional V reduces to the simple average utility each act has yielded in the past. Formally,

Corollary A2.3: *The following two statements are equivalent*:

(i) *for every p and* $E, \geq \cdot = \geq_{p,E}$ *satisfies A1′–A4′, A6 and A7*;
(ii) *for every p and E, for all* $a \neq b, x \in F_a, y \in F_b,$

$$x \geq_{p,E} y \Leftrightarrow \frac{\sum_{q \in H_a} x(q)}{|H_a|} \geq \frac{\sum_{q \in H_b} y(q)}{|H_b|}.$$

References

Allais, M. (1953), "Le Comportement de L'Homme Rationel devant le Risque: Critique des Postulates et Axioms de l'Ecole Americaine," *Econometrica*, **XXI**: 503–546.

Anscombe, F. J. and R. J. Aumann (1963), "A Definition of Subjective Probability," *The Annals of Mathematics and Statistics*, **XXXIV**: 199–205.

Bewley, T. (1986), "Knightian Decision Theory: Part I," published as "Knightian Decision Theory: Part I", *Decisions in Economics and Finance*, **25**: 79–110.

Camerer, C. and M. Weber (1992), "Recent Developments in Modeling Preferences: Uncertainty and Ambiguity," *Journal of Risk and Uncertainty*, **V**: 325–370.

Cross, J. G. (1983), *A Theory of Adaptive Economic Behavior*, New York: Cambridge University Press.

de Finetti, B. (1937), "La Prevision: Ses Lois Logiques, Ses Sources Subjectives," *Annales de l'Institute Henri Poincare*, **VII**: 1–68; translated by H. E. Kyburg in Kyburg and Smokler (1964).

Ellsberg, D. (1961), "Risk, Ambiguity, and the Savage Axioms," *Quarterly Journal of Economics*, **LXXV**: 643–669.

Falkenhainer, B., K. D. Forbus and D. Gentner (1989), "The Structure-Mapping Engine: Algorithmic Example," *Artificial Intelligence*, **XLI**: 1–63.

Gick, M. L. and K. J. Holyoak (1980), "Analogical Problem Solving," *Cognitive Psychology*, **XII**: 306–355.

Gick, M. L. and K. J. Holyoak (1983), "Schema Induction and Analogical Transfer," *Cognitive Psychology*, **XV**: 1–38.

Gilboa, I. and D. Schmeidler (1989), "Maxmin Expected Utility with a Non-Unique Prior," *Journal of Mathematical Economics*, **XVIII**: 141–153.

Gilboa, I. and D. Schmeidler (1993a), "Case-Based Consumer Theory," published as "Cumulative Utility Consumer Theory", *International Economic Review*, **38** (1997): 737–761.

Gilboa, I. and D. Schmeidler (1994a), "Act Similarity in Case-Based Decision Theory," published as *Economic Theory*, **9** (1997): 47–61. Reprinted as Chapter 2 of this volume.

Gilboa, I. and D. Schmeidler (1994b), "Case-Based Knowledge Representation," published as "Case-Based Knowledge and Introduction", *IEEE Transactions on Systems, Man, and Cybernetics*, **30** (2000): 85–95.

Gilboa, I. and D. Schmeidler (forthcoming), "Case-Based Optimization," *Games and Economic Behavior*, **15** (1996): 1–26.

Gittins, J. C. (1979), "Bandit Processes and Dynamic Allocation Indices," *Journal of the Royal Statistical Society, B*, **XLI**: 148–164.

Harless, D. and C. Camerer (1994), "The Utility of Generalized Expected Utility Theories," *Econometrica*, **LXII**: 1251–1289.

Hume, D., (1966), *Enquiry into the Human Understanding* (Oxford: Clarendon Press, 1748, 2nd ed.)

Keynes, J. M. (1921), *A Treatise on Probability*, London: Macmillian and Co.

Knight, F. H. (1921), *Risk, Uncertainty, and Profit*, Boston, New York: Houghton Mifflin.

Machina, M. (1987), "Choice Under Uncertainty: Problems Solved and Unsolved," *Economic Perspectives*, **I**: 121–154.

March, J. G. and H. A. Simon (1958), *Organizations*, New York: John Wiley and Sons.

Matsui, A. (1993), "Expected Utility Theory and Case-Based Reasoning," mimeo.

Ramsey, F. P. (1931), "Truth and Probability," in *The Foundations of Mathematics and Other Logical Essays*, London: Kegan Paul; and New York: Harcourt, Brace and Co.

Riesbeck, C. K. and R. C. Schank (1989), *Inside Case-Based Reasoning*, Hillsdale, NJ: Lawrence Erlbaum Associates.

Savage, L. J. (1954), *The Foundations of Statistics*, New York: John Wiley and Sons.

Schank, R. C. (1986), *Explanation Patterns: Understanding Mechanically and Creatively*, Hillsdale, NJ: Lawrence Erlbaum Associates.

Schmeidler, D. (1989), "Subjective Probability and Expected Utility without Additivity," *Econometrica*, **LVTI**: 571–587.

Selten, R. (1978), "The Chain-Store Paradox," *Theory and Decision*, **IX**: 127–158.

Simon, H. A. (1957), *Models of Man*, New York: John Wiley and Sons.

Tversky, A., "Features of Similarity," *Psychological Review*, **LXXXIV**: 327–352.

von Neumann, J. and O. Morgenstern (1944), *Theory of Games and Economic Behavior* Princeton, NJ: Princeton University Press.

Chapter 2

Act Similarity in Case-Based Decision Theory*

Itzhak Gilboa[†] and David Schmeidler[‡]

Reprinted from *Economic Theory*, *9* (1997): 47–61.

Case-Based Decision Theory (CBDT) postulates that decision making under uncertainty is based on analogies to past cases. In its original
version, it suggests that each of the available acts is ranked according
to its own performance in similar decision problems encountered in
the past.

The purpose of this paper is to extend CBDT to deal with cases in
which the evaluation of an act may also depend on past performance
of different, but similar acts. To this end we provide a behavioral
axiomatic definition of the similarity function over problem-act pairs
(and not over problem pairs alone, as in the original model).

We propose a model in which preferences are context-dependent.
For each conceivable history of outcomes (to be thought of as the
"context" of decision) there is a preference order over acts. If these
context-dependent preference relations satisfy our consistency-
across-contexts axioms, there is an essentially unique similarity

---

*We are grateful to Akihiko Matsui for the discussions that motivated this work. We also thank
Enriqueta Aragones, Roger Myerson, Zvika Neeman, Ariel Rubinstein, Peyton Young, and an
anonymous referee for their comments. Partial financial support from the Alfred Sloan Foundation
is gratefully acknowledged.
†KGSM-MEDS, Northwestern University, Leverone Hall, Evanston, IL 60208, USA.
‡Tel Aviv University and Ohio State University, Department of Economics, Arps Hall, 1945 N.
High St., Columbus, OH 43210, USA.

function that represents these preferences via the (generalized) CBDT functional.

JEL classification numbers: D80, D81, D83

2.1. Introduction

2.1.1. *Motivation*

"Case-Based Decision Theory" (CBDT) suggests that people making decisions under uncertainty tend to choose acts that performed well in similar decision situations in the past. More specifically, the theory in its original version assumes that a decision maker has "cases" in her memory, each of which is a triple (q, a, r), where q is a decision problem, a is the act chosen in it, and r is the result that was obtained. CBDT assumes a utility function over the set of results, u, and a similarity function over the set of problems, s, such that, given a memory (i.e., a set of cases) M and a decision problem p, each act a is evaluated according to the weighted sum

$$(*) \qquad U(a) = U_{p,M}(a) = \sum_{(q,a,r)\in M} s(p, q)u(r).$$

While it stands to reason that past performance of an act would affect the act's evaluation in current problems, it is not necessarily the case that past performance is the only relevant factor in the evaluation process. Specifically, an act's desirability may be affected by the performance of other, *similar* acts. For instance, suppose that a hypothetical decision maker, Ms. A, is looking for a house to buy. One of her options is to purchase a house in a neighborhood where she owned a house in the past. Ms. A hasn't lived in the same house she is now considering, yet it seems unavoidable that her past experience with a close, and probably similar house would influence her evaluation of the new one. Similarly, consider a decision maker, Mr. B, who tries to decide whether or not to buy a new product in the supermarket. He has never purchased this product in the past, but he has consumed similar products by the same producer. Again, we would expect Mr. B's decision to depend on his experience with similar acts.

More generally, a decision maker is often faced with new acts, that is, acts that her memory contains no information about their past performance. According to the CBDT formulation (Gilboa and Schmeidler, 1995), the valuation index attached to these acts is the default value zero. As in the examples above, this application of CBDT is not very plausible. Correspondingly, it may lead to counter-intuitive behavioral predictions. For instance, it would suggest that Ms. A will be as likely to buy the house in the neighborhood she knows as a house in a neighborhood she does not know. Similarly, Mr. B's decision will be predicted to buy the new product based on B's aspiration level alone, without distinguishing among products by the record of their producers.

Act similarity effects are especially pronounced in economic problems involving a continuous parameter. For instance, the decision whether or not to "Offer to sell at price p" for a specific value p, would likely be affected by the results of previous offers to sell with different but close values of p. Generally, if there are infinitely many acts available to the decision maker, it is always the case that most of them are "new" to her. However, she will typically infer something about these "new" acts from the performance of other acts she has actually tried. While a straightforward application of CBDT to economic models with an infinite set of acts may result in counterintuitive and unrealistic predictions, the introduction of act similarity may improve these predictions, Furthermore, some of the results obtained in Gilboa and Schmeidler (1993a) and Gilboa and Schmeidler (1993c) for a finite set of acts are likely to have natural extensions to cases with infinitely many acts, provided some notion of similarity over the latter.

The need for modeling act-similarity may sometimes be obviated by redefining "acts" and "problems." For instance, Ms. A's acts may be simply "To Buy" and "Not to Buy," where each possible purchase is modeled as a separate decision problem.

However, such a model is hardly very intuitive, especially when many acts are considered simultaneously. It is more natural to explicitly model a similarity function between acts. Moreover, in many cases the similarity function is most naturally defined on problem-act pairs. For example, "Driving on the left in New York" may be similar to "Driving on the right in London;" "Buying when the price is low," may be more similar to "Selling when the price is high" than to "Selling when the price is low," and so forth.

In short, we would like to have a model in which the similarity function s is defined on such pairs, and — again, given a memory M and a decision problem p, — each act a is evaluated according to the weighted sum

$$(\bullet) \qquad U(a) = U_{p,M}(a) = \sum_{(q,b,r)\in M} s((p,a),(q,b))u(r).$$

Observe that a case (q, b, r) in memory may be viewed as a pair $((q, b), r)$, where the problem-act pair (q, b) is a single entity, describing the circumstances from which the outcome r resulted. That is, when past cases are considered, the distinction between problems and acts is immaterial. Indeed, it may also be fuzzy in the decision maker's memory. By contrast, when evaluating currently available acts, this distinction is both clearer and more important: the "problem" refers to the given circumstances, which are not under the decision maker's control, whereas the various "acts" describe alternative choices.

2.1.2. *Axiomatization*

In this paper we provide an axiomatization of the decision rule $(\bullet)$. The axiomatization is different in nature from those typically found in the literature, as well as from the axiomatization of $(*)$ we present in Gilboa and Schmeidler (1995). In the latter, an act is identified with its own past history. That is, each act can be thought of as a function, associating an outcome with each problem in which this act was chosen. Thus the mathematical structure we use there resembles that of de Finetti (1937) and, to an extent, also that of Savage (1954) and Anscombe and Aumann (1963). Namely, the objects of choice are vectors of outcomes. True, in the case-based model any two distinct acts are defined on disjoint domains, whereas in the other models all acts have the same domain. Yet the spirit of the axioms in Gilboa and Schmeidler (1995) is similar to the classical models of expected utility theory.

By contrast, when one takes into account the effect of act similarity, one can no longer identify an act with its own history of outcomes. The latter does not summarize all the information relevant to the evaluation of the act. Rather, the outcome of any act in any problem may, a-priori, affect the evaluation of any other act in the current decision problem.

We therefore use a model in which preferences are context-dependent. For each conceivable history of outcomes (to be thought of as the "context" of decision) there is a preference order over acts. Our axioms relate these preference relations given different contexts. A decision maker whose preferences across contexts is "consistent" enough to satisfy our axioms can be ascribed an essentially-unique problem-act similarity function such that her preferences are repre-sented by ($\bullet$).

The axiomatization of ($\bullet$) may be viewed as a recipe for indirect elicitation of similarity by a direct elicitation of preferences between acts in hypothetical contexts. The decision maker facing a problem with a set of available (feasible) acts A_p has one 'true' memory $M = ((q_i, b_i, r_i))_{i=1}^n$. Our implicit assumption is that her considerations in selecting an act from A_p follow ($\bullet$). That is, in evaluating an act, say a, she assigns weights to consequences in her memory, the $(u(r_i))_{i=1}^n$, by their similarities to the act being evaluated, $(s((p,a),(q_i b_i)))_{i=1}^n$, correspondingly. When faced with a hypothetical memory $M' = ((q_i, b_i, r_i'))_{i=1}^n$ she ranks the alternatives using, we assume, *the same* similarities. Indeed, when using a hypothetical memory to elicit preferences, only the results are changed and the problems and acts of the "true" memory are left unchanged.

There is a restrictive but simplifying assumption in our axiomatiza-tion. The results are expressed in "utiles," and the utility function is not derived simultaneously with the similarity function. In Section 2.3 the possibility to generalize the present work in this and other directions in presented, together with additional remarks. In the next section we describe our formal model, axioms and results. The proofs are relegated to Section 2.4.

Before concluding this section, it should be mentioned that Matsui (1993) also provides an axiomatization of CBDT with problem-act similarity. His model allows the similarity function to depend on past cases' outcomes as well. The axiomatic derivation of Matsui presupposes a binary relation on functions that assign an outcome to triples of the form (past case, act, act). From a mathematical viewpoint, his model and results resemble those of Gilboa and Schmeidler (1995), but the objects of choice he uses are naturally more complex entities than the "act profiles" used in the latter. There appears to be no simple relationship between Matsui's results and those of the present paper. While both papers axiomatize similar functionals, they employ rather different frameworks.

2.2. The Axiomatization and the Theorem

We start with the model's primitives, following Gilboa and Schmeidler (1995).

Let P be a non-empty set *of problems*. Let A be a non-empty set of *acts*. For each $p \in P$ there is a non-empty subset $A_p \subseteq A$ of acts available at p. Let R be a set of *outcomes* or *results*. The set of *cases* is $C \equiv P \times A \times R$. A *memory* is a finite subset $M \subseteq C$. Without loss of generality we assume that for every memory M, if $m = (p, a, r)$, $m' = (p', a', r') \in M$ and $m \neq m'$ then $p \neq p'$. As mentioned in the introduction, we assume (explicitly) that $R = \Re$ and, implicitly, that the utility function is the identity.

To state our axioms the following notation will prove useful: given a memory M, denote its projection on $P \times A$ by E. That is,

$$E = E(M) = \{(q, a) | \exists\, r \in R,\, (q, a, r) \in M\}$$

is the set of problem-act pairs recalled. We will also use the projection of M (or of E) on P, denoted by H. That is,

$$H = H(M) = H(E) = \{q \in P | \exists\, a \in A,\, r \in R,\ \text{s.t.}\ (q, a, r) \notin M\}$$

is the set of problems recalled.

For every memory M and every problem $p \notin H$ we assume a preference relation over acts, $\geq_{p,M} \subseteq A_p \times A_p$. Our main result derives the numerical representation for a *given set* E and a given *problem*

$p \notin H$. Let us therefore assume that E and p are given. Every memory M with $E(M) = E$ may be identified with the results it associates with the problem-act pairs, i.e., with a function $x = x(M) \in \mathfrak{R}^E$. An element $x \in \mathfrak{R}^E$ specifies the history of results, or the *context* of the decision problem p. Denoting $n = |E|$, we abuse notation and identify $\mathfrak{R}^E$ with $\mathfrak{R}^n$. Thus a relation $\geq_{p,M}$ over A_p may be thought of as a relation $\geq_x$. Moreover, we will assume that $\geq_x$ is defined for every $x \in \mathfrak{R}^n$. We define $>_x$ and $\approx_x$ to be the asymmetric and symmetric parts of $\geq_x$, as usual.

We will use the following axioms:

A1 Order: For every $x \in \mathfrak{R}^n$, $\geq_x$ is complete and transitive on A_p.

A2 Continuity: For every $\{x_k\}_{k \geq 1} \subseteq \mathfrak{R}^n$ and $x \in \mathfrak{R}^n$, and every $a, b \in A_p$, if $x_k \to x$ (in the standard topology on $\mathfrak{R}^n$) and $a \geq_{x_k} b$ for all $k \geq 1$, then $a \geq_x b$.

A3 Additivity: For every $x, y \in \mathfrak{R}^n$ and every $a, b \in A_p$, if $a >_x b$ and $a \geq_y b$, then $a >_{x+y} b$.

A4 Neutrality: For every $a, b \in A_p$, $a \approx_0 b$, $(0 = (0, \ldots, 0) \in \mathfrak{R}^n)$.

Axiom 1 is probably the most standard of all. It simply requires that, given any conceivable context, the decision maker's preference relation over acts is a weak order. Axioms 2–4 are new in the sense that they are formulated in terms of contexts, rather than in terms of acts. However, at least Axioms 2 and 3 cannot fail to remind the reader of standard axioms in the classical approach: A2 requires that preferences be continuous *in the space of contexts*. A3 states that preferences be additive in this space. That is, if both contexts x and y suggest that a is preferred to b, and at least one of the preferences is strict, then also the "sum" context $x + y$ suggests that a is strictly preferred to b. The logic of this axiom is that a context may be thought of as "evidence" in favor of one act or another. Thus, if both x and y "lend support" to choosing a over b, then so should the "accumulated evidence" $x + y$. For instance, assume that $E = \{(q_1, c), (q_2, d)\}$ and that $x = (1, 1)$ and $y = (0, -1)$. Assume that $a >_x b$. Presumably, this is so because (p, a) is, on average, more similar to (q_1, c) and to

(q_2, d) than is (p, b). Further assume that $a >_y b$. This is probably due to the fact that (p, b) is more similar to (q_2, d) than (p, a) is, hence act d's failure colors the decision maker's impression of b more negatively than that of a. It seems natural that for the context $x + y = (1, 0)$ it will be true that $a >_{x+y} b$. Indeed, the act c, which (in the corresponding problem) was more similar to a succeeded, and that which was more similar to b resulted in a neutral outcome. Thus a is preferred to b.

Needless to say, A3 is one of the main axioms, and it carries most of the responsibility for the additive functional we end up with. Naturally, it cannot be any more plausible than the additive functional itself, and there are reasonable examples in which additivity fails. (See Gilboa and Schmeidler, 1993b.) However, the main role of the axiomatization here is to relate the theoretical construct, "problem-act similarity," to observable preferences. Hence we do not attempt to present A3 as a "canon of rationality." While we believe it is a sensible requirement in some cases, we concede it may fail in others.

The utility level 0 in Axiom 4 represents a neutral result. Intuitively, it makes the decision maker neither sad nor happy. We refer to this utility as the "aspiration level." If all problem-act pairs in the decision maker's memory resulted in this neutral utility level, she cannot use her memory to differentiate between available acts. Any act, whether close to or remote from previous acts, cannot be expected to perform better than any other act. (Note that if every act in memory resulted in the same positive utility, then one would prefer an act that is, on average, closer to past acts, to a more remote act. In this case strict preferences between some acts are expected.)

Axioms 1–4 are easily seen to be necessary for the functional form we would like to derive. By contrast, the next axiom we introduce is not. While the theorem we present is an equivalence theorem, it characterizes a more restricted class of preferences than the decision rule discussed in the introduction, namely those preferences satisfying Axiom 5 as well. This axiom should be viewed merely as a technical requirement. It states that preferences are "diverse" in the following sense: for any four acts, there is a conceivable context that would distinguish among all four of them.

A5 Diversity: For every distinct $a, b, c, d \in A_p$ there exists $x \in \mathfrak{R}^n$ such that $a >_x b >_x c >_x d$.

(Observe that A5 is trivially satisfied when $|A_p| < 4$.)

Note that, specifically, A5 rules out preferences according to which acts c and d are always "between" a and b. This may be particularly restrictive for some applications. For instance, consider acts that are linearly ordered, say, they are parameterized by quantity. In this case it may well be the case that "Sell 100 shares" is preferred to "Sell 300 shares," or vice versa — but that in both cases, "Sell 200 shares" is ranked in between the two. Yet (in the presence of at least four acts), this is precluded by A5. Therefore there is certainly room to study more general axiom systems. In the next section we discuss Axiom 5 in more detail and provide examples to show that Axioms 1–4 alone cannot guarantee the desired result.

Our main result can now be formulated.

Theorem: *Let there be given E and p as above. Then the following two statements are equivalent:*

(i) *$\{\geq_x\}_{x \in \mathfrak{R}^n}$ satisfy A1–A5;*
(ii) *There are vectors $\{s^a\}_{a \in A_p}, s^a \in \mathfrak{R}^n$, such that:*
for every $x \in \mathfrak{R}^n$ and every $a, b \in A_p$,

$$(**) \qquad a \geq_x b \quad \text{iff} \quad \sum_{i=1}^{n} s_i^a x_i \geq \sum_{i=1}^{n} s_i^b x_i,$$

and, for every distinct $a, b, c, d \in A_p$, the vectors $(s^a - s^b), (s^b - s^c)$ and $(s^c - s^d)$ are linearly independent.

Furthermore, in this case, if $|A_p| \geq 4$, the vectors $\{s^a\}_{a \in A_p}$ are unique in the following sense: if $\{s^a\}_{a \in A_p}$ and $\{\hat{s}^a\}_{a \in A_p}$ both satisfy $(**)$, then there are a scalar $\alpha > 0$ and a vector $w \in \mathfrak{R}^n$ such that for all $a \in A_p$, $\hat{s}^a = \alpha s^a + w$.

We remind the reader that $\mathfrak{R}^n$ is used as a proxy for $\mathfrak{R}^E$. Thus the vectors $\{s_{a \in A_p}^a\}$ provided by the theorem can also be thought of as functions from E to $\mathfrak{R}$. Furthermore, these can be viewed as defining similarity on problem-act pairs. Specifically, the theorem implies that under A1–A5, there exists a similarity function

$$s_E: \{(p, a) | p \notin H(E), \ a \notin A_p\} \times E \to \mathfrak{R}$$

defined by

$$s_E((p, a), (q, b)) = s^a((q, b)),$$

such that for every p, M with $E(M) = E$ and $p \notin H(M)$ the functional

$$U_{p,M}(a) = \sum_{(q,b,r) \in M} s_E((p, a), (q, b)) u(r)$$

represents $\geq_{p,M}$ on A_p.

Since the formulation in ($\bullet$) did not include explicitly the dependence of the similarity function on the pairs of problem-act recalled, E, the following question naturally arises: What additional condition (or "axiom") we need to impose so that the similarity between pairs (p, a) and $(q, b) \in E$ would be independent of E? The answer is deferred to the next section.

2.3. Remarks

Remark 1 (Aspiration level adjustment): The decision rule ($\bullet$) is generally not invariant under shifts of the utility function. The utility level zero has been interpreted as the "aspiration level," and different aspiration levels would lead to different choices. Moreover, a decision maker's aspiration level need not be constant over time. If we adopt a cognitive interpretation of this behaviorally-defined concept, it is plausible that the payoff people "count on" getting in a given problem depends on the payoff they experienced in the past. Correspondingly, the way in which aspiration levels are adjusted may have important behavioral implications. For instance, in Gilboa and Schmeidler (1993c) we show that an aspiration level adjustment rule that satisfies certain conditions will asymptotically lead to optimal choices in a repeated problem. More generally, there is a need for a theory of aspiration level adjustments.

We do not offer here any such theory. Yet our main theorem axiomatizes a class of decision rules that includes both ($\bullet$), and variations thereof allowing the aspiration level to change over time, as long as it is a linear function of the payoffs received in the past. Specifically, consider a functional

$$U_{p,M}(a) = \sum_{(q,b,r)\in M} s_E((p,a),(q,b))[u(r) - h(x)]$$

where $h(x)$ is linear, i.e., $h(x) = \sum_{i=1}^{n} \alpha_i x_i$ for some $\alpha_i \in \Re$, $(i = 1, \ldots, n)$, and x is the vector of payoff, as above. It is easily verified that in this case one can re-define the similarity function so that the above expression reduces to that axiomatized in the theorem. In other words, as long as we allow the similarity function to depend on E, the functional form we axiomatize is general enough to describe at least a certain class of aspiration level adjustment rules.

Remark 2 (Insufficiency of A1–4): As mentioned above, only A1–4 are necessary for the numerical representation we axiomatize. One may wonder whether we can make do without A5 altogether. We note here that this is not the case, i.e., that axioms 1–4 are not sufficient for the numerical representation ($\bullet$). We prove this by two counter-examples in the next section. The first is combinatorial in nature, and uses only four acts. The second is based on cardinality of the set of acts. Specifically, note that axioms A1–4 do not restrict the act set in terms of cardinality, topology, and so forth. Hence, as this example shows, A1–4 are compatible with a lexicographic ordering of a large set of acts, and therefore cannot suffice for a numerical representation.

Remark 3 (Memory independence): The condition guaranteeing that the similarity function is independent of memory will be expressed here in terms of the similarity function provided by the representation theorem. Alternatively, one may use the language of the original data, namely the preference orders $\geq_{p,M} \subseteq A_p \times A_p$. However, it may make the following condition even more cumbersome, with no obvious theoretical advantage.[1]

---

[1] Since the similarity function is almost uniquely determined by the main theorem, it is already translated to observable data. Note that the condition that follows does not depend on the specific choice of the similarity function provided by the theorem.

For every $p, q, q' \in P$, every $a, b, c, a', b', c' \in A_p$ and every E^1 and E^2 such that $(q, b), (q', b') \in E^1, E^2$, and $p \notin H(E^1), H(E^2)$, the following holds:

$$\frac{s_{E^1}((p, a'), (q', b')) - s_{E^1}((p, c'), (q', b'))}{s_{E^1}((p, a), (q, b)) - s_{E^1}((p, c), (q, b))}$$
$$= \frac{s_{E^2}((p, a'), (q', b')) - s_{E^2}((p, c'), (q', b'))}{s_{E^2}((p, a), (q, b)) - s_{E^2}((p, c), (q, b))}$$

whenever the denominators do not vanish. It is further assumed that if one denominator equals zero, so does the other. If the condition holds, then $s_{E^1} = s_{E^2}$.

The proof follows that of the corresponding result — Theorem 2 — in Gilboa and Schmeidler (1995).

Remark 4: For some purposes, one might be interested in a model with an infinite memory, and a functional that is a similarity-weighted integral over it. Since all our arguments are based on duality (or separation) results, the axiomatic derivation of such a similarity measure, together with the corresponding functional, is relatively straightforward.

Remark 5: From a theoretical point of view, the concept of "utility," as that of "similarity," should also be related to observable choices. Specifically, one would like to have an axiomatic derivation of both concepts simultaneously. One way to obtain such a derivation would be to first derive a notion of comparison of utility differences from choices among acts, and then to require that our additivity axiom hold with respect to these differences. Using a cardinal utility function that represents the difference comparisons as in Alt (1936), we may then proceed with our proof in the utility space. Although we do not provide such an axiomatization here, we point out a possible way to infer utility differences comparisons from act preference data.

Note that in the decision rule (•), the pair of similarity and utility functions (s, u) is equivalent to the pair $(-s, -u)$. Intuitively, observing a preference of act a over b given a history of act c, it is possible that c was a success, and a (in the current problem) is more similar to it (in the recalled problem) than b is, but also that c was a failure, and a is *less* similar to it than is b.

One may therefore assume a given preference order over the set of outcomes R. Using it, a qualitative similarity order on *pairs* of problem-act pairs is defined. One may then proceed to define a binary relation on pairs of results, to be interpreted as strength of preferences. Thus, $(r, r') \geq (t, t')$ if for all $a, b \in A_p$, and all M_1, M_2, M_3, M_4 with $E(M_i) = E, i = 1, \ldots, 4$: If M_1 and M_2 as well as M_3 and M_4 (viewed as functions from E to R) differ only on $(q, c), (\bar{q}, \bar{c}) \in E$; $(q, c, t), (\bar{q}, \bar{c}, s) \in M_1, (q, c, t'), (\bar{q}, \bar{c}, s') \in M_2$, $(q, c, r), (\bar{q}, \bar{c}, s) \in M_3$ and $(q, c, r'), (\bar{q}, \bar{c}, s') \in M_4$; (q, c) is more similar to (p, a) than to (p, b), and the outcome r is preferred to r'; and $a \approx_{M_1} b, a \approx_{M_2} b$ and $a \approx_{M_3} b$, then $a \leq_{M_4} b$. One then imposes a few axioms on this binary relation on $R \times R$ that guarantee a cardinal utility representation.

Remark 6: The axiomatization we propose here assumes that preference relations are given for memories that differ from the actual one in the results obtained (but not in the problem-act pairs encountered). It was pointed out to us by Peyton Young and by Roger Myerson that one may also provide an axiomatization of our decision rule assuming that preferences are given for memories that are derived from the actual one by replication of cases. That is, such memories would have a different number of cases than the actual one, but each of these cases will be identical to a real case. The formal structure of such a derivation follows closely the derivations of scoring rules by Young (1975) and by Myerson (1993).

Specifically, for a given (actual) memory M, let the set of conceivable memories be $\mathbf{N}^M$, where $\mathbf{N}$ stands for the non-negative integers. Summation of elements in $\mathbf{N}^M$, as well as multiplication by an integer, are interpreted pointwise. Let $\phi: \mathbf{N}^M \twoheadrightarrow A$ be a non-empty-valued correspondence. We interpret ϕ as the observable choice correspondence, selecting the "best" acts in A for every conceivable memory. We seek axioms on ϕ that will be equivalent to the existence of "weights" $w(c, a)$ for every case $c \in M$ and every act $a \in A$, such that, for every conceivable memory $\Theta \in N^M$,

$$(\diamond) \qquad \phi(\Theta) = \arg\max_{a \in A} \left[\sum_{c \in M} \Theta(c) w(c, a) \right].$$

That is, $w(c, a)$ is the weight assigned to case c in the evaluation of act a in the present problem. In our model, $w(c, a) = s((p, a), (q, b)) u(r)$ where $c = (q, b, r)$ and p is the problem at hand. However, $(\diamond)$ need not be interpreted as a product of a similarity and a utility function.

The Young-Myerson axioms would be:

Additivity: For all $\Theta, \Xi \in N^M$ with $\phi(\Theta) \cap \phi(\Xi) \neq \varnothing, \phi(\Theta + \Xi) \subseteq \phi(\Theta) \cap \phi(\Xi)$.

(Archimedean) Continuity: For all $\Theta, \Xi \in N^M$ there is a K such that for all $n \geq K$, $\phi(n\Theta + \Xi) \subseteq \phi(\Theta)$.

Diversity: For every non-empty $B \subseteq A$, there is a $\Theta \in N^M$ such that $\phi(\Theta) = B$.

This model has the conceptual advantage that all conceivable memories, for which preferences are assumed to be given, are obtained from the actual one by replicating actual cases. On the other hand, for the very same reason such a model cannot differentiate between the effects of past problems, of the acts chosen at them, and of the results that followed. In particular, one cannot obtain a separate axiomatic derivation of the similarity function and of the utility function; rather, only their product is observable.

2.4. Proofs

We split the proof into three parts: (i) implies (ii), the opposite direction, and the uniqueness of the representation in part (ii). Finally, we prove Remark 2.

Part 1: (i) implies (ii)

To outline the proof we will first state its main 3 lemmata.

Lemma 1:   *There are vectors $\{s^{ab}\}_{a,b \in A_p}$, $s^{ab} \in \Re^n$, such that*

(i) $X^{ab} \equiv \{x \in \Re^n | a \geq_x b\} = \{x \in \Re^n | s^{ab} \cdot x \geq 0\};$

(ii) $\Upsilon^{ab} \equiv \{x \in \mathfrak{R}^n | a >_x b\} = \{x \in \mathfrak{R}^n | s^{ab} \cdot x > 0\}$;
   In particular, $\Upsilon^{ab} = \varnothing$ *iff* $s^{ab} = 0$;
(iii) $-s^{ab} = s^{ba}$;
(iv) s^{ab} *satisfying* (i), *and* (ii), *is unique up to multiplication by a positive number.*

The proof uses Axioms 1–4 to show that the sets X^{ab} and Υ^{ab} are convex and a separation theorem can therefore be applied.

Lemma 2: *For every three acts,* $f, g, h \in A_p$, *and the corresponding vectors* s^{fg}, s^{gh}, s^{fh} *from Lemma 1 there are* $\alpha, \beta \geq 0$ *such that*:

(i) $\alpha s^{fg} + \beta s^{gh} = s^{fh}$.
(ii) *Moreover, if the acts are distinct, and* $|A_p| \geq 4$, *then* α, β *from* (i) *are unique.*

The proof of (i) uses the linear programming duality result. In the proof of (ii) Axiom 5 is used for the first time; it also is required for the proof of the following lemma.

Lemma 3: *Suppose that* $|A_p| \geq 4$, *then there are vectors* $\{s^{ab}\}_{a,b \in A_p}$, $s^{ab} \in \mathfrak{R}^n$ *such that*: (i)–(iii) *of Lemma 1 hold, and for any three acts,* $f, g, h \in A_p$, *the Jacobi identity* $s^{fg} + s^{gh} = s^{fh}$ *holds.*

Note that Lemma 3, unlike Lemma 2, guarantees the possibility to rescale *simultaneously all* the s^{ab}-s from Lemma 1 such that the Jacobi identity will hold on A_p.

Lemma 3 is proved by double induction; it is transfinite if A_p is uncountably infinite. Before proving the lemmata, we will show how Part (i) is derived from Lemma 3.

Proof of Part (i): First let us point out, omitting details, that for the case $|A_p| \geq 2$ Lemma 1 implies Part (i), and for the case $|A_p| = 3$, Lemma 2(i) implies Part (i). Hence we restrict attention to the case $|A_p| \geq 4$.

Choose an arbitrary act, say e in A_p. Define $s^e = 0$, and for any other act, a, define $s^a = s^{ae}$, where the s^{ae}-s are from Lemma 3.

Given $x \in \mathfrak{R}^n$ and $a, b \in A_p$ we have:

$$a \geq_x b \Leftrightarrow s^{ab} \cdot x \geq 0 \Leftrightarrow (s^{ae} + s^{eb}) \cdot x \geq 0 \Leftrightarrow (s^{ae} - s^{be}) \cdot x \geq 0$$

$$\Leftrightarrow s^a \cdot x - s^b \cdot x \geq 0 \Leftrightarrow s^a \cdot x \geq s^b \cdot x$$

The first implication follows from Lemma 1(i), the second from the Jacobi identity of Lemma 3, the third from Lemma 1(iii), and the fourth from the definition of the s^a-s. Hence, $(**)$ of the theorem has been proved.

Given four distinct acts $a, b, c, d \in A_p$, suppose by way of negation, that the vectors $(s^a - s^b), (s^b - s^c)$ and $(s^c - s^d)$ are linearly dependent. That is,

$$\alpha(s^a - s^b) + \beta(s^b - s^c) + \gamma(s^c - s^d) = 0,$$

where one coefficient is positive and at least another one is nonnegative. As an example set $\alpha < 0$, $\beta > 0$, $\gamma \geq 0$. Then,

$$s^a - s^b = \left(-\frac{\beta}{\alpha}\right)(s^b - s^c) + \left(-\frac{\gamma}{\alpha}\right)(s^c - s^d).$$

Here, applying $(**)$, for any $x \in \mathfrak{R}^n$: if $b >_x c$ and $c >_x d$, then $a >_x b$. This precludes the ranking $b >_x c >_x d >_x a$ (for any x), the existence of which is guaranteed by A5. Such a contradiction can be found for any of the remaining possible combinations of signs of the coefficients α, β and γ. $\diamond$

Proof of Lemma 1: We first state and prove four claims that are immediate implications of the axioms.

Claim 1:   *Reflection:* $a \geq_x b \Rightarrow b \geq_{-x} a$

Proof. Otherwise, by A1, $a >_{-x} b$. This, together with the antecedent of the claim and A3 contradict A4. $\diamond$

Claim 2:   *Reflection (continued):* $a >_x b \Rightarrow b >_{-x} a$

Proof. Otherwise Claim 1 is contradicted. $\diamond$

Claim 3:   *Additivity* (continued): $a >_x b$ and $a \geq_y b$ imply $a \geq_{x+y} b$.

Proof. Otherwise, by A1 and Claim 2, $a >_{-(x+y)} b$. Since $-(x+y) + y = x$, we get by A3; $a >_{-x} b$. In view of Claim 1, this contradicts $a \geq_x b$.

$\diamond$

Claim 4: *Homogeneity*: For $\lambda > 0$, $a \geq_x b$ iff $a \geq_{\lambda x} b$.

Proof. Prove first for λ an integer, then for λ a rational number and complete by A2.

Completion of the Proof of Lemma 1: Applying Axioms 1–4 and Claims 1–4 we get that X^{ab} and Υ^{ab} are convex cones with vertex at the origin. Also X^{ab} is closed, Υ^{ab} is open and $X^{ba} = -X^{ab}$, $\Upsilon^{ba} = -\Upsilon^{ab}$.

We will now construct the s^{ab}-s for all unordered pairs and singletons $\{\{a, b\}|a, b \in A_p\}$. If Υ^{ba}, the complement of X^{ab}, is empty, then define $s^{ab} = 0 = s^{ba}$. Clearly, (i), (ii), (iii), hold for this case. If Υ^{ba} is non-empty then the origin is a boundary point of X^{ab}. By a supporting hyperplane argument there is a vector $s^{ab} \neq 0$ so that (i) holds. Define $s^{ba} = -s^{ab}$; thus, (iii) holds. Obviously (ii) holds too. By Claims 1 and 2 (Reflection), conditions (i) and (ii) also hold for (b, a). (iv) is implied by (i).

$\diamond$

Proof of Lemma 2:

(i) Given three vectors s^{fg}, s^{gh}, s^{fh} define the following LP problem:

$$(P) \qquad \underset{x \in \Re^n}{\text{Min }} s^{fh} \cdot x$$
$$\text{s.t.} \quad s^{fg} \cdot x \geq 0$$
$$s^{gh} \cdot x \geq 0.$$

Consider also its dual problem (with a degenerate objective function),

$$(D) \qquad \underset{\alpha, \beta}{\text{Max }} 0\alpha + 0\beta (\equiv 0)$$
$$\text{s.t.} \quad \alpha s^{fg} + \beta s^{gh} = s^{fh}$$
$$\alpha, \beta \geq 0.$$

Transitivity of $\geq_x$ implies that (P) is bounded: every x that is feasible for (P) satisfies $f \geq_x g$ and $g \geq_x h$. Hence it also satisfies

$f \geq_x h$, which implies that $s^{fh} \cdot x \geq 0$. Since (P) is bounded, (D) is feasible. It follows that there are $\alpha, \beta \geq 0$ such that $\alpha s^{fg} + \beta s^{gh} = s^{fh}$.

$$\Diamond$$

(ii) Assume by way of negation, that for some distinct $f, g, h \in A_p$: $\alpha s^{fg} + \beta s^{gh} = s^{fh} = \gamma s^{fg} + \delta s^{gh}$, or $(\alpha - \gamma) s^{fg} = (\delta - \beta) s^{gh}$, where $(\alpha, \beta) \neq (\gamma, \delta)$. By Lemma 1, this implies that either $X^{fg} = X^{gh}$ or $X^{fg} = -X^{gh}$. The first case precludes the existence of x for which $f >_x h >_x g$ and the second case precludes the existence of x such that $f >_x g >_x h$. In both cases A5 is violated. $\Diamond$

Proof of Lemma 3: Assume that A_p is (well-) ordered. Let acts i, j and k be the first three acts in this order. Let s^{ik} be a vector from Lemma 1, and apply Lemma 2 to the three acts, i, j, k. Define $\hat{s}^{ij} = \alpha s^{ij}$ and $\hat{s}^{jk} = \beta s^{jk}$. (Recall that α and β are unique, given s^{ik}). Supplement these and later definitions by $s^{aa} = 0, \forall a \in A_p$, and by condition (iii) of Lemma 1. Clearly, the conclusions of Lemma 3 hold for all $f, g, h \in \{i, j, k\}$. For simplicity of presentation we rename the vectors, $\hat{s}^{ij}$ and $\hat{s}^{jk}$ by s^{ij} and s^{jk}, respectively.

Continue by induction where the induction hypothesis is stated as follows: Suppose that for some two acts, say c and d, with c preceding d, the vectors s^{ab} satisfying the conclusion of Lemma 3 have been defined for all pairs of acts a, b preceding d as well as for pairs of acts a, d and d, a with a preceding c. Suppose also that s^{dc} has not yet been defined.

If there is no such a d, then the proof of the Lemma has been completed. If there is no such a c, then the main induction step has been completed and we move to the act immediately following d in the order. If $c = i$, then we start the secondary induction step: To define s^{di} and s^{dj} we apply Lemma 2, as above, to i, d, j, using s^{ij} as given. Obviously the conclusions of the Lemma hold on the extended domain.

We present now the induction's main step. To define s^{dc} apply Lemma 2 to the three acts i, d, c with s^{ic} from the induction hypothesis. We get a uniquely defined s^{dc}, but we get also a new definition of s^{id}, to be denoted by λs^{id}. I.e., the Jacobi identity from Lemma 2 in the present notation is:

$$\lambda s^{id} + s^{dc} = s^{ic}$$

On the other hand, for any act preceding c, say a, we have by the induction hypothesis,

$$s^{id} + s^{da} = s^{ia}$$

Applying once again Lemma 2 to the three acts a, d, c we get,

$$\vartheta s^{ad} + \eta s^{dc} = s^{ac}.$$

Subtracting the last two equalities from the previous one we get,

$$(\lambda - 1)s^{id} + (1 - \eta)s^{dc} + (-\vartheta - 1)s^{ad} = s^{ic} - s^{ia} - s^{ac} = 0$$

The right side equality follows from the Jacobi identity for the three acts i, a, c by the induction hypothesis.

As a conclusion we have that either the vectors s^{id}, s^{dc} and s^{ad} are linearly dependent or all three coefficients are zero. In the latter case it has been shown that $\lambda = 1$ and adding the vector s^{ad} (and $s^{da} = -s^{ad}$ and $s^{dd} = 0$) preserves the Jacobi identity and other conclusions of Lemma 3 on the extended domain. I.e., the induction step has been proved in this case, and we can pass to the immediate follower of c (in the order), if it differs from d.

In the linear dependence case we get a contradiction to Axiom 5. Indeed, it implies that a vector in one of the pairs $\{\{s^{id}, s^{di}\}, \{s^{dc}, s^{cd}\},$ $\{s^{ad}, s^{da}\}\}$ can be represented as nonnegative linear combination of two vectors, one from each of the remaining pairs. Any such representation leads to a contradiction, as in the Proof of Part (i) above. For example, if s^{dc} is a nonnegative linear combination of s^{di} and s^{ad}, then for no $x : c >_x a >_x d >_x i$. ◇

Part 2: (ii) implies (i)

It is straightforward to verify that if $\{\geq_x\}_{x \in \Re^n}$ are representable by $\{s^a\}_{a \in A_p}$ as in (**), they have to satisfy Axioms 1–4. We will therefore only prove that this representation — coupled with the linear independence condition — imply Axiom 5.

Assume, then, that some quadruple of distinct acts $a, b, c, d \in A_p$ is given, and that the vectors $(s^a - s^b), (s^b - s^c)$ and $(s^c - s^d)$ are linearly independent. We will prove that there exists $x \in \Re^n$ such that $a >_x b >_x c >_x d$. Because of the linear independence, the following linear system has a solution:

$$(s^a - s^b) \cdot x = 1$$
$$(s^b - s^c) \cdot x = 1$$
$$(s^c - s^d) \cdot x = 1.$$

The desired order is implied by (∗∗).   ◇

Part 3: Uniqueness

Suppose that $|A_p| \geq 4$ and that $\{s^a\}_{a \in A_p}$, and $\{\hat{s}^a\}_{a \in A_p}$, both satisfy (∗∗), and we wish to show that there are a scalar $\alpha > 0$ and a vector $w \in \Re^n$ such that for all $a \in A_p$, $\hat{s}^a = \alpha s_a + w$. Recall that, for $a \neq b$, $s^a \neq s^b$ and $\hat{s}^a \neq \hat{s}^b$ by A5.

Choose $a \neq b\,(a, b \in A_p)$. From Lemma 1(iv) there exists a unique $\alpha > 0$ such that $(\hat{s}^a - \hat{s}^b) = \alpha(s^a - s^b)$. Define $w = \hat{s}^a - \alpha s^a$.

We now wish to show that, given $c \in A_p : \hat{s}^c = \alpha s^c + w$. Again, from Lemma 1(iv) there are unique $\gamma, \delta > 0$ such that

$$(\hat{s}^c - \hat{s}^a) = \gamma(s^c - s^a)$$

and

$$(\hat{s}^b - \hat{s}^c) = \delta(s^b - s^c).$$

Summing up these two with $(\hat{s}^a - \hat{s}^b) = \alpha(s^a - s^b)$, we get

$$\alpha(s^a - s^b) + \gamma(s^c - s^a) + \delta(s^b - s^c) = 0.$$

Thus

$$(\alpha - \delta)(s^a - s^b) + (\delta - \gamma)(s^a - s^c) = 0.$$

Since $(s^a - s^b)$ and $(s^b - s^c)$ are linearly independent (condition (ii) of the Theorem), so are $(s^a - s^b)$ and $(s^a - s^c)$, and we conclude that $\alpha = \gamma = \delta$. This implies $\hat{s}^c = \alpha(s^c - s^a) + \hat{s}^a = \alpha s^c + w$.   ◇

This completes the proof of the theorem.

Proof of Remark 2: We will now show that A1–4 are not sufficient for (•) to hold. As mentioned above, we provide two counter-examples. Each highlights a different aspect of A5.

Example 1: Let $A_p = \{a, b, c, d\}$ and $n = 3$. Define the vectors

$$s^{ab} = (-1, 1, 0) \quad s^{ac} = (0, -1, 1); \quad s^{ad} = (1, 0, -1);$$
$$s^{bc} = (2, -3, 1); \quad s^{cd} = (1, 2, -3); \quad s^{bd} = (3, -1, -2),$$

and $s^{uv} = -s^{vu}$ and $s^{uu} = 0$ for $u, v \in A_p$.

For $u, v \in A_p$ and $x \in \Re^3$ define, $u \geq_x v$ iff $s^{uv} \cdot x \geq 0$.

We claim that $\{\geq_x\}_{x \in \Re^3}$ satisfy A1–4 and yet they cannot be represented as in (•). With the exception of transitivity, A1–4 follow immediately from the definition of $\{\geq_x\}_{x \in \Re^3}$, and Lemma 1. Transitivity is verified triple by triple. We now show that the numerical representation (•) does not hold.

Assume by way of negation that there are vectors $\{s^u\}_{u \in A_p}$ representing $\{\geq_x\}_{x \in \Re^3}$. Define $\hat{s}^{uv} = s^u - s^v$. By Lemma 2, for every $u, v \in A_p$ there exists a coefficient $\lambda^{uv} > 0$ such that $\hat{s}^{uv} = \lambda^{uv} s^{uv}$. Without loss of generality we may assume $\lambda^{bc} = 1$. Since s^{bc} and s^{cd} are linearly independent, we also get $\lambda^{cd} = \lambda^{db} = 1$.

Considering the equation $\lambda^{ab} s^{ab} + \lambda^{ca} s^{ca} = s^{cb}$, observe that it has a unique solution with $\lambda^{ab} = 2$. (Where uniqueness follows from linear independence of s^{ab} and s^{ca}). Hence we have $\hat{s}^{ab} = (-2, 2, 0)$. By a similar token, the equation $\lambda^{da} s^{da} + \lambda^{ab} s^{ab} = s^{db}$ also has a unique solution in which $\lambda^{ab} = 1$. Thus we also have $\hat{s}^{ab} = (-1, 1, 0)$, a contradiction.

Example 2: Let $A_p = [0, 1]^2$ and let $\geq_l$, be the lexicographic order on it. For any given $n \geq 1$, define $\{\geq_x\}_{x \in \Re^n}$ as follows:

$$\text{if } \sum_{i=1}^{n} x_i > 0, \quad a >_x b \quad \text{iff } a >_l b;$$

$$\text{if } \sum_{i=1}^{n} x_i = 0, \quad a \approx_x b \quad \text{for all } a, b \in A_p;$$

and

$$\text{if } \sum_{i=1}^{n} x_i < 0, \quad b >_x a \quad \text{iff } a >_l b.$$

Thus, for every $x \in \Re^n$, $\geq_x$ is one of (i) $\geq_l$; (ii) $\geq_l^{-1}$; or (iii) the trivial relation (according to which any two acts are equivalent). Hence $\leq_x$ satisfies Al. It can also be verified that A2–A4 are satisfied by $\{\geq_x\}_{x \in \Re^n}$. However, one would not expect to obtain a representation as in (∗∗), since it would imply a numerical representation of $\geq_l$, as well.

We therefore conclude that A5, which is also used in our proof for the finite case, implicitly bounds the cardinality of the set of acts A_p. Specifically, $|A_p| \geq \aleph$ since there cannot be more than a continuum of independent vectors in $\Re^n$. ◇

References

Alt, F. (1936), On the measurability of utility. In: John S. Chipman, Leonid Hurwicz, Marcel K. Richter, Hugo F. Sonnenschein (eds.) Preferences, Utility, and demand, A. Minnesota Symposium, pp. 424–431, Chap. 20. New York, Chicago, San Francisco, Atlanta: Harcourt Brace Jovanovich, Inc., [Über die Messbarkeit des Nutzens. Zeitschrift fuer Nationaloekonomie 7, 161–169, Abstract (in English): On the Measurability of Utility. Translation by Siegfried Schach, revised and corrected by Dr. Alt]

de Finetti, B. (1937), "La Prevision: Ses Lois Logiques, Ses Sources Subjectives", *Annales De l'Institute Henri Poincare*, 7: 1–68.

Gilboa, I. and D. Schmeidler (1995), "Case-Based Decision Theory", *The Quarterly Journal of Economics*, **110**: 605–639. Reprinted as Chapter 1 of this volume.

Gilboa, I. and D. Schmeidler (1993a), *Case-Based Consumer Theory*, mimeo. Northwestern University, revised 1994. Published as "Cumulative Utility Consumer Theory", *International Economic Review*, **38** (1997): 737–761.

Gilboa, I and D. Schmeidler (1993b), *Case-Based Knowledge Representation*, mimeo. Northwestern University, revised 1994. Published as "Case-Based Knowledge and Introduction", *IEEE Transactions on Systems, Man, and Cybernetics*, **30** (2000): 85–95.

Gilboa, I. and D. Schmeidler (forthcoming), "Case-Based Optimization", *Games and Economic Behavior*, **15** (1996): 1–26.

March, J. G. and H. A. Simon (1958), Organizations. New York: Wiley.

Matsui, A. (1993), *Expected Utility Theory and Case-Based Reasoning*, mimeo. University of Pennsylvania.

Myerson, R. B. (forthcoming), Axiomatic Derivation of Scoring Rules Without the Ordering Assumption. *Social Choice and Welfare*, 1995, **12**(1), pp. 59–74.

Ramsey, F. P. (1964), Truth and Probability. In: *The Foundations of Mathematics and Other Logical Essays*. New York: Harcourt Brace Jovanovich, Inc. 1931. Reprinted in Kyburg and Smokier.

Savage, L. J. (1972), *The Foundations of Statistics*, 2nd edn. New York: John Wiley and Sons.

Simon, A. (1957), *Models of Man*. New York: Wiley.

Young, P. H. (1975), Social Choice Scoring Function. *SIAM Journal of Applied Mathematics*, **28**: 824–838.

Chapter 3

Cognitive Foundations of Probability

Itzhak Gilboa* and David Schmeidler[†]

Reprinted from *Mathematics of Operations Research, 27* (2002): 68–81.

Prediction is based on past cases. We assume that a predictor can rank eventualities according to their plausibility given any memory that consists of repetitions of past cases. In a companion paper, we show that under mild consistency requirements, these rankings can be represented by numerical functions, such that the function corresponding to each eventuality is linear in the number of case repetitions. In this paper we extend the analysis to rankings of events. Our main result is that a cancellation condition à la de Finetti implies that these functions are additive with respect to union of disjoint sets. If the set of past cases coincides with the set of possible eventualities, natural conditions are equivalent to ranking events by their empirical frequencies. More generally, our results may describe how individuals form probabilistic beliefs given cases that are only partially pertinent to the prediction problem at hand, and how this subjective measure of pertinence can be derived from likelihood rankings.

3.1. Introduction

The Bayesian approach holds that, when facing uncertainty, one should form a prior and, given new information, update it according to Bayes rule. It relies on sound axiomatic foundations: Ramsey (1931), de

---

*Berglas School of Economics, Tel Aviv University, 69978 Tel Aviv, Israel; igilboa@ post.tau.ac.il
[†]School of Mathematics, Tel Aviv University, 69978 Tel Aviv, Israel; schmeid@ post.tau.ac.il

Finetti (1937), and Savage (1954) argue that Bayesian expected utility maximization is the only normatively acceptable decision rule, and that in-principle-observable preferences can uniquely define a prior. Probabilities have also been axiomatically derived from qualitative plausibility judgments, where the latter are modeled as binary relations (see Kraft *et al.*, 1959; Krantz *et al.*, 1971; Fine, 1973; Fishburn, 1986) or as propositions (see Fagin *et al.*, 1990; Fagin and Halpern, 1994; Aumann, 1995; Heifetz and Mongin, 1999). (The idea of postulating qualitative probabilities as a basis for numeric probabilities originates with de Finetti and constitutes a main step in Savage's derivation of subjective probabilities.) But these axiomatic derivations do not explicitly model the information based on which a prior is formed. Further, they do not attempt to provide an account of this cognitive process. Thus, the axiomatizations of Bayesian beliefs and of Bayesian expected utility maximization may convince one that one would like to be Bayesian, but they do not provide the self-help tools that are needed to become a practising Bayesian.

The goal of this paper is to model explicitly the link between factual knowledge and derived beliefs. A special case of such a model should be a frequentist approach: When observing an experiment that is repeated under (seemingly) identical conditions, one may use past empirical frequencies as probabilities of future occurrences. But we also wish to discuss situations that are not repeated under precisely the same conditions. For instance, a physician encounters a new patient who is not identical to any past patient she has treated, but may be similar to some. How is she to form a prior probability over the outcome of various treatments?

In a companion paper (Gilboa and Schmeidler, 1999) we suggest and axiomatize a rule for ranking of possible eventualities: Each past case c and each possible eventuality x are assigned a number $v(x, c)$, and, given a collection of cases M, eventuality x is considered more likely than eventuality y if and only if

$$\sum_{c \in M} v(x, c) > \sum_{c \in M} v(y, c).$$

This rule can be viewed as a description of a cognitive process by which the predictor decides which of two eventualities is more likely, but it falls short of generating a prior over all possible events. It does not assume that the eventualities have any logical or algebraic structure. Further, the axiomatic derivation demands that, for every four eventualities, every permutation is a possible ranking (for an appropriately chosen M). This condition is counterintuitive if eventualities are endowed with some additional structure. Specifically, if y is logically derived from x, one cannot expect x to be strictly more likely than y.

In this paper we deal with the case in which the objects of the likelihood rankings are events in a given algebra. We first show how, in this set-up, one may appropriately weaken the axioms of Gilboa and Schmeidler (1999) to obtain a result for the ranking of pairs events, none of which is included in the other. We then proceed to impose an additional condition of event cancellation à la de Finetti. Our main result is that this condition is equivalent to the condition that the functions $v(\cdot, c)$, for each case c, are additive with respect to the union of disjoint sets. It therefore describes the way that a predictor who is committed to this cancellation condition may form probabilistic beliefs over events given any possible memory.

We then proceed to test our model in the benchmark example of frequentism. That is, we assume that each case observed in the past can only be one of the possible eventualities in the problem at hand. Under this structural assumption it is natural to state two additional assumptions on plausibility rankings, which are shown to be equivalent to frequentism, namely, to ranking events by their empirical frequencies.

While it is reassuring to know that frequentism is a special case of our model, we consider it a conceptually simple problem. The more interesting problems, as in the example of the physician above, are those in which each past case is pertinent to the present problem to a certain subjective degree. Our results show how one may form probabilistic beliefs based on partially relevant information. Conversely, they also show how qualitative "at least as plausible as" comparisons may be used to elicit the subjective similarity judgments, and when these can be assumed additive with respect to set union.

The rest of this paper is organized as follows. Section 2 contains the main results. Subsection 2.1 quotes relevant results from a companion paper. Subsection 2.2 adapts the representation result to an algebra of events, while § 2.3 contains the result about additivity of the set functions $v(\cdot, c)$. Section 3 deals with the situation in which the set of past cases coincides with the set of possible eventualities and shows how frequentism then follows as a special case of our approach. Finally, an appendix contains all proofs.

3.2. Main Results

3.2.1. *States*

In this subsection we describe a result that is proven in a companion paper, and which is the basis of the analysis that follows. Consider a prediction problem, in which one is asked to rank *eventualities* in a nonempty set X. For concreteness, one may think of these eventualities as states of the world throughout this subsection. But the axioms that we use here do not presuppose that the eventualities to be ranked are mutually exclusive or exhaustive. Indeed, the result reported in this subsection will be used later for various collections of events.

The predictor is equipped with knowledge of cases, facts, observations, or stories. Let M be a finite and nonempty set of *cases*, representing the predictor's knowledge. The person (or the machine) who is supposed to come up with predictions is assumed to have a well-defined "at least as likely as" relation on X that presumably relies on M. Hence, for a different collection of cases the predictor may have a different "at least as likely as" relation. We assume that such a relation is given not only for the actual state of knowledge, but also for all hypothetical ones that are generated from it by replication of cases.

Formally, consider the set of repetitions of cases $\mathbb{J} = \mathbb{Z}_+^M = \{I \,|\, I : M \rightarrow \mathbb{Z}_+\}$, where $\mathbb{Z}_+$ denotes the nonnegative integers. For simplicity, we will refer to elements of $\mathbb{J}$ as *memories*. We assume that for every $I \in \mathbb{J}$ the predictor has a binary relation "at least as likely as" $\succsim_I$ on X (i.e., $\succsim_I \subseteq X \times X$).

Algebraic operations on $\mathbb{J}$ are performed pointwise. We define $\succ_I$ and $\approx_I$ to be the asymmetric and symmetric parts of $\succsim_I$, as usual.

We will use the following axioms:

Axiom 1: Order.   *For every $I \in \mathbb{J}$ $\succsim_I$ is complete and transitive on X.*

Axiom 2: Combination.   *For every $I, J \in \mathbb{J}$ and every $x, y \in X$, if $x \succsim_I y \, (x \succ_I y)$ and $x \succsim_J y$, then $x \succsim_{I+J} y \, (x \succ_{I+J} y)$.*

Axiom 3: Archimedeanity.   *For every $I, J \in \mathbb{J}$ and every $x, y \in X$, if $x \succ_I y$, then there exists $k \in N$ such that $x \succ_{kI+J} y$.*

Observe that in the presence of Axiom 2, Axiom 3 also implies that for every $I, J \in \mathbb{J}$ and every $x, y \in X$, if $x \succ_I y$, then there exists $l \in \mathbb{N}$ such that for all $k \geq l, x \succ_{kI+J} y$.

Axiom 1 simply requires that, given any conceivable memory, the decision maker's preference relation over acts is a weak order. Axiom 2 states that if eventuality x is more plausible than eventuality y given two disjoint memories, x should also be more plausible than y given the combination of these memories. In our set-up, combination (or concatenation) of memories takes the form of adding the number of repetitions of each case in the two memories. Axiom 3 is a continuity, or an Archimedean axiom. It states that if, given the memory I, the predictor believes that eventuality x is strictly more plausible than y, then no matter what her ranking is for another memory, J, there is a number of repetitions of I that is large enough to overwhelm the ranking induced by J.

We also need a diversity axiom that is not necessary for the functional form we would like to derive. While the theorem we present is an equivalence theorem, it characterizes a more restricted class of plausibility rankings than those discussed in the introduction. Specifically, we require that for any four eventualities, there is a memory that would distinguish among all four of them.

Axiom 4: Diversity.   *For every list (x, y, z, w) of distinct elements of X there exists $I \in \mathbb{J}$ such that $x \succ_I y \succ_I z \succ_I w$. If $|X| < 4$, then for any strict ordering of the elements of X there exists $I \in \mathbb{J}$ such that $\succ_I$ is that ordering.*

Finally, we need the following definition: A matrix of real numbers is called *diversified* if no row in it is dominated by an affine combination of three (or fewer) other rows in it. Formally:

Definition: A matrix $v: X \times Y \to \mathbb{R}$, where $|X| \geq 4$, is *diversified* if there are no distinct four elements $x, y, z, w \in X$ and $\lambda, \mu, \theta \in \mathbb{R}$ with $\lambda + \mu + \theta = 1$ such that $v(x, \cdot) \leq \lambda v(y, \cdot) + \mu v(z, \cdot) + \theta v(w, \cdot)$. If $|X| < 4$, v is *diversified* if no row in v is dominated by an affine combination of the others.

We now quote a result of a previous work which will be used in this paper.

Theorem 2.1 (Gilboa and Schmeidler, 1999, 2001): *Let* X, M, *and* $\{\succsim_I\}_{I \in \mathbb{J}}$ *be given as above. Then the following two statements are equivalent:*

(i) $\{\succsim_I\}_{I \in \mathbb{J}}$ *satisfy Axioms 1–4;*
(ii) *There is a diversified matrix* $v: X \times M \to \mathbb{R}$ *such that:*

$$(**) \quad \begin{cases} \text{for every } I \in \mathbb{J} \text{ and every } x, y \in X, \\ x \succsim_I y \quad \text{iff} \quad \sum_{c \in M} I(c) v(x, c) \geq \sum_{c \in M} I(c) v(y, c) \end{cases}$$

Furthermore, in this case the matrix v *is unique in the following sense:* v *and* u *both satisfy* $(**)$ *iff there are a scalar* $\lambda > 0$ *and a matrix* $\beta: X \times M \to \mathbb{R}$ *with identical rows (i.e., with constant columns) such that* $u = \lambda v + \beta$.

This theorem has several applications mentioned in Gilboa-Schmeidler (1999, revised version 2001). In particular, it can be viewed as axiomatizing kernel methods for estimation of density functions, as well as for classification problems. The theorem can also be interpreted as an axiomatization of maximum likelihood estimation: Assume that X is a set of theories or general rules that one is to rank according to plausibility given memory. Axioms 1–4 appear reasonable for this case, and one can derive the representation $(**)$. If we shift the weights $v(x, c)$ so that they are all negative, they can be interpreted as logarithms of the conditional probability of case c given theory x.

Thus, Theorem 2.1 can be viewed as an axiomatization of ranking theories or probability distributions, based on the likelihood function, together with a derivation of the conditional probabilities used in the likelihood function.

3.2.2. *Events*

The predictions discussed in § 3.2.1 are abstract eventualities lacking any logical or algebraic structure. It is natural to ask how similarity-based ranking of prediction relates to basic logical or set operations. In an attempt to address this question, we focus here on the case in which the alternatives to be ranked are events.

Let Ω be a state space. Let Σ be an algebra of events on Ω. Assume that Σ contains all singletons. Assume further that Ω contains at least five states. We assume an "at least as likely as" relation between events in Σ that are not included in each other. More precisely, two events $A, B \in \Sigma$ are said to be *nonincluded* if $A \backslash B, BA \neq \varnothing$. We assume that only such pairs are ranked. In particular, we are interested only in proper nonempty subsets of Ω in Σ. For reasons that will be clarified in the proof, it is convenient to rule out of the discussion all proper subsets of Ω whose complement is a singleton. We therefore focus on $\Sigma' = \{A \in \Sigma | A \neq \varnothing \text{ and } |A^c| > 1\}$.

The application of Theorem 2.1 to the case of events is not immediate, because we only assume ranking between nonincluded events, and because the diversity axiom would require some modifications. We start by restating the first three axioms for the case at hand.

Axiom 1*: Order. *For every $I \in \mathbb{J}$, for every pair of nonincluded events $A, B \in \Sigma', A \succsim_I B$ or $B \succsim_I A$. Further, if $A, B, C \in \Sigma'$ are pairwise nonincluded, then $A \succsim_I B$ and $B \succsim_I C$ imply $A \succsim_I C$.*

Axiom 2*: Combination. *For every $I, J \in \mathbb{J}$ and every pair of nonincluded events $A, B \in \Sigma'$, if $A \succsim_I B(A \succ_I B)$ and $A \succsim_J B$, then $A \succsim_{I+J} B(A \succ_{I+J} B)$.*

Axiom 3*: Archimedeanity. *For every $I, J \in \mathbb{J}$ and every pair of nonincluded events $A, B \in \Sigma'$, if $A \succ_I B$, then there exists $k \in N$ such that $A \succ_{kI+J} B$.*

Next we turn to the diversity axiom used in § 3.2.1. Observe that, as stated, it cannot hold for any list of four events. First, one cannot expect an event A to be strictly more plausible than an event B if $A \subseteq B$. Second, if one ranks plausibility according to a probability measure, and if $1_A + 1_B \leq 1_C + 1_D$ or $1_A + 1_C \leq 1_B + 1_D$ (where 1_E denotes the indicator function of $E \in \Sigma$), one cannot expect any memory to induce the ranking $A \succ_I B \succ_I C \succ_I D$. The proposition that follows shows that the exceptions above are the only ones.

Proposition 2.2: *Suppose that* (A, B, C, D) *are four events in a measurable space* (Ω, Σ). *Then there exists a probability measure P on* Σ *such that* $P(A) > P(B) > P(C) > P(D)$ *iff*

(i) *no event in the list* (A, B, C, D) *is a subset of a follower in the list; and*

(ii) *neither* $1_A + 1_B \leq 1_C + 1_D$ *nor* $1_A + 1_C \leq 1_B + 1_D$.

It follows that for a quadruple of pairwise nonincluding events, Condition (ii) is necessary and sufficient for the existence of a measure that strictly ranks the four events in the given order. The proposition above motivates the following definitions: $(A, B, C, D) \in \Sigma'^4$ is a list of *orderly differentiated* events if no event in the list is a subset of a follower in the list, and neither $1_A + 1_B \leq 1_C + 1_D$ nor $1_A + 1_C \leq 1_B + 1_D$. The events $\{A, B, C, D\}$ are *properly differentiated* if every permutation thereof generates a list of orderly differentiated events. In order not to rule out rankings that agree with probability measures, we will restrict the requirement of diversity as follows:

Axiom 4*: Restricted Diversity.   *For every list of orderly differentiated events* (A, B, C, D), *there exists* $I \in \mathbb{J}$ *such that* $A \succ_I B \succ_I C \succ_I D$.

We can now state the following.

Theorem 2.3:   *Under the structural assumptions above, the following two statements are equivalent:*

(i) $\{\succsim_I\}_{I \in \mathbb{J}}$ *satisfy A1*–A4*;*

(ii) *There is a diversified matrix $v: \Sigma' \times M \to \mathbb{R}$ such that*:

$(**)$
$$\begin{cases} \text{for every } I \in \mathbb{J} \text{ and every pair of nonincluding} \\ \qquad\qquad \text{events } A, B \in \Sigma', \\ A \succsim_I B \quad \text{iff} \quad \sum_{c \in M} I(c)v(A,c) \geq \sum_{c \in M} I(c)v(B,c), \end{cases}$$

Furthermore, in this case the matrix v is unique in the following sense: v and w both satisfy $(**)$ iff there are a scalar $\lambda > 0$ and a matrix $u: \Sigma' \times M \to \mathbb{R}$ with identical rows (i.e., with constant columns) such that $w = \lambda v + u$.

3.2.3. *Additivity*

Theorem 2.3 states that a method that ranks events by their likelihood, given any possible repetition of known cases, has to be equivalent to a numerical ranking where the number attached to each event is a linear function of the numbers of case repetitions. One naturally wonders what it would take to make these numbers probabilities. That is, when is there a probability measure μ_c for each case c, such that memory I induces the same ranking of events as the measure $\sum_{c \in M} I(c)\mu_c$?

Obviously, a necessary condition for a representation by additive measures is that, for every $I \in J, \succsim_I$ satisfies de Finetti's cancellation axiom: for every three events A, B, C such that $(A \cup B) \cap C = \varnothing$, we have $A \succsim_I B \Leftrightarrow A \cup C \succsim_I B \cup C$. A key result is that, if we impose this condition (restricted to nonincluded events) on top of the conditions of Theorem 2.3, the resulting matrix v can be normalized so that it is additive in events, namely, so that $v(A \cup B, \cdot) = v(A, \cdot) + v(B, \cdot)$ whenever $A \cap B = \varnothing$. Moreover, in this case we obtain uniqueness of the representation up to multiplication by a positive constant. Formally, we introduce the following axiom:

Axiom 5: Cancellation. *For every $I \in \mathbb{J}$, and for every three pairwise nonincluded events A, B, C such that $(A \cup B) \cap C = \varnothing$, we have $A \succsim_I B \Leftrightarrow A \cup C \succsim_I B \cup C$.*

Theorem 2.4:   *Under the structural assumptions above, the following two statements are equivalent*:

(i) $\{\succsim_I\}_{I \in \mathbb{J}}$ *satisfies* $A1^* - A4^*$ *and* $A5$;
(ii) *There are finite, signed, and finitely additive measures* $\{\mu_c\}_{c \in M}$ *such that*:

$$(**) \quad \begin{cases} \textit{for every } I \in \mathbb{J} \textit{ and every pair of nonincluded} \\[1mm] \textit{events } A, B \in \Sigma', \\[1mm] A \succsim_I B \quad \textit{iff} \quad \sum_{c \in M} I(c)\mu_c(A) \geq \sum_{c \in M} I(c)\mu_c(B), \end{cases}$$

and, for every list of orderly differentiated events (A, B, C, D), *the vectors* $(\mu_c(A))_c, (\mu_c(B))_c, (\mu_c(C))_c,$ *and* $(\mu_c(D))_c$, *define a diversified matrix.*

Furthermore, in this case the measures $\{\mu_c\}_{c \in M}$ *are unique up to multiplication by a positive number.*

Remark 2.5:   Theorem 2.4 may not hold if Ω contains less than five states.

The statement (and proof) of Theorem 2.4 does not restrict Ω to be finite. Yet, the restricted diversity axiom may do so. For instance, it is easy to see that if the measures μ_c are nonnegative, Axiom 4* can be satisfied only if Ω is, indeed, finite. However, one may have versions of the theorems that allow infinite Ω (say, with an infinite set of cases).

Theorem 2.4 only guarantees representation by signed measures. Indeed, since we only use comparisons of pairwise nonincluded events, the data $\{\succsim_I\}_{I \in \mathbb{J}}$ do not imply that likelihood rankings are monotone with respect to set inclusion. One may require that, for each $I \in \mathbb{J}, \succsim_I$ be a *qualitative probability* according to de Finetti (1931), namely that:

(i) $\succsim_I$ is complete and transitive on Σ;
(ii) for every three events A, B, C such that $(A \cup B) \cap C = \emptyset$, we have
$A \succsim_I B \Leftrightarrow A \cup C \succsim_I B \cup C$;
(iii) for every event $A, A \succsim_I \emptyset$;
(iv) $\Omega \succ_I \emptyset$.

This condition strengthens both Axiom 1* and Axiom 5. One may conjecture that imposing it would yield a representation such as in $(**)$

of Theorem 2.4 for *all* pairs of events. As stated, the answer cannot be in the affirmative because any numerical representation by $\{\mu_c\}_{c\in M}$ would yield $A \approx_0 B$ for all $A, B \in \Sigma$ (where $0 \in \mathbb{J}$ denotes the memory in which all cases appear zero times), contradicting (iv). A more natural condition to impose is, therefore,

Axiom 1′: Qualitative Probability. *For every $I \in \mathbb{J}, \succsim_I$ satisfies* (i)–(iii) *of the definition above, and if $I \neq 0, \succsim_I$ also satisfies* (iv).

Yet, even with this weakening the conjecture is false.

Remark 2.6: Assume that $\{\succsim_I\}_{I\in\mathbb{J}}$ satisfy Axioms 1′, 2*, 3*, and 4*. It is possible that the signed measures $\{\mu_c\}_{c\in M}$ obtained in Theorem 2.4 fail to be nonnegative.

3.3. Frequentism

The framework of § 3.2 does not assume any formal relationship between past cases and states of the world. Indeed, one of the strengths of the approach outlined above is that any such relationships may be inferred from plausibility rankings given various memories, rather than assumed a priori. Still, an interesting special case that is also an important test case is the situation where memory consists only of past occurrences of the same states that are now possible. For instance, one may be asked to rank the possible outcomes of a roll of a die based on empirical frequencies of these outcomes in past rolls of the same die. It would be reassuring to know that our approach is compatible with frequentism, i.e., that the numerical rankings derived in § 3.2.3 may boil down to relative empirical frequencies in this case.

Assume, then, that $M = \Omega = \{1, \ldots, n\}$, where $I \in \mathbb{J}$ is interpreted as the empirical frequencies of the possible outcomes. Assume that the relations in $\{\succsim_I\}_{I\in\mathbb{J}}$ are qualitative probability relations (as defined by Axiom 1′). We impose two additional assumptions. The first is a symmetry axiom, stating that the names of the outcomes are immaterial. The second is a specificity axiom, requiring that an outcome that has never been observed does not increase the plausibility of events containing it. For the symmetry axiom, we introduce the following notation: let $\pi \colon \{1, \ldots, n\} \to \{1, \ldots, n\}$ be a permutation. For $I \in \mathbb{J}$, define $I \circ \pi \in \mathbb{J}$ by $I \circ \pi(c) = I(\pi(c))$ and for $A \in \Sigma$ let $A_\pi \in \Sigma$ be defined by $A_\pi = \{\pi^{-1}(i) | i \in A\}$. Using this notation,

Axiom 6: Symmetry.   *For every permutation π, every $I \in \mathbb{J}$, and every $A, B \in \Sigma, A \succsim_{I \circ \pi} B$, iff $A_\pi \succsim_I B_\pi$.*

Next we introduce:

Axiom 7: Specificity.   *Assume that for $j \in \Omega$ and $I \in \mathbb{J}, I(j) = 0$. For every $A, B \in \Sigma$, such that $B \succsim_I A$, we also have $B \succsim_I A \cup \{j\}$.*

Theorem 3.1:   *Assume that $n \geq 5$ and that $\{\succsim_I\}_{I \in \mathbb{J}}$ satisfies Axioms $1', 2^*, 3^*, 4^*, 6$, and 7. Let $\{\mu_i\}$ be the measures provided by Theorem 2.4. Then $(\mu_i(\{j\}))_{1 \leq i, j \leq n}$ is the identity matrix, up to multiplication by a positive number.*

When $n < 5$ Theorem 2.4 does not provide a representation of $\{\succsim_I\}_{I \in \mathbb{J}}$ by measures μ_i. Yet, we can simply assume a representation along the lines of Theorem 2.4 and obtain a similar result:

Remark 3.2:   Let $\{\succsim_I\}_{I \in \mathbb{J}}$ be a collection of binary relations on $\Sigma = 2^\Omega$, and let $\{\mu_i\}$ be nonnegative measures on Ω such that for every $I \in \mathbb{J}$ and every $A, B \in \Sigma$,

$$A \succsim_I B \quad \text{iff} \quad \sum_{c \in M} I(c)\mu_c(A) \geq \sum_{c \in M} I(c)\mu_c(B).$$

Assume that $\{\succsim_I\}_{I \in \mathbb{J}}$ satisfies Axioms $1', 6$, and 7. Then $(\mu_i(\{j\}))_{1 \leq i, j \leq n}$, is the identity matrix, up to multiplication by a positive number.

Observe that Axioms 6 and 7 are obviously necessary for the two results.

Appendix: Proofs and Related Analysis

Proof of Proposition 2.2:

Let there be given four events A, B, C, D in a measurable space (Ω, Σ). It suffices to consider the minimal algebra containing these events, which is finite. Assume that the atoms in this algebra are $\Omega' = \{1, \ldots, n\}$. It is easy to see that there exists a probability measure P on Σ such that $P(A) > P(B) > P(C) > P(D)$ iff the following LP problem is feasible:

$$(P) \quad \min_{x \in \mathbb{R}_+^M} 0 \cdot x$$
$$\text{s.t.} \qquad (1_A - 1_B) \cdot x \geq 1$$
$$(1_B - 1_C) \cdot x \geq 1$$
$$(1_C - 1_D) \cdot x \geq 1.$$

Problem (P) is feasible iff its dual is bounded. The dual is

$$(D) \quad \max \quad \alpha + \beta + \gamma$$
$$\text{s.t.} \quad \alpha(1_A - 1_B) + \beta(1_B - 1_C) + \gamma(1_C - 1_D) \leq 0 \qquad (*)$$
$$\alpha, \beta, \gamma \geq 0.$$

Clearly, (D) is bounded iff it is bounded by zero, and this is the case iff its only feasible point is $\alpha, \beta, \gamma, = 0$.

We now prove that Conditions (i) and (ii) imply these equalities. Assume, by way of negation, that $(*)$ holds and let $\alpha, \beta, \gamma > 0$ but not all three are equal to zero.

Observe first that if exactly one of α, β, and γ is positive, it follows that at least one of the inclusions; $A \subseteq B, B \subseteq C$, or $C \subseteq D$ holds, in contradiction to (i).

Next assume that exactly two of α, β, and γ are positive. Suppose first that $\beta = 0$. Condition (ii) implies that for some k, $1_A(k) + 1_C(k) > 1_B(k) + 1_D(k)$. Since the indicator vectors take only the values 1 or 0, there are five possibilities in which this inequality may be satisfied: $1 + 1 > 1 + 0, 0 + 1, 0 + 0$ or $1 + 0, 0 + 1 > 0 + 0$. Evaluating $(*)$ at k for these five possibilities leads to one of the following: $\alpha - \alpha + \gamma \leq 0, \alpha + \gamma - \gamma \leq 0, \alpha + \gamma \leq 0, \alpha \leq 0$, or $\gamma \leq 0$. All these inequalities are inconsistent with positivity of α and γ, so this case is ruled out.

If $\gamma = 0$, consider $i \in A\backslash C$ (whose existence follows from (i)). Now (∗) implies $\alpha \leq 0$ or $\alpha - \beta + \beta \leq 0$. Both contradict positivity of α. The last case is $\alpha = 0$. Let $l \in B\backslash D$ (again, such a state exists by (i)). In this case (∗) implies $\beta \leq 0$ or $\beta - \beta + \gamma \leq 0-$ a contradiction to $\beta, \gamma > 0$.

So we are left with the case where all three, α, β, and γ, are positive. Since the first inequality in (ii) does not hold, there exists i for which $1_A(i) + 1_B(i) > 1_C(i) + 1_D(i)$. We consider the five possibilities, as above: $1+1 > 1+0, 0+1, 0+0$ or $1+0, 0+1 > 0+0$. Substituting the corresponding values in (∗) we get that at least one of the following five inequalities holds: $\alpha - \alpha + \beta - \beta + \gamma \leq 0, \alpha - \alpha + \beta - \gamma \leq 0, \alpha - \alpha + \beta \leq 0$, $\alpha \leq 0, -\alpha + \beta \leq 0$. The possibility of α, β, and γ leaves only $\beta \leq \alpha$ or $\beta \leq \gamma$.

By (i), there is a state $j \in A\backslash D$. Again, using (∗), we get one of the following four inequalities: $\alpha \leq 0, \alpha - \alpha + \beta \leq 0, \alpha - \beta + \gamma \leq 0$, and $\alpha - \alpha + \beta - \beta + \gamma 0$, depending on whether j belongs to B and to C or not. Three of the inequalities are directly inconsistent with positivity of α, β, and γ. The fourth, namely, $\alpha + \gamma \leq \beta$, is also inconsistent with positivity when coupled with either of $\beta \leq \alpha$ or $\beta \leq \gamma$ obtained previously. This concludes the proof that (i) and (ii) suffice for the existence of a probability measure P such that $P(A) > P(B) > P(C) > P(D)$. The necessity of (i) and (ii) is obvious. □

Proof of Theorem 2.3:

We will construct the numerical representation by "patching" together numerical representations for subsets of events that are properly differentiated. In doing so, a few auxiliary results will be of help. We start by the following definition. Suppose that for a subset of events $\Delta \subseteq \Sigma'$ there is a matrix $v^\Delta : \Delta \times M \to \mathbb{R}$. Let Δ' be a subset of Δ. We say that v^Δ *ranks* Δ' if for every nonincluded $E, F \in \Delta'$, and every $I \in \mathbb{J}$,

$$E \succsim_I F \quad \textit{iff} \quad \sum_{c \in M} I(c)v^\Delta(E, c) \geq \sum_{c \in M} I(c)v^\Delta(F, c).$$

To extend a numerical representation v^Δ to a larger set Δ, we would like to know that such a representation is unique on relatively small subsets Δ'. For instance, when we consider triples of pairwise nonincluded events A, B, C, it would be nice to know that a function v that ranks $\{A, B, C\}$ is unique as in Theorem 2.1. For this one would need to have a diversity axiom for triples of events, namely, that for any permutation thereof there exists an $I \in \mathbb{J}$ such that $\succ_I$ agrees with the given permutation. One would expect this to follow from the seemingly more powerful diversity assumption Axiom 4*, stated for all quadruples of orderly differentiated events. However, not every triple of pairwise nonincluded events can be complemented to a quadruple of orderly differentiated events. Consider, for instance, $n = 5, A = \{1, 2, 3\}, B = \{4\}, C = \{5\}$. These are pairwise nonincluded, but there is no event D that is pairwise nonincluded with respect to all of them.

This case is anomalous enough to deserve a definition. We say that three events $A, B, C \in \Sigma'$ *form an all-but-two partition* if two of them are singletons and the third is the complement of the (union of the) first two. We now state

Lemma 1:   *Let $A, B, C \in \Sigma'$ be three pairwise nonincluded events that do not form an all-but-two partition. Then there exists $I \in \mathbb{J}$ such that $A \succ_I B \succ_I C$.*

Proof.   First observe that since A, B, C are pairwise nonincluded, there exist probability measures P on Ω such that $P(A) > P(B) > P(C)$. We will shortly prove that there exists an event D that is nonincluded with respect to each of $A, B,$ and C. We can choose a probability P such that $P(A) > P(B) > P(C)$ and that $P(D)$ differs from each of $\{P(A), P(B), P(C)\}$. This would mean, by Proposition 2.2, that one of the four lists, $\{(D, A, B, C), (A, D, B, C), (A, B, D, C), (A, B, C, D)\}$, is orderly differentiated. We can then use the diversity axiom for that list to deduce the desired result.

We therefore wish to prove that there exists an event D that is nonincluded with respect to each of $A, B,$ and C. If there exists a state $i \in (A \cup B \cup C)^c$, then choosing $D = \{i\}$ will do. Assume, then, that $A \cup B \cup C = \Omega$. Next, if $A, B,$ and C are pairwise disjoint, then, because they do not form an all-but-two partition, it has to be the case that at least two of them contain more than one element. Assume without loss of generality that $\{i,j\} \subseteq A$ and that $\{k,l\} \subseteq B$. In this case, $D = \{i,k\}$ is noninlcuded with respect to each of A, B, and C.

We now deal with the cases where $A \cup B \cup C = \Omega$ and the three events are not pairwise disjoint. Assume that one of them is disjoint from the other two, say, $A \cap (B \cup C) = \emptyset$. Let $i \in B \backslash C$ and $j \in C \backslash B$. Since not all three events are disjoint, $B \cap C \neq \emptyset$, and it follows that $D = \{i,j\}$ is noninincluded with respect to all three.

We therefore assume that $A \cup B \cup C = \Omega$ and that each event intersects the union of the other two. Observe that this implies that none of A, B, or C is a singleton. Assume that one of them is not contained in the union of the other two, say, $A \backslash (B \cup C) = \emptyset$. Then there exists $i \in A \backslash (B \cup C)$. Choose $j \in B \backslash A$ and let $D = \{i,j\}$.

Finally, we are left with the case where $A \cup B \cup C = \Omega$ and where each event is contained in the union of the other two. Hence, every state in Ω is included in at least two of $\{A, B, C\}$. In this case $A^c = (B \cap C) \backslash A$; $B^c = (A \cap C) \backslash B$; and $C^c = (A \cap B) \backslash C$. Since A, B, C are in Σ', each of these pairwise disjoint events includes at least two elements. Construct D by selecting one of each. □

Lemma 2: *Let there be given two subsets $\sigma, \tau \subseteq \Sigma'$ with corresponding matrices $v^\sigma : \sigma \times M \to \mathbb{R}$ and $v^\tau : \tau \times M \to \mathbb{R}$. Assume that there are three pairwise nonincluded events $A, B, C \in \sigma \cap \tau$ that do not form an all-but-two partition, and that both v^σ and v^τ rank $\{A, B, C\}$. If, for all $c \in M, v^\sigma(A, c) = v^\tau(A, c)$ and $v^\sigma(B, c) = v^\tau(B, c)$, then also $v^\sigma(C, c) = v^\tau(C, c)$ for all $c \in M$.*

Proof. In view of the previous lemma, the triple $\{A, B, C\}$ satisfies the conditions of Theorem 2.1. The conclusion follows from the uniqueness result of the theorem. □

Lemma 3: *Let there be given a subset of events $\Delta \subseteq \Sigma'$ and a matrix $v^\Delta : \Delta \times M \to \mathbb{R}$. Assume that $A, B, C, D \in \Delta$ are properly differentiated. If v^Δ ranks $\{A, C, D\}$ and $\{B, C, D\}$, then it also ranks $\{A, B\}$.*

Proof. Since A, B, C, D are properly differentiated, we can apply Theorem 2.1 to $\sigma \equiv \{A, B, C, D\}$, and conclude that there exists a matrix v^σ that ranks σ. Without loss of generality we may assume that $v^\sigma(C, c) = v^\Delta(C, c)$ and $v^\sigma(D, c) = v^\Delta(D, c)$ for all $c \in M$. Since A, B, C, D are properly differentiated, we know that none of $\{A, C, D\}, \{B, C, D\}$ forms an all-but-two partition. The previous remark therefore states that for all $c \in M, v^\sigma(A, C) = v^\Delta(A, C)$ and $v^\sigma(B, c) = v^\Delta(B, c)$ also hold. Since v^σ ranks $\{A, B\}$, so does v^Δ. $\qquad\square$

We now turn to the construction of a matrix $v^{\Sigma'}$ that ranks Σ'. The strategy is as follows: we start by constructing a representation for all pairs. (Recall that Σ includes all singletons, and therefore also all pairs.) We then extend it to singletons. Next we show that it can be extended to all events in Σ'.

Define $\Delta_2 = \{A \in \Sigma' \mid |A| = 2\}$.

Lemma 4: *There exists $v^{\Delta_2} : \Delta_2 \times M \to \mathbb{R}$ that ranks Δ_2.*

Proof. Choose an element of Ω, and call it 1. We first consider $\Delta_2^1 = \{\{1, i\} \mid i \neq 1\}$. Any four events in Δ_2^1 are properly differentiated, and Theorem 2.1 can be applied to obtain a representation $v^{\Delta_2^1} : \Delta_2^1 \times M \to \mathbb{R}$ that ranks Δ_2^1.

Next we wish to extend $v^{\Delta_2^1}$ to v^{Δ_2} on all of Δ_2. Consider $A = \{i, j\}$ where $i, j \neq 1$. Any four events in $\Delta_2^1 \cup \{A\}$ are properly differentiated. Hence Theorem 2.1 offers a unique definition of $v^{\Delta_2}(A, c)$ for $c \in M$.

We claim that v^{Δ_2} ranks Δ_2. Let there be given $A, B \in \Delta_2$. Notice that if at least one of them is in Δ_2^1, we already know that Δ_2^1 ranks $\{A, B\}$. Assume, then, that $A, B, \in \Delta_2 \backslash \Delta_2^1$. Distinguish between two cases: (i) $A \cap B = \varnothing$, and (ii) $A \cap B \neq \varnothing$.

In Case (i) we have $A = \{i, j\}, B = \{k, l\}$ where i, j, k, l are distinct and differ from 1. Observe that any distinct four events out of $\Delta_2^1 \cup \{A, B\}$ are properly differentiated. It follows that there exists a matrix defined on $(\Delta_2^1 \cup \{A, B\}) \times M$ that ranks $\Delta_2^1 \cup \{A, B\}$, and that it coincides with our definition of v^{Δ_2}.

In Case (ii) we have $A = \{i, j\}, B = \{i, k\}$ where i, j, k are distinct and differ from 1. Observe that $\{\{1, i\}, \{1, j\}, \{i, j\}, \{i, k\}\}$ are properly differentiated. By Lemma 3, since v^{Δ_2} ranks $\{\{1, i\}, \{1, j\}, \{i, j\}\}$ and $\{\{1, i\}, \{1, j\}, \{i, k\}\}$, it also ranks $\{\{i, j\}, \{i, k\}\}$. $\qquad\square$

Our next step is to extend v^{Δ_2} to singletons. Let $\bar{\Delta}_2 = \{A \in \Sigma' \mid |A| \leq 2\}$.

Lemma 5:   *There exists $v^{\bar{\Delta}_2} \colon \bar{\Delta}_2 \times M \to R$ that ranks $\bar{\Delta}_2$.*

Proof. Let $v^{\bar{\Delta}_2}$ equal v^{Δ_2} on all pairs. We now extend it to all singletons, and then show that this extension indeed ranks $\bar{\Delta}_2$. Let there be given $i \in \Omega$. Choose distinct $j, k, l \neq i$. There is a unique definition of $v^{\bar{\Delta}_2}(\{i\}, c)$ (for $c \in M$) such that $v^{\bar{\Delta}_2}$ ranks $\{\{i\}, \{j, k\}, \{j, l\}\}$. We first claim that $v^{\bar{\Delta}_2}$ thus defined ranks $\Delta_2 \cup \{\{i\}\}$. Indeed, for any event $B \in \Delta_2$ that differs from $\{j, k\}, \{j, l\}, \{B, \{j, k\}, \{j, l\}\}$ are pairwise nonincluded, and they do not form an all-but-two partition. Hence, $v^{\bar{\Delta}_2}$ ranks $\{B, \{j, k\}, \{j, l\}\}$. Further, if $i \notin B$, then $\{B, \{i\}, \{j, k\}, \{j, l\}\}$ are also properly differentiated, and, by Lemma 3, $v^{\bar{\Delta}_2}$ also ranks $\{B, \{i\}\}$.

Let this be the definition of $v^{\bar{\Delta}_2}(\{i\}, c)$ for each $i \in \Omega$. We need to show that for every distinct $i, j \in \Omega, v^{\bar{\Delta}_2}$ ranks $\{\{j\}, \{i\}\}$. Since $|\Omega| \geq 5$, there are two distinct $C, D \in \Delta_2$ that are disjoint from $\{i, j\}$. Thus $\{\{i\}, \{j\}, C, D\}$ are properly differentiated, while v^{Δ_2} ranks both $\{\{i\}, C, D\}$ and $\{\{j\}, C, D\}$, which completes the proof. $\square$

The following combinatorial lemma will prove useful.

Lemma 6:   *Let A and B be two nonincluded events in Σ'. Then there are C and D in Σ', with $|C|, |D| = 2$, such that $\{A, B, C, D\}$ are properly differentiated.*

Proof. Assume without loss of generality that $|A| \geq |B|$, and distinguish between two cases:

Case 1. $|A \backslash B| \geq 2$.

Case 2. $|A \backslash B| = |B \backslash A| = 1$.

In Case 1 set $i, j \in A \backslash B, i \neq j$, and $k \in B \backslash A$. Define $C = \{i, k\}$ and $D = \{j, k\}$. By direct verification one can check that the conclusion of the lemma holds.

Next consider Case 2. Since $|\Omega \backslash A|, |\Omega \backslash B| \geq 2$, there is a state $p \in \Omega \backslash (A \cup B)$. Assume first that there also is a state $q \neq p$, such that $q \in \Omega \backslash (A \cup B)$. Let $i \in A \backslash B$, and $k \in B \backslash A$. If A and B are singletons, there exists $j \in \Omega \backslash \{i, k, p, q\}$ and we can choose $C = \{j, p\}$

and $D = \{j, q\}$. Otherwise (namely, $A \cap B \neq \emptyset$) defining $C = \{i, p\}$ and $D = \{k, q\}$ results in the desired conclusion.

We are now left with Case 2 under the additional restriction that $|\Omega \backslash (A \cup B)| = 1$. Since $|\Omega| \geq 5$, we know that $|A \cap B| \geq 2$. Define $C = \{p, k\}$ and $D = \{p, l\}$, where $k \neq l, k, l \in A \cap B$. Once again, direct verification completes the proof. $\qquad\square$

Completion of the Proof of Theorem 2.3:

We now proceed to define $v = v^{\Sigma'} : \Sigma' \times M \rightarrow \mathbb{R}$ that ranks Σ', as an extension of v^{Δ_2}. Let there be given an event $A \in \Sigma'$ with $|A| > 2$. Let i be an element in Ω that is not included in A, and let j, k be two elements that are in A. Since $\{A, \{i, j\}, \{i, k\}\}$ are pairwise nonincluded and they do not form an all-but-two partition, Theorem 2.1 applies to them and offers a unique definition of $v(A, c)$ (for all $c \in M$) such that v ranks $\{A, \{i, j\}, \{i, k\}\}$.

We now wish to show that v thus defined ranks $\{A, B\}$ for all nonincluded $A, B \in \Sigma'$. If $|A|, |B| \geq 2$, the result follows from the definition of v as an extension of v^{Δ_2}. Assume, then, that $|A| > 2$. We split the proof into three parts according to the number of elements in B.

First assume that $|B| = 2$. Recall that i is an element in Ω that is not included in A, and that j, k are two elements that are in A. There are four cases to check, according to whether $i \in B$ and whether A and B are disjoint. In all cases, direct verification shows that $\{A, B, \{i, j\}\}$, $\{i, k\}$ are properly differentiated, and Lemma 3 implies that v ranks $\{A, B\}$.

Next assume that $|B| = 1$, i.e., that $B = \{l\}$ where $l \notin A$. Choose $s \notin A \cup \{l\}$. Since j, k are in $A, \{A, B, \{s, j\}, \{s, k\}\}$ are properly differentiated, and Lemma 3 implies that v ranks $\{A, \{1\}\}$.

Finally, assume $|B| > 2$. By Lemma 6 there are C and D in Σ', with $|C|, |D| = 2$, such that $\{A, B, C, D\}$ are properly differentiated. We already know that v ranks $\{A, C, D\}$ and $\{B, C, D\}$, since C and D are pairs. Again, Lemma 3 implies that v ranks $\{A, B\}$.

It is easy to see that, if for some list of orderly differentiated events (A, B, C, D), the vectors $(\mu_c(A)), (\mu_c(B))_c, (\mu_c(C))_c$, and $(\mu_c(D))_c$

form a matrix that is not diversified, then A4* is violated. This was also proven in detail in Gilboa and Schmeidler (1999).   □

The proof of sufficiency and of uniqueness are as in Gilboa and Schmeidler (1999).   □

Proof of Theorem 2.4:

The fact that (ii) implies (i) is immediate. We will first show that (i) implies (ii) and the uniqueness result for the case of a finite algebra.

Assume, then, that $\Omega = \{1, \ldots, n\}$ (recall that $n \geq 5$) and that $\Sigma = 2^{\Omega}$, and let $\hat{v}$ be the matrix provided by Theorem 2.3. Set $w(\cdot) = \hat{v}(\{1,2\}, \cdot) - \hat{v}(\{1\}, \cdot) - \hat{v}(\{2\}, \cdot)$, and define a matrix v by $v(A, \cdot) = \hat{v}(A, \cdot) + w(\cdot)$ so that $v(\{1,2\}, \cdot) = v(\{1\}, \cdot) + v(\{2\}, \cdot)$. We wish to show that for this v, $v(A, \cdot) = \sum_{i \in A} v(\{i\}, \cdot)$ for every $A \in \Sigma'$. Observe that if we find such a v, it is unique up to positive multiplication, since a shift by a vector w can preserve additivity only if $w = 0$.

Some notation may prove useful. We will use event superscripts to denote rows in the matrix. Thus, v^A denotes the vector $v(A, \cdot)$. Also, for $A, B, \in \Sigma'$, define $v^{A,B} = v^A - v^B$. Note that for any three pairwise nonincluded events A, B, C, we have the *Jacobi identity* $v^{A,C} = v^{A,B} + v^{B,C}$. A key observation is the following

Lemma 1:   *For every three pairwise nonincluded events A, B, C such that $(A \cup B) \cap C = \emptyset$, there exists a unique $\lambda > 0$ such that $v^{A \cup C, B \cup C} = \lambda v^{A,B}$.*

Proof. By the cancellation axiom we know that the set of memories I for which $A \cup C \succeq_I B \cup C$ is precisely the same set for which $A \succeq_I B$. The conclusion follows from the uniqueness result of Theorem 2.1 applied to the events A, B.   □

Next we show that, when we focus on a singleton $C = \{i\}$, the coefficient λ does not depend on the sets A, B:

Lemma 2:   *For every $i \in \Omega$ there exists a unique $\lambda_i > 0$ such that, for every nonincluded $A, B \in \Sigma'$ such that $i \notin A \cup B$ and $|A|, |B| < n - 2$, $v^{A \cup \{i\}, B \cup \{i\}} = \lambda_i v^{A,B}$.*

Proof. Consider three events A, B, D that are pairwise nonincluded, none of which includes i, and none of which has more than $n - 3$ elements. We know that

$$v^{A \cup \{i\}, B \cup \{i\}} = v^{A \cup \{i\}, D \cup \{i\}} + v^{D \cup \{i\}, B \cup \{i\}}.$$

Applying Lemma 1 to each of the three elements above we obtain numbers $\lambda, \mu, \eta > 0$ such that

$$v^{A \cup \{i\}, B \cup \{i\}} = \lambda v^{A,B}$$

$$v^{A \cup \{i\}, D \cup \{i\}} = \mu v^{A,D}$$

$$v^{D \cup \{i\}, B \cup \{i\}} = \eta v^{D,B}.$$

Hence,

$$\lambda v^{A,B} = \mu v^{A,D} + \eta v^{D,B},$$

or

$$v^{A,B} = \frac{\mu}{\lambda} v^{A,D} + \frac{\eta}{\lambda} v^{D,B}.$$

However, since we also have the Jacobi identity

$$v^{A,B} = v^{A,D} + v^{D,B},$$

the restricted diversity axiom implies that $\lambda = \mu = \eta$.

Applying this result for the case that A, B, D are singletons, we conclude that there exists $\lambda_i > 0$ such that the conclusion holds for all singletons A, B. Next, for every $A \in \Sigma'$ such that $i \notin A$ and $|A| < n-2$, and every $j \notin A \cup \{i\}$, we also get $v^{A \cup \{i\}, \{i,j\}} = \lambda_i v^{A, \{j\}}$. Similarly, consider such a set A, and choose $j, k \notin A \cup \{i\}$ and $l \in A$. Define $B = \{l, k\}$ and $D = \{j\}$. We obtain that $v^{A \cup \{i\}, \{i,k,l\}} = \lambda_i v^{A, \{k,l\}}$. Finally, consider two nonincluded events $A, B \in \Sigma'$ such that $i \notin A \cup B$ and $|A|, |B| < n - 2$. Choose $l \in A \backslash B$ and $k \in B \backslash A$. Define $D = \{l, k\}$ and apply the result above to these three events. The desired result follows. □

Lemma 3: *There exists a unique $\lambda > 0$ such that $\lambda = \lambda_i$ for all $i \in \Omega$ (where λ_i is the coefficient defined by Lemma 2). Further, this λ satisfies,* *for every distinct $i, j, k, l \in \Omega$,*

$$v^{\{i,j\},\{k,l\}} = \lambda v^{\{i\},\{k\}} + \lambda v^{\{j\},\{l\}}.$$

Proof. By the Jacobi identity and Lemma 2,

$$v^{\{i,j\},\{k,l\}} = v^{\{i,j\},\{k,j\}} + v^{\{k,j\},\{k,l\}} = \lambda_j v^{\{i\},\{k\}} + \lambda_k v^{\{j\},\{l\}}.$$

Similarly,

$$v^{\{i,j\},\{k,l\}} = v^{\{i,j\},\{l,j\}} + v^{\{l,j\},\{k,l\}} = \lambda_j v^{\{i\},\{l\}} + \lambda_l v^{\{j\},\{k\}};$$

hence,

$$\lambda_j v^{\{i\},\{k\}} + \lambda_k v^{\{j\},\{l\}} = \lambda_j v^{\{i\},\{l\}} + \lambda_l v^{\{j\},\{k\}}$$

or

$$\lambda_j v^{\{i\},\{k\}} = \lambda_j v^{\{i\},\{l\}} + \lambda_k v^{\{l\},\{j\}} + \lambda_l v^{\{j\},\{k\}}.$$

That is,

$$v^{\{i\},\{k\}} = v^{\{i\},\{l\}} + \frac{\lambda_k}{\lambda_j} v^{\{l\},\{j\}} + \frac{\lambda_l}{\lambda_j} v^{\{j\},\{k\}}.$$

But the Jacobi identity also implies

$$v^{\{i\},\{k\}} = v^{\{i\},\{l\}} + v^{\{l\},\{j\}} + v^{\{j\},\{k\}}.$$

and, coupled with the diversity axiom this means that $\lambda_j = \lambda_k = \lambda_l$. $\square$

Until the end of the proof we reserve the symbol λ to the coefficient defined by Lemma 3.

Lemma 4:   *For every distinct $i \in \Omega \setminus \{1, 2\}$,*

$$v^{\{1,i\}} = v^{\{1\}} + (1 - \lambda) v^{\{2\}} + \lambda v^{\{i\}}$$
$$v^{\{2,i\}} = (1 - \lambda) v^{\{1\}} + v^{\{2\}} + \lambda v^{\{i\}}.$$

Proof. By symmetry between 1 and 2, it suffices to prove the first equation. Since $v^{A,B} = v^A - v^B$ (applied to $A = \{1, i\}$ and $B = \{1, 2\}$)

we get

$$v^{\{1,i\}} = v^{\{1,i\},\{1,2\}} + v^{\{1,2\}}$$

By Lemma 3

$$= \lambda v^{\{i\},\{2\}} + v^{\{1,2\}}$$

and, using the fact $v^{\{1,2\}} = v^{\{1\}} + v^{\{2\}}$ and $v^{A,B} = v^A - v^B$ (applied to $A = \{i\}$ and $B = \{2\}$)

$$= \lambda v^{\{i\}} - \lambda v^{\{2\}} + v^{\{1\}} + v^{\{2\}}$$

$$= v^{\{1\}} + (1 - \lambda)v^{\{2\}} + \lambda v^{\{i\}}. \qquad \square$$

Lemma 5: *For every distinct $i, j \in \Omega \backslash \{1, 2\}$,*

$$v^{\{i,j\}} = (1 - \lambda)v^{\{1\}} + (1 - \lambda)v^{\{2\}} + \lambda v^{\{i\}} + \lambda v^{\{j\}}.$$

Proof. Since $v^{A,B} = v^A - v^B$ (applied to $A = \{i, j\}$ and $B = \{1, 2\}$) we get

$$v^{\{i,j\}} = v^{\{i,j\},\{1,2\}} + v^{\{1,2\}}.$$

Using the Jacobi identity,

$$v^{\{i,j\}} = v^{\{i,j\},\{1,i\}} + v^{\{1,i\},\{1,2\}} + v^{\{1,2\}}.$$

By Lemma 3 the right-hand side equals

$$\lambda v^{\{j\},\{1\}} + \lambda v^{\{i\},\{2\}} + v^{\{1,2\}}.$$

Using the facts $v^{\{1,2\}} = v^{\{1\}} + v^{\{2\}}$ and $v^{A,B} = v^A - v^B$, we get

$$v^{\{i,j\}} = \lambda v^{\{j\}} - \lambda v^{\{1\}} + \lambda v^{\{i\}} - \lambda v^{\{2\}} + v^{\{1\}} + v^{\{2\}}$$

$$= (1 - \lambda)v^{\{1\}} + (1 - \lambda)v^{\{2\}} + \lambda v^{\{i\}} + \lambda v^{\{j\}}. \qquad \square$$

Lemma 6: $\lambda = 1.$

Proof. Consider the set $\{3, 4, 5\}$. By definition of $v^{A,B}$,

$$v^{\{3,4,5\}} = v^{\{3,4,5\},\{2,5\}} + v^{\{2,5\}}.$$

By Lemma 3 the right-hand side equals

$$\lambda v^{\{3,4\},\{2\}} v^{\{2,5\}} = \lambda v^{\{3,4\}} - \lambda v^{\{2\}} + v^{\{2,5\}},$$

where the last equality follows from $v^{A,B} = v^A - v^B$.

Using Lemma 5 for the set $\{3, 4\}$ and Lemma 4 for the set $\{2, 5\}$, we get

$$\begin{aligned}
v^{\{3,4,5\}} &= \lambda[(1 - \lambda)v^{\{1\}} + (1 - \lambda)v^{\{2\}} + \lambda v^{\{3\}} + \lambda v^{\{4\}}] - \lambda v^{\{2\}} \\
&\quad + (1 - \lambda)v^{\{1\}} + v^{\{2\}} + \lambda v^{\{5\}} \\
&= (1 - \lambda^2)v^{\{1\}} + (1 - \lambda^2)v^{\{2\}} + \lambda^2 v^{\{3\}} + \lambda^2 v^{\{4\}} + \lambda v^{\{5\}}
\end{aligned}$$

By symmetry between 4 and 5, we also get

$$v^{\{3,4,5\}} = (1 - \lambda^2)v^{\{1\}} + (1 - \lambda^2)v^{\{2\}} + \lambda^2 v^{\{3\}} + \lambda v^{\{4\}} + \lambda^2 v^{\{5\}}.$$

Equating the two, it has to be the case that

$$\lambda v^{\{4\}} + v^{\{5\}} = v^{\{4\}} + \lambda v^{\{5\}}$$

or

$$(1 - \lambda)v^{\{4\},\{5\}} = 0.$$

By the diversity axiom, $v^{\{4\},\{5\}} \neq 0$, hence, the conclusion follows. $\square$

Lemma 7:   *For every $i, j \in \Omega$, $v^{\{i,j\}} = v^{\{i\}} + v^{\{j\}}$.*

Proof. Use Lemmata 4, 5, and 6. $\square$

Lemma 8:   *For every $A \in \Sigma'$, $v^A = \Sigma_{i \in A} v^{\{i\}}$.*

Proof. By induction on $|A|$. We already know that the lemma holds for $|A| \leq 2$. Assume it is true for $|A| \leq k$ and consider a set B with

$|B| = k + 1 < n - 1$. Choose $i \in B$ and $j \notin B$. Since B and $\{i, j\}$ are nonincluded, we may write

$$v^B = v^{B, \{i, j\}} + v^{\{i, j\}}.$$

Denoting $A = B \backslash \{i\}$, we can write $v^{B, \{i, j\}} = V^{A \cup \{i\}, \{j\} \cup \{i\}}$ where $i \notin A \cup \{j\}$. Using Lemmata 3 and 6,

$$v^{B, \{i, j\}} = v^{A \cup \{i\}, \{j\} \cup \{i\}} = v^{A, \{j\}}.$$

Plugging this into the first equation we get

$$v^B = v^{A, \{j\}} + v^{\{i, j\}} = v^A - v^{\{j\}} + v^{\{i, j\}}$$

and, using the induction hypothesis,

$$= \sum_{k \in A} v^{\{k\}} - v^{\{j\}} + v^{\{i\}} + v^{\{j\}} = \sum_{k \in B} v^{\{k\}}.$$

This completes the proof of Theorem 2.4 for the finite case. ☐

We now turn to the case of an infinite Ω. Choose an element of Ω, say, 1. Consider all the finite subalgebras Σ_0 of Σ that include $\{1\}$ and that have at least five atoms. For each such Σ_0 there exists a matrix $v^{\Sigma_0} : \Sigma_0' \times M \rightarrow R$ that ranks Σ_0' and that satisfies additivity with respect to the union of disjoint sets. Further, such a matrix is unique up to multiplication by a positive number. Choose one such Σ_0 and a corresponding v^{Σ_0} for it . Let Σ_1 be another subalgebra of Σ that includes $\{1\}$ and that has at least five atoms. Let v^{Σ_1} be a matrix that ranks Σ_1'. Let Σ_2 be the minimal algebra containing both Σ_0 and Σ_1. Applying the result to the (finite) subalgebra Σ_2, there exists v^{Σ_2} that ranks Σ_2'. Since v^{Σ_2} ranks both Σ_0' and $\Sigma_1', (v^{\Sigma_2}(\{1\}, c))_{c \in M}$ differs from both $(v^{\Sigma_0}(\{1\}, c))_{c \in M}$ and $(v^{\Sigma_1}(\{1\}, c))_{c \in M}$ by a positive multiplicative number. This implies that the latter two vectors also differ by a positive multiplicative number, and that there is a unique v^{Σ_1} that ranks Σ_1' and agrees with v^{Σ_0} on the row of $\{1\}$. (Observe that this row in not the 0 vector due to the diversity condition.)

Given an event $A \in \Sigma'$, choose a finite subalgebra Σ_1 as above that includes A, and define $(v(A, c))_{c \in M}$ by the unique v^{Σ_1} identified above. By similar considerations one concludes that this definition of $(v(A, c))_{c \in M}$ does not depend on the choice of Σ_1 that includes A. Hence, v is well defined. Finally, we wish to show that if $A, B, A \cup B \in \Sigma'$, and $A \cap B \neq \emptyset$, We have

$$v(A \cup B, c) = v(A, c) + v(B, c) \quad \text{for all } c \in M.$$

Choose a finite subalgebra Σ_1 as above that includes both A and B, and observe that v coincides with $v^{\Sigma'_1}$ on Σ'_1, where $v^{\Sigma'_1}$ satisfies this equality by Lemma 8.   □

Proof of Remark 2.5: Consider $M = \Omega = \{1, 2, 3, 4\}$ and define $\{\succsim_I\}_I$ by the matrix v given by the table below (where empty entries denote zeroes). It is straightforward to check that $\{\succsim_I\}_I$ satisfies Axiom 5 for all triples of events, yet the matrix v is not additive in events.

$v^A(c)$		Case c			
		1	2	3	4
	1	2			
	2		2		
	3			2	
	4				2
Event A	1,2	2	2		
	1,3	2	1	1	
	1,4	2	1		1
	2,3	1	2	1	
	2,4	1	2		1
	3,4	1	1	1	1

Proof of Remark 2.6: Consider $M = \Omega = \{1, 2, 3, 4, 5\}$ and define $\{\succsim_I\}_I$ as follows: if $A \subseteq B$, then $A \preceq_I B$ for every $I \in J$. If A is a proper subset of B, then $A \prec_I B$ for every $I \neq 0$. If A and B are nonincluded, define $\{\succsim_I\}_I$ by the signed measures $\{\mu_c\}_{c \in M}$ given by the table below (where empty entries denote zeros):

$$\mu_c(\{i\})$$

Case c

State i		1	2	3	4	5
	1	−1	1	1	1	1
	2	2	1			
	3	4		1		
	4	8			1	
	5	16				1

One may verify that $\{\gtrsim_I\}_I$ are qualitative probability relations satisfying Axioms $2^* - 4^*$, even though $\mu_1(\{1\}) < 0$.

Proof of Theorem 3.1: We first show:

Lemma 9: *The symmetry axiom implies that there are two numbers, $a > b \geq 0$, such that $\mu_i(\{j\}) = b$ if $i \neq j$ and $\mu_i(\{i\}) = a$.*

Observe that this condition is also equivalent to the symmetry axiom.

Proof of Lemma 1:

Claim 1: For every $i \leq n$ there exists a number $b_i \in \mathbb{R}$, such that $\mu_i(\{j\}) = b_i$ for all $j \neq i$.

Proof. Consider the memory $I = 1_{\{i\}} \in \mathbb{J}$ and a permutation π that swaps only two states $j, k \neq i$. Since $I = I \circ \pi$, the symmetry axiom implies that $\{j\} \approx_I \{k\}$, hence $\mu_i(\{j\}) = \mu_i(\{k\})$. $\quad\square$

We denote $a_i = \mu_i(\{i\})$.

Claim 2: *For every $i, j, \leq n, a_j - b_j = a_i - b_i$.*

Proof. Consider $I = 1_{\{i,j\}}, \in \mathbb{J}$ and a permutation π that swaps only i, j. Since $I = I \circ \pi$, the symmetry axiom implies that $\{i\} \approx_I \{j\}$, hence $a_j + b_i = b_j + a_i$, or $a_j - b_j = a_i - b_i$. $\quad\square$

Claim 3: *For every distinct $i, j, k \leq n, \{k\} \succ_I \{i,j\}$ for $I = 1_{\{k\}} \in \mathbb{J}$.*

Proof. By the diversity axiom we know that there exists a memory $I \in \mathbb{J}$ such that $\{k\} \succ_I \{i,j\}$. The combination axiom implies that there exists $s \leq n$ such that $\{k\} \succ_I \{i,j\}$ for $I = 1_{\{s\}}$. We claim that this can only hold for $s = k$. Indeed, first observe that if $s \notin \{i, j, k\}$, then for $I = l_{\{S\}}$ we also get, by symmetry, $\{i\} \succ_I \{j, k\}$. But, since $\succsim_I$ is a qualitative probability, $\{k\} \succ_I \{i, j\}$ implies $\{j, k\} \succ_I, \{i\}$, a contradiction. Next assume that $s = i$. By similar reasoning, $\{k\} \succ_I \{i, j\}$ implies $\{i, k\} \succ_I \{i, j\}$, but this implies $\{k\} \succ_I \{j\}$, which contradicts symmetry (as in Claim 1). $s = j$ is similarly excluded, and the conclusion follows. $\quad\square$

Claim 4:   *For every $i, j \leq n, a_i = a_j > 0$.*

Proof. Let there be given distinct $i, j, k \leq n$. Consider, for every two nonnegative integers m, l, the momory $I = I(m, l) = m1_{\{j\}} + l1_{\{k\}} \in \mathbb{J}$, and the permuation π that swaps only i, j. Thus, $I \circ \pi = m1_{\{i\}} + l1_{\{k\}}$. It follow from the symmetry axiom that $\{i, j\} \succ_I \{k\}$ iff $\{i, j\} \succ_{I \circ \pi} \{k\}$. This means that, for every $m, l \geq 0$,

$$ma_i + mb_i + 2lb_k \geq mb_i + la_k \quad \text{iff} \quad ma_j + mb_j + 2lb_k \geq mb_j + la_k$$

or

$$ma_i \geq l(a_k - 2b_k) \quad \text{iff} \quad ma_j \geq l(a_k - 2b_k). \qquad (*)$$

Further, we argue that $a_i, a_j, (a_k - 2b_k) > 0$. First, observe that by Claim 3 (corresponding to $m = 0, l = 1$) it has to be the case that $(a_k - 2b_k) > 0$. Also, Claim 3 implies that $\{i\} \succ_I \{j, k\} for I = 1_{\{i\}}$ (corresponding to $m = 1, l = 0$), and $\{i, j\} \succ_I \{k\}$ follows by monotonicity of $\succsim_I$ with respect to set inclusion. Hence $a_i > 0$. Similarly, $a_j > 0$ has to hold as well. The desired result now follows from $(*)$. $\quad\square$

Combining Claims 1, 2, and 4, we conclude that there are two numbers, $a, b \in \mathbb{R}$, such that $\mu_i(\{j\}) = b$ if $i \neq j$ and $\mu_i(\{i\}) = a$. Furthermore, we know that $a > 0$.

Claim 5: $a > b$.

Proof. As in the proof of Claim 4, $(a - 2b) > 0$ and this suffices since $a > 0$. $\quad\square$

Claim 6: $b \geq 0$.

Proof. Let $i, j, k \leq n$ be distinct. Consider $I = 1_{\{i, j\}}$. We know that $\{i\} \approx_I \{j\}$, hence $\{i, k\} \succsim_I \{j\}$. Hence $a + 3b \geq a + b$. $\quad\square$

This completes the proof of Lemma 1. It remains to show that the specificity axiom implies that $b = 0$ as well. To see this, consider again the proof of Claim 6 above, and observe that if $b > 0$, it has to be the case that $\{i, k\} \succ_I \{j\}$, where $I = l_{\{i,j\}}$. But this is a contradiction to the specificity axiom, because $\{i\} \approx_I \{j\}$ and $I(k) = 0$. □

Proof of Remark 3.2: Assume that for some $i, j \in \Omega$, $\mu_i(\{j\}) < 0$. This means that for $I = l_{\{i\}}$ we have $\varnothing \succ_I \{j\}$, contradicting A1'. Hence μ_i are nonnegative. Next we consider $i \neq j$ and show that $\mu_i(\{j\}) = 0$. Assume, to the contrary, that $\mu_i(\{j\}) > 0$ for $i \neq j$. In this case, for $I = l_{\{i\}}$ we have $\varnothing \succsim_I \varnothing$ and $\{j\} \succ_I \varnothing$ while $I(j) = 0$, contradicting A7. Also, since by A1', $\Omega \succ_I \varnothing$ for $I = l_{\{i\}}$, it follows that $\mu_i(\{i\}) > 0$ for all $i \in \Omega$. Finally, consider $I = l_{\{i,j\}}$ for $i \neq j$. By A6, $\{i\} \approx_I \{j\}$. Hence $\mu_i(\{i\}) = \mu_j(\{j\})$. □

Acknowledgments

The authors wish to thank Didier Dubois, Roger Myerson, Peter Wakker, and Peyton Young for conversations and comments that motivated and influenced this work. The authors also thank an anonymous referee for helpful comments. This work was partially supported by ISF Grant No. 790/00.

References

Aumann, R. J. (1995), Interactive epistemology, Discussion paper no. 67, Center for Rationality and Interactive Decision Theory, The Hebrew University of Jerusalem, Jerusalem, Israel.

Fagin, R. and J. Y. Halpern (1994), Reasoning about knowledge and probability. *Journal of the Association for Computing Machinery*, **41**: 340–367.

—, —, N. Megiddo (1990), A logic for reasoning about probabilities. *Information and Computation*, **87**: 78–128.

Fine, T. (1973), *Theories of Probability*, Academic Press, New York.

de Finetti, B. (1931), Sul significato soggettivo della probabilita. *Fundamenta Mathematicae*, **17**: 298–329.

—. (1937), La prevision: Ses lois logiques, ses sources subjectives. *Annales de l'Institute Henri Poincare*, **7**: 1–68.

Fishburn, P. C. (1986), The axioms of subjective probability. *Statistical Science,* **1**: 335–358.

Gilboa, I. and D. Schmeidler (1999), Inductive inference: An axiomatic approach. Working paper no. 29–99, Foerder Institute for Economic Research, Tel Aviv University, Tel Aviv, Israel. Revised, 2001. Published in *Econometrica,* **71** (2003): 1–26. Reprinted as Chapter 4 in this volume.

—, —. (2001), *A Theory of Case-Based Decisions.* Cambridge University Press, Cambridge, U.K.

Heifetz, A. and P. Mongin (1999), Probability logic for type spaces. *Games Econom. Behavior.* Published as *Games and Economic Behavior,* **35**: 31–53 (2008).

Kraft, C. H., J. W. Pratt and A. Seidenberg (1959), *Annals of Mathematical Statistics,* **30**: 408–419.

Krantz, D. H., D. R. Luce, P. Suppes and A. Tversky (1971), *Foundations of Measurement I,* Academic Press, New York.

Ramsey, F. P. (1931), Truth and probability. *The Foundation of Mathematics and Other Logical Essays,* Harcourt, Brace and Co., New York.

Savage, L. J. (1954) *Foundations of Statistics,* John Wiley and Sons, New York.

Chapter 4

Inductive Inference: An Axiomatic Approach*

Itzhak Gilboa[†] and David Schmeidler[‡]

Reprinted from *Econometrica*, *71* (2003): 1–26.

A predictor is asked to rank eventualities according to their plausibility, based on past cases. We assume that she can form a ranking given any memory that consists of finitely many past cases. Mild consistency requirements on these rankings imply that they have a numerical representation via a matrix assigning numbers to eventuality-case pairs, as follows. Given a memory, each eventuality is ranked according to the sum of the numbers in its row, over cases in memory. The number attached to an eventuality-case pair can be interpreted as the degree of support that the past case lends to the plausibility of the eventuality. Special instances of this result may be viewed as axiomatizing kernel methods for estimation of densities and for classification problems. Interpreting the same result for rankings of theories or hypotheses, rather than of specific eventualities, it is shown that one may ascribe to the predictor subjective conditional probabilities of cases given theories, such that her rankings of theories agree with rankings by the likelihood functions.

---

*We wish to thank Yoav Binyamini, Didier Dubois, Drew Fudenberg, John Geanakoplos, Bruno Jullien, Edi Karni, Simon Kasif, Daniel Lehmann, Sujoy Mukerji, Roger Myerson, Klaus Nehring, Ariel Rubinstein, Lidror Troyanski, Peter Wakker, Peyton Young, and three anonymous referees for the discussions that motivated this work, as well as for comments and references.
[†]The Eitan Berglas School of Economics, Tel-Aviv University, Tel-Aviv 69978, ISRAEL; The Recanati School of Business, Tel-Aviv University; Cowles Foundation, Yale University; igilboa@post.tau.ac.il.
[‡]Department of Statistics, Tel-Aviv 69978, ISRAEL; Recanati School of Business, Tel-Aviv University and Department of Economics, The Ohio State University. schmeid@post.tau.ac.il.

Keywords: Case-based reasoning; case-based decision theory; prediction; maximum likelihood; kernel functions; kernel classification.

4.1. Introduction

Prediction is based on past cases. As Hume (1748) argued, "From causes which appear similar we expect similar effects. This is the sum of all our experimental conclusions." Over the past decades Hume's approach has found re-incarnations in the artificial intelligence literature as reasoning by analogies, reasoning by similarities, or case-based reasoning. (See Schank (1986) and Riesbeck and Schank (1989).) Many authors accept the view that analogies, or similarities to past cases hold the key to human reasoning. Moreover, the literature on machine learning and pattern recognition deals with using past cases, or observations, for predicting or classifying new data. (See, for instance, Forsyth and Rada (1986) and Devroye, Gyorfi, and Lugosi (1996).) But how should past cases be used? How does, and how should one resolve conflicts between different analogies? To address these questions, let us first consider a few examples.

Example 1:　A die is rolled over and over again. One has to predict the outcome of the next roll. As far as the predictor can tell, all rolls were made under identical conditions. Also, the predictor does not know of any a-priori reason to consider any outcome more likely than any other. The most reasonable prediction seems to be the mode of the empirical distribution, namely, the outcome that has appeared most often in the past. Moreover, empirical frequencies suggest a plausibility ranking of all possible outcomes, and not just a choice of the most plausible ones.[1]

Example 2:　A physician is asked by a patient if she predicts that a surgery will succeed in his case. The physician knows whether the procedure succeeded in most cases in the past, but she will be quick to remind her patient that every human body is unique. Indeed, the

---

[1]The term "likelihood" in the context of a binary relation, "at least as likely as", has been used by de Finetti (1937) and by Savage (1954). It should not be confused with "likelihood" in the context of likelihood functions, also used in the sequel. At this point we use "likelihood" and "plausibility" informally and interchangeably.

physician knows that the statistics she read included patients who varied in terms of age, gender, medical condition, and so forth. It would therefore be too naive of her to quote statistics as if the empirical frequencies were all that mattered. On the other hand, if the physician considers only past cases of patients that are identical to hers, she will probably end up with an empty database.

Example 3: An expert on international relations is asked to predict the outcome of the conflict in the Middle East. She is expected to draw on her vast knowledge of past cases, coupled with her astute analysis thereof, in forming her prediction. As in Example 2, the expert has a lot of information she can use, but she cannot quote even a single case that was identical to the situation at hand. Moreover, as opposed to Example 2, even the possible eventualities are not identical to outcomes that occurred in past cases.

We seek a theory of prediction that will permit the predictor to make use of available information, where different past cases might have differential relevance to the prediction problem. Specifically, we consider a prediction problem for which a set of possible eventualities is given. This set may or may not be an exhaustive list of all conceivable eventualities. We do not model the process by which such a set is generated. Rather, we assume the set given and restrict attention to the problem of qualitative ranking of its elements according to their likelihood.

The prediction rule Consider the following prediction rule for Example 2. The physician considers all known cases of successful surgery. She uses her subjective judgment to evaluate the similarity of each of these cases to the patient she is treating, and she adds them up. She then does the same for unsuccessful treatments. Her prediction is the outcome with the larger aggregate similarity value. This generalizes frequentist ranking to a "fuzzy sample": in both examples, likelihood of an outcome is measured by summation over cases in which it occurred. Whereas in Example 1 the weight attached to each past case is 1, in this example this weight varies according to the physician's subjective assessment of similarity of the relevant cases. Rather than a dichotomous distinction between data points that do and those that do not belong to the sample, each data point belongs to the sample to a certain degree, say, between 0 and 1.

The prediction rule we propose can also be applied to Example 3 as follows. For each possible outcome of the conflict in the Middle East, and for each past case, the expert is asked to assess a number, measuring the degree of support that the case lends to this outcome. Adding up these numbers, for all known cases and for each outcome, yields a numerical representation of the likelihood ranking. Thus, our prediction rule can be applied also when there is no structural relationship between past cases and future eventualities.

Formally, let M denote the set of known cases. For each $c \in M$ and each eventuality x, let $v(x, c) \in \mathbb{R}$ denote the degree of support that case c lends to eventuality x. Then the prediction rule ranks eventuality x as more likely than eventuality y if and only if

$$\sum_{c \in M} v(x, c) > \sum_{c \in M} v(y, c). \tag{1}$$

Axiomatization The main goal of this paper is to axiomatize this rule. We assume that a predictor has a ranking of possible eventualities given any possible memory (or database). A memory consists of a finite set of past cases, or stories. The predictor need not envision all possible memories. She might have a rule, or an algorithm that generates a ranking (in finite time) for each possible memory. We only rely on qualitative plausibility rankings, and do not assume that the predictor can quantify them in a meaningful way. Cases are not assumed to have any particular structure. However, we do assume that for every case there are arbitrarily many other cases that are deemed equivalent to it by the predictor (for the prediction problem at hand). For instance, if the physician in Example 2 focuses on five parameters of the patient in making her prediction, we can imagine that she has seen arbitrarily many patients with particular values of the five parameters. The equivalence relation on cases induces an equivalence relation on memories (of equal sizes), and the latter allows us to consider replication of memories, that is, the disjoint union of several pairwise equivalent memories.

Our main assumption is that prediction satisfies a *combination axiom*. Roughly, it states that if an eventuality x is more likely than an eventuality y given two possible disjoint memories, then x is more likely than y also given their union. For example, assume that the patient in Example 2 consults two physicians, who were trained in the same medical school but who have been working in different hospitals since graduation. Thus, the physicians can be thought of as having disjoint databases on which they can base their prediction, while sharing the inductive algorithm. Assume next that both physicians find that success is more likely than failure in the case at hand. Should the patient ask them to share their databases and re-consider their predictions? If the inductive algorithm that the physicians use satisfies the combination axiom, the answer is negative.

We also assume that the predictor's ranking is *Archimedean* in the following sense: if a database M renders eventuality x more likely than eventuality y, then for every other database N there is a sufficiently large number of replications of M, such that, when these memories are added to N, they will make eventuality x more likely than eventuality y. Finally, we need an assumption of *diversity*, stating that any list of four eventualities may be ranked, for some conceivable database, from top to bottom. Together, these assumptions necessitate that prediction be made according to the rule suggested by the formula (1) above. Moreover, we show that the function v in (1) is essentially unique.

This result can be interpreted in several ways. From a descriptive viewpoint, one may argue that experts' predictions tend to be consistent as required by our axioms (of which the combination is the most important), and that they can therefore be represented as aggregate similarity-based predictions. From a normative viewpoint, our result can be interpreted as suggesting the aggregate similarity-based predictions as the only way to satisfy our consistency axioms. In both approaches, one may attempt to measure similarities using the likelihood rankings given various databases.

Observe that we assume no a priori conceptual relationship between cases and eventualities. Such relationships, which may exist in the predictor's mind, will be revealed by her plausibility rankings. Further, even if cases and eventualities are formally related (as in Example 2), we do not assume that a numerical measure of distance, or of similarity is given in the data.

Our decision rule generalizes several well-known statistical methods, apart from ranking eventualities by their empirical frequencies. Kernel methods for estimation of a density function, as well as for classification problems, are special case of our rule. If the objects that are ranked by plausibility are general theories, rather than specific eventualities, our rule can be viewed as ranking theories according to their likelihood function. In particular, these established statistical methods satisfy our combination axiom. This may be taken as an argument for this axiom. Conversely, our result can be used to axiomatize these statistical methods in their respective set-ups.

Methodological remarks The Bayesian approach (Ramsey (1931), de Finetti (1937), and Savage (1954)) holds that all prediction problems should be dealt with by a prior subjective probability that is updated in light of new information via Bayes' rule. This requires that the predictor have a prior probability over a space that is large enough to describe all conceivable new information. We find that in certain examples (as above) this assumption is not cognitively plausible. By contrast, the prediction rule (1) requires the evaluation of support weights only for cases that were actually encountered. For an extensive methodological discussion, see Gilboa and Schmeidler (2001).

Since the early days of probability theory, the concept of probability serves a dual role: one relating to empirical frequencies, and the other — to quantification of subjective beliefs or opinions. (See Hacking (1975).) The Bayesian approach offers a unification of these roles employing the concept of a subjective prior probability. Our approach may also be viewed as an attempt to unify the notions of empirical frequencies and subjective opinions. Whereas the axiomatic derivations of de Finetti (1937) and Savage (1954) treat the process of the generation of a prior as a black box, our rule aims to make a preliminary step towards the modeling of this process.

Thus, our approach is complementary to the Bayesian approach at two levels: first, it may offer an alternative model of prediction, when the information available to the predictor is not easily translated to the language of a prior probability. Second, our approach may describe how a prior is generated. (See also Gilboa and Schmeidler (2002).)

The rest of this paper is organized as follows. Section 2 presents the formal model and the main results. Section 3 discusses the relationship to kernel methods and to maximum likelihood rankings. Section 4 contains a critical discussion of the axioms, attempting to outline their scope of application. Finally, Section 5 briefly discusses alternative interpretations of the model, and, in particular, relates it to case-based decision theory. Proofs are relegated to the appendix.

4.2. Model and Result

4.2.1. *Framework*

The primitives of our model consist of two non-empty sets X and $\mathbb{C}$. We interpret X as the set of all conceivable *eventualities* in a given prediction problem, p, whereas $\mathbb{C}$ represents the set of all conceivable *cases*. To simplify notation, we suppress the prediction problem p whenever possible. The predictor is equipped with a finite set of cases $M \subset \mathbb{C}$, her *memory*, and her task is to rank the eventualities by a binary relation, "at least as likely as".

While evaluating likelihoods, it is insightful not only to know what has happened, but also to take into account what could have happened. The predictor is therefore assumed to have a well-defined "at least as likely as" relation on X for many other collections of cases in addition to M itself. Let $\mathbb{M}$ be the set of finite subsets of $\mathbb{C}$. For every $M \in \mathbb{M}$, we denote the predictor's "at least as likely as" relation by $\succsim_M \subset X \times X$.

Two cases c and d are *equivalent*, denoted $c \sim d$, if, for every $M \in \mathbb{M}$ such that $c, d \notin M$, $\succsim_{M \cup \{c\}} = \succsim_{M \cup \{d\}}$. To justify the term, we note the following.

Observation: $\sim$ is an equivalence relation.

Note that equivalence of cases is a subjective notion: cases are equivalent if, in the eyes of the predictor, they affect likelihood rankings in the same way. Further, the notion of equivalence is also context-dependent: two cases c and d are equivalent as far as a specific prediction problem is concerned.

We extend the definition of equivalence to memories as follows. Two memories $M_1, M_2 \in \mathbb{M}$ are equivalent, denoted $M_1 \sim M_2$, if there is a bijection $f : M_1 \to M_2$ such that $c \sim f(c)$ for all $c \in M_1$. Observe that memory equivalence is also an equivalence relation. It also follows that, if $M_1 \sim M_2$, then, for every $N \in \mathbb{M}$ such that $N \cap (M_1 \cup M_2) = \varnothing$, $\succsim_{N \cup M_1} = \succsim_{N \cup M_2}$.

Throughout the discussion, we impose the following structural assumption.

Richness Assumption: For every case $c \in \mathbb{C}$, there are infinitely many cases $d \in \mathbb{C}$ such that $c \sim d$.

A note on nomenclature: the main result of this paper is interpreted as a representation of a prediction rule. Accordingly, we refer to a "predictor" who may be a person, an organization, or a machine. However, the result may and will be interpreted in other ways as well. Instead of ranking eventualities one may rank *decisions, acts,* or a more neutral term, *alternatives.* Cases, the elements of $\mathbb{C}$, may also be called *observations* or *facts.* A memory M in $\mathbb{M}$ represents the predictor's *knowledge* and will be referred to also as a *database.*

4.2.2. Axioms

We will use the four axioms stated below. In their formalization let $\succ_M$ and $\approx_M$ denote the asymmetric and symmetric parts of $\succsim_M$, as usual. $\succsim_M$ is *complete* if $x \succsim_M y$ or $y \succsim_M x$ for all $x, y \in X$.

A1 Order: For every $M \in \mathbb{M}$, $\succsim_M$ is complete and transitive on X.

A2 Combination: For every disjoint $M, N \in \mathbb{M}$ and every $x, y \in X$, if $x \succsim_M y$ ($x \succ_M y$) and $x \succsim_N y$, then $x \succsim_{M \cup N} y$ ($x \succ_{M \cup N} y$).

A3 Archimedean Axiom: For every disjoint $M, N \in \mathbb{M}$ and every $x, y \in X$, if $x \succ_M y$, then there exists $l \in \mathbb{N}$ such that for any l-list $(M_i)_{i=1}^l$ of pairwise disjoint M_i's in $\mathbb{M}$, where for all $i \leq l$, $M_i \sim M$ and $M_i \cap N = \varnothing$, $x \succ_{M_1 \cup \cdots \cup M_l \cup N} y$ holds.

Axiom 1 simply requires that, given any conceivable memory, the predictor's likelihood relation over eventualities is a weak order. Axiom 2 states that if eventuality x is more plausible than eventuality y given two disjoint memories, x should also be more plausible than y given the union of these memories. Axiom 3 is states that if, given the memory M, the predictor believes that eventuality x is strictly more plausible than y, then, no matter what is her ranking for another memory, N, there is a number of "repetitions" of M that is large enough to overwhelm the ranking induced by N.

Finally, we need a diversity axiom. It is not necessary for representation of likelihood relations by summation of real numbers. Theorem 1 below is an equivalence theorem, characterizing precisely which matrices of real numbers will satisfy this axiom.

A4 Diversity: For every list (x, y, z, w) of distinct elements of X there exists $M \in \mathbb{M}$ such that $x \succ_M y \succ_M z \succ_M w$. If $|X| < 4$, then for any strict ordering of the elements of X there exists $M \in \mathbb{M}$ such that $\succ_M$ is that ordering.

4.2.3. Results

For clarity of exposition, we first state the sufficiency result.

Theorem 1 Part I — Sufficiency: *Let there be given* X, $\mathbb{C}$, *and* $\{\succsim_M\}_{M \in \mathbb{M}}$ *satisfying the richness assumption as above. Then (i) implies (ii(a)):*

(i) $\{\succsim_M\}_{M \in \mathbb{M}}$ *satisfy A1–A4;*
(ii(a)) *There is a matrix* $v : X \times \mathbb{C} \to \mathbb{R}$ *such that:*

$$\begin{cases} \text{for every } M \in \mathbb{M} \text{ and every } x, y \in X, \\ x \succsim_M y \quad \text{iff} \quad \sum_{c \in M} v(x, c) \geq \sum_{c \in M} v(y, c), \end{cases} \tag{2}$$

In other words, axioms A1–A4 imply that $\{\succsim_M\}_{M \in \mathbb{M}}$ follow our prediction rule for an appropriate choice of the matrix v. Not all of these axioms are, however, necessary for the representation to obtain. Indeed, the axioms imply special properties of the representing matrix v. First, it can be chosen in such a way that equivalent cases are attached identical columns. Second, every four rows of the matrix satisfy an additional condition. Existence of a matrix v satisfying these two properties together with (2) does imply axioms A1–A4. Before stating the necessity part of theorem, we present two additional definitions.

Definition: A matrix $v : X \times \mathbb{C} \to \mathbb{R}$ *respects case equivalence* (with respect to $\{\succsim_M\}_{M \in \mathbb{M}}$) if for every $c, d \in \mathbb{C}$, $c \sim d$ iff $v(\cdot, c) = v(\cdot, d)$.

When no confusion is likely to arise, we will suppress the relations $\{\succsim_M\}_{M \in \mathbb{M}}$ and will simply say that "v respects case equivalence".

The following definition applies to real-values matrices in general. It will be used for the matrix $v : X \times \mathbb{C} \to \mathbb{R}$ in the statement of the theorem, but also for another matrix in the proof. It defines a matrix to be diversified if no row in it is dominated by an affine combination of any other three (or less) rows. Thus, if v is diversified, no row in it dominates another. Indeed, the property of diversification can be viewed as a generalization of this condition.

Definition: A matrix $v : X \times \Upsilon \to \mathbb{R}$, where $|X| \geq 4$, is *diversified* if there are no distinct four elements $x, y, z, w \in X$ and $\lambda, \mu, \theta \in \mathbb{R}$ with $\lambda + \mu + \theta = 1$ such that $v(x, \cdot) \leq \lambda v(y, \cdot) + \mu v(z, \cdot) + \theta v(w, \cdot)$. If $|X| < 4$, v is diversified if no row in v is dominated by an affine combination of the others.

We can finally state

Theorem 1 Part II — Necessity:

  (i)   *also implies*
  (ii(b)) *the matrix v is diversified; and*
  (ii(c)) *the matrix v respects case equivalence.*

  Conversely, (ii(a,b,c)) *implies* (i).

Theorem 1 Part III — Uniqueness: *If* (i) [*or* (ii)] *hold, the matrix v is unique in the following sense: v and u both satisfy* (2) *and respect case equivalence iff there are a scalar $\lambda > 0$ and a matrix $\beta : X \times \mathbb{C} \to \mathbb{R}$ with identical rows (i.e., with constant columns), that respects case equivalence, such that $u = \lambda v + \beta$.*

Observe that, by the richness assumption, $\mathbb{C}$ is infinite, and therefore the matrix v has infinitely many columns. Moreover, the theorem does not restrict the cardinality of X, and thus v may also have infinitely many rows.

Given any real matrix of order $|X| \times |\mathbb{C}|$, one can define for every $M \in \mathbb{M}$ a weak order on X through (2). It is easy to see that it will satisfy

A1 and A2. If the matrix also respects case equivalence, A3 will also be satisfied. However, these conditions do not imply A4. For example, A4 will be violated if a row in the matrix dominates another row. Since A4 is not necessary for a representation by a matrix v via (2) (even if it respects case equivalence), one may wonder whether it can be dropped. The answer is given by the following.

Proposition: Axioms A1, A2, and A3 do not imply the existence of a matrix v that satisfies (2).

Some remarks on cardinality are in order. Axiom A4 can only hold if the set of types, $\mathbb{T} = \mathbb{C}/\sim$, is large enough relatively to X. For instance, if there are two distinct eventualities, the diversity axiom requires that there be at least two different types of cases. However, six types suffice for X to have the cardinality of the continuum.[2]

Finally, one may wonder whether (2) implies that v respects case equivalence. The negative answer is given below.

Remark: Condition (2) does not imply that v respects case equivalence.

4.3. Related Statistical Methods

4.3.1. *Kernel estimation of a density function*

Assume that Z is a continuous random variable taking values in $\mathbb{R}^m$. Having observed a finite sample $(z_i)_{i \leq n}$, one is asked to estimate the density function of Z. Kernel estimation (see Akaike (1954), Rosenblatt (1956), Parzen (1962), Silverman (1986), and Scott (1992) for a survey) suggests the following. Choose a (so-called "kernel") function $k : \mathbb{R}^m \times \mathbb{R}^m \to \mathbb{R}_+$ with the following properties: (i) $k(z, y)$ is a non-increasing function of $\|z - y\|$; (ii) for every $z \in \mathbb{R}^m$, $\int_{\mathbb{R}^m} k(z, y) dy = 1$.[3] Given the sample $(z_i)_{i \leq n}$, estimate the density function by $f(y|z_1, \ldots, z_n) \equiv \frac{1}{n} \sum_{i \leq n} k(z_i, y)$.

---

[2] The proof is omitted for brevity's sake.
[3] More generally, the kernel may be a function of transformed coordinates. The following discussion does not depend on assumptions (i) and (ii) and they are retained merely for concreteness.

Consider the estimated function f as a measure of likelihood: $f(y) > f(w)$ is interpreted as saying that a small neighborhood around y is more likely than the corresponding neighborhood around w. With this interpretation, kernel estimation is clearly a special case of our prediction rule, with $v(y, z) = \frac{1}{n} k(z, y)$. Observe that kernel estimation presupposes a notion of distance on $\mathbb{R}^m$, whereas our theorem derives the function v from qualitative rankings alone.

4.3.2. *Kernel classification*

Kernel methods are also used for classification problems. Assume that a classifier is confronted with a data point $y \in \mathbb{R}^m$, and it is asked to guess to which member of a finite set A it belongs. The classifier is equipped with a set of examples $M \subset \mathbb{R}^m \times A$. Each example (x, a) consist of a data point $x \in \mathbb{R}^m$, with a known classification $a \in A$. Kernel classification methods would adopt a kernel function as above, and, given the point y, would guess that y belongs to a class $a \in A$ that maximizes the sum of $k(x, y)$ over all x's in memory that were classified as a.

Our general framework can accommodate classification problems as well. As opposed to kernel estimation, one is not asked to rank (neighborhoods of) points in $\mathbb{R}^m$, but, *given* such a point, to rank classes in A. Assume a point $y \in \mathbb{R}^m$ is given, and, for a case $(x, a) \in M$, define $v_y(b, (x, a)) = k(x, y) 1_{a=b}$ (where $1_{a=b}$ is 1 if $a = b$ and zero otherwise). Clearly, the ranking defined by v_y boils down to the ranking defined by kernel classification.

As above, this axiomatization can be viewed as a normative justification of kernel methods, and also as a way to elicit the "appropriate" kernel function from qualitative ranking data. Again, our approach does not assume that a kernel function is given, but derives such a function together with the kernel classification rule.

A popular alternative to kernel classification methods is offered by nearest neighbor methods. (See Fix and Hodges (1951, 1952), Royall (1966), Cover and Hart (1967), Stone (1977), and Devroye, Gyorfi, and Lugosi (1996).) It is easily verified that nearest neighbor approaches do not satisfy the Archimedean axiom. Moreover, for $k > 1$ a majority vote among the k-nearest neighbors violates the combination axiom. Thus, our axioms offer a normative justification for preferring kernel methods to nearest neighbor methods.

4.3.3. *Maximum likelihood ranking*

Our model can also be interpreted as referring to ranking of theories or hypotheses given a set of observations. The axioms we formulated apply to this case as well. In particular, our main requirements are that theories be ranked by a weak order for every memory, and that, if theory x is more plausible than theory y given each of two disjoint memories, x should also be more plausible than y given the union of these memories.

Assume, therefore, that Theorem 1 holds. Suppose that, for each case c, $v(x, c)$ is bounded from above. (This is the case, for instance, if there are only finitely many theories to be ranked.) Choose a representation v where $v(x, c) < 0$ for every theory x and case c. Define $p(c|x) = \exp(v(x, c))$, so that $\log(p(c|x)) = v(x, c)$.

Our result states that, for every two theories x, y,

$$x \succsim_M y \quad \text{iff} \quad \sum_{c \in M} v(x, c) \geq \sum_{c \in M} v(y, c),$$

which is equivalent to

$$\exp\left(\sum_{c \in M} v(x, c)\right) \geq \exp\left(\sum_{c \in M} v(y, c)\right) \quad \text{or to}$$

$$\prod_{c \in M} p(c|x) \geq \prod_{c \in M} p(c|y).$$

In other words, if a predictor ranks theories in accordance with A1–A4, there exist conditional probabilities $p(c|x)$, for every case c and theory x, such that the predictor ranks theories as if by their likelihood functions, under the implicit assumption that the cases were stochastically independent.[4] On the one hand, this result can

---

[4]We do not assume that the cases that have been observed (M) constitute an exhaustive state space. Correspondingly, there is no requirement that the sum of conditional probabilities $\sum_{c \in M} p(c|x)$ be the same for all x.

be viewed as a normative justification of the likelihood rule: any method of ranking theories that is not equivalent to ranking by likelihood (for *some* conditional probabilities $p(c|x)$) has to violate one of our axioms. On the other hand, our result can be descriptively interpreted, saying that likelihood rankings of theories are rather prevalent. One need not consciously assign conditional probabilities $p(c|x)$ for every case c given every theory x, and one need not know probability calculus in order to generate predictions in accordance with the likelihood criterion. Rather, whenever one satisfies our axioms, one may be ascribed conditional probabilities $p(c|x)$ such that one's predictions are in accordance with the resulting likelihood functions. Thus, relatively mild consistency requirements imply that one predicts *as if* by likelihood functions.

Finally, our result may be used to elicit the subjective conditional probabilities $p(c|x)$ of a predictor, given her qualitative rankings of theories. However, our uniqueness result is somewhat limited. In particular, for every case c one may choose a positive constant β_c and multiply $p(c|x)$ by β_c for all theories x, resulting in the same likelihood rankings. Similarly, one may choose a positive number α and raise all probabilities $\{p(c|x)\}_{c,x}$ to the power of α, again without changing the observed ranking of theories given possible memories. Thus there will generally be more than one set of conditional probabilities $\{p(c|x)\}_{c,x}$ that are consistent with $\{\succsim_M\}_{M\in\mathbb{M}}$.

The likelihood function relies on independence across cases. Conceptually, stochastic independence follows from two assumptions in our model. First, we have defined $\{\succsim_M\}_{M\in\mathbb{M}}$ where each M is a set. This implicitly assumes that only the number of repetitions of cases, and not their order, matters. This structural assumption is reminiscent of de Finetti's exchangeability condition (though the latter is defined in a more elaborate probabilistic model). Second, our combination axiom also has a flavor of independence. In particular, it rules out situations in which past occurrences of a case make future occurrences of the same case less likely.[5]

---

[5] See the clause "mis-specified case" in the next section.

4.4. Discussion of the Axioms

The rule we axiomatize generalizes rankings by empirical frequencies. Moreover, the previous section shows that it also generalizes several well-known statistical techniques. It follows that there is a wide range of applications for which this rule, and the axioms it satisfies, are plausible.

But there are applications in which the axioms do not appear compelling. We discuss here several examples, trying to delineate the scope of applicability of the axioms, and to identify certain classes of situations in which they may not apply.

In the following discussion we do not dwell on the first axiom, namely, that likelihood relations are weak orders. This axiom and its limitations have been extensively discussed in decision theory, and there seems to be no special arguments for or against it in our specific context.

We also have little to add to the discussion of the diversity axiom. While it does not appear to pose conceptual difficulties, there are no fundamental reasons to insist on its validity. One may well be interested in other assumptions that would allow a representation as in (2) by a matrix v that is not necessarily diversified.

The Archimedean axiom is violated when a single case may outweigh any number of repetitions of other cases. For instance, a physician may find a single observation, taken from the patient she is currently treating, more relevant than any number of observations taken from other patients.[6] In the context of ranking theories, it is possible that a single case c constitutes a direct refutation of a theory x. If another theory y was not refuted by any case in memory, a single occurrence of case c will render theory x less plausible than theory y regardless of the number of occurrences of other cases, even if these lend more support to x than to y.[7] In such a situation, one would like to assign conditional probability of zero to case c given theory x, or, equivalently, to set $v(x, c) = -\infty$. Since this is beyond the scope of the present model, one may drop the Archimedean axiom and seek representations by non-standard numbers.

---

[6]Indeed, the nearest neighbor approach to classification problems violates the Archimedean axiom.

[7]This example is due to Peyton Young.

We now turn to the combination axiom. As is obvious from the additive formula in (2), our rule implicitly presupposes that the weight of evidence derived from a given case does not depend on other cases. It follows that the combination axiom is likely to fail whenever this "separability" property does not hold. We discuss here several examples of this type. We begin with those in which re-definition of the primitives of the model resolves the difficulty. Examples we find more fundamental are discussed later.

Mis-specified cases Consider a cat, say Lucifer, who every so often dies and then may or may not resurrect. Suppose that, throughout history, many other cats have been observed to resurrect exactly eight times. If Lucifer had died and resurrected four times, and now died for the fifth time, we'd expect him to resurrect again. But if we double the number of cases, implying that we are now observing the ninth death, we would not expect Lucifer to be with us again. Thus, one may argue, the combination axiom does not seem to be very compelling.

Obviously, this example assumes that all of Lucifer's deaths are equivalent. While this may be a reasonable assumption of a naive observer, the cat connoisseur will be careful enough to distinguish "first death" from "second death", and so forth. Thus, this example suggests that one has to be careful in the definition of a "case" (and of case equivalence) before applying the combination axiom.

Mis-specified theories Suppose that one wishes to determine whether a coin is biased. A memory with 1,000 repetitions of "Head", as well as a memory with 1,000 repetitions of "Tail" both suggest that the coin is indeed biased, while their union suggests that it is not. Observe that this example hinges on the fact that two rather different theories, namely, "the coin is biased toward Tail" and "the coin is biased toward Head" are lumped together as "the coin is biased". If one were to specify the theories more fully, the combination axiom would hold.[8]

---

[8]Observe that if one were to use the maximum likelihood principle, one would have to specify a likelihood function. This exercise would highlight the fact that "the coin is biased" is not a fully specified theory. However, this does not imply that only theories that are given as conditional distributions are sufficiently specified to satisfy the combination axiom.

Theories about patterns A related class of examples deal with concepts that describe, or are defined by patterns, sequences, or sets of cases. Assume that a single case consists of 100 tosses of a coin. A complex sequence of 100 tosses may lend support to the hypothesis that the coin generates random sequences. But many repetitions of the very same sequence would undermine this hypothesis. Observe that "the coin generates random sequences" is a statement about *sequences* of cases. Similarly, statements such as "The weather always surprises" or "History repeats itself" are about sequences of cases, and are therefore likely to generate violations of the combination axiom.

Second-order induction An important class of examples in which we should expect the combination axiom to be violated, for descriptive and normative purposes alike, involves learning of the similarity function. For instance, assume that one database contains but one case, in which Mary chose restaurant x over y.[9] One is asked to predict what John's decision would be. Having no other information, one is likely to assume some similarity of tastes between John and Mary and to find it more plausible that John would prefer x to y as well. Next assume that in a second database there are no observed choices (by anyone) between x and y. Hence, based on this database alone, it would appear equally likely that John would choose x as that he would y. Assume further that this database does contain many choices between other pairs of restaurants, and it turns out that John and Mary consistently choose different restaurants. When combining the two databases, it makes sense to predict that John would choose y over x.

This is an instance in which the similarity function is learned from cases. Linear aggregation of cases by fixed weights embodies learning *by* a similarity function. But it does not describe how this function *itself* is learned. In Gilboa and Schmeidler (2001) we call this process "second-order induction" and show that the additive formula cannot capture such a process.

Combinations of inductive and deductive reasoning Another important class of examples in which the combination axiom is not very reasonable consists of prediction problems in which some structure

---

[9]This is a variant of an example by Sujoy Mukerji.

is given. Consider a simple regression problem where a variable x is used to predict another variable y. Does the method of ordinary least squares satisfy our axioms? The answer depends on the unit of analysis. If we consider the regression equation $y = \alpha + \beta x + \varepsilon$ and attempt to estimate the values of α and β given a sample $M = \{(x_i, y_i)\}_{i \leq n}$, the answer is in the affirmative. Consider, for instance, α. Let a, a' be two real numbers interpreted as estimates of α. Define $a \succsim_M a'$ if a has a higher value of the likelihood function given $\{(x_i, y_i)\}_{i \leq n}$ than does a'. This implies that $\succsim_M$ satisfies the combination axiom. Since the least squares estimator a is a maximum likelihood estimator of the parameter α (under the standard assumptions of regression analysis), choosing the estimate a is consistent with choosing a $\succsim_M$-maximizer.

Assume now that the units of analysis are the particular values of y_p for a new value of x_p. That is, rather than accepting the regression model $y = \alpha + \beta x + \varepsilon$ and asking what are the values of α and β, suppose that one is asked to predict (formulate $\succsim_M$) directly on potential values of y_p. The regression estimates a, b define a density function for y_p (a normal distribution centered around the value $a + bx_p$). This density function can be used to define $\succsim_M$, but these relations will generally not satisfy the combination axiom.

The reason is that the regression model is structured enough to allow some deductive reasoning. In ranking the plausibility of values of y for a given value of x, one makes two steps. First, one uses inductive reasoning to obtain estimates of the parameters a and b. Then, espousing a belief in the linear model, one uses these estimates to rank values of y by their plausibility. This second step involves deductive reasoning, exploiting the particular structure of the model. While the combination axiom is rather plausible for the first, inductive step, there is no reason for it to hold also for the entire inductive-deductive process.

To consider another example, assume that a coin is about to be tossed in an i.i.d. manner. The parameter of the coin is not known, but one knows probability rules that allow one to infer likelihood rankings of outcomes given any value of the unknown parameter. Again, when one engages in inference about the unknown parameter, one performs only inductive reasoning, and the combination axiom seems plausible. But when one is asked about particular outcomes, one uses inductive reasoning as well as deductive reasoning. In these cases, the combination axiom is too crude.[10]

In conclusion, there are classes of counterexamples to our axioms that result from under-specification of cases, of eventualities, or of memories. There are others that are more fundamental. Among these, two seem to deserve special attention. First, there are situations where second-order induction is involved, and the similarity function itself is learned. Indeed, our model deals with accumulated evidence but does not capture the emergence of new insights. Second, there are problems where some theoretical structure is assumed, and it can be used for deductive inferences. Our model captures some forms of inductive reasoning, but does not provide a full account of inferential processes involving a combination of inductive and deductive reasoning.

4.5. Other Interpretations

Decisions Theorem 1 can also have other interpretations. In particular, the objects to be ranked may be possible acts, with the interpretation of ranking as preferences. In this case, $v(x, c)$ denotes the support that case c lends to the choice of act x. The decision rule that results generalizes most of the decision rules of case-based decision theory (Gilboa and Schmeidler (2001)), as well as expected utility maximization, if beliefs are generated from cases in an additive way (see Gilboa and Schmeidler (2002)). Gilboa, Schmeidler, and Wakker (1999) apply this theorem, as well as an alternative approach, to axiomatize a theory of case-based decisions in which both the similarity function between problem-act pairs and the utility function of outcomes are derived from preferences. This model generalizes Gilboa and Schmeidler (1997), in which the utility function is assumed given and only the similarity function is derived from observed preferences.

Probabilities The main contribution of Gilboa and Schmeidler (2002) is to generalize the scope of prediction from eventualities

---

[10]We have received several counterexamples to the combination axiom that are, in our view, of this nature. In particular, we would like to thank Bruno Jullien, Klaus Nehring, and Ariel Rubinstein.

to events. That is, in that paper we assume that the objects to be ranked belong to an algebra of subsets of a given set. Additional assumptions are imposed so that similarity values are additive with respect to the union of disjoint sets. Further, it is shown that ranking by empirical frequencies can also be axiomatically characterized in this set-up. Finally, tying the derivation of probabilities with expected utility maximization, one obtains a characterization of subjective expected utility maximization in face of uncertainty. As opposed to the behavioral axiomatic derivations of de Finetti (1937) and Savage (1954), which infer beliefs from decisions, this axiomatic derivation follows a presumed cognitive path leading from belief to decision.

Appendix: Proofs

Proof of Observation:
It is obvious that $\sim$ is reflexive and symmetric. To show that it is transitive, assume that $c \sim d$ and $d \sim e$ for distinct c, d, e. Let M be such that $c, e \notin M$. If $d \notin M$, then $\succsim_{M \cup \{c\}} = \succsim_{M \cup \{d\}}$ by $c \sim d$ and $\succsim_{M \cup \{d\}} = \succsim_{M \cup \{e\}}$ by $d \sim e$, and $\succsim_{M \cup \{c\}} = \succsim_{M \cup \{e\}}$ follows. If $d \in M$, define $N = M \backslash \{d\}$. Since $c, d \notin N \cup \{e\}$, $c \sim d$ implies $\succsim_{N \cup \{e\} \cup \{c\}} = \succsim_{N \cup \{e\} \cup \{d\}}$. Similarly, since $d, e \notin N \cup \{c\}$, $d \sim e$ implies $\succsim_{N \cup \{c\} \cup \{d\}} = \succsim_{N \cup \{c\} \cup \{e\}}$. It follows that $\succsim_{M \cup \{c\}} = \succsim_{N \cup \{c,d\}} = \succsim_{N \cup \{c,e\}} = \succsim_{N \cup \{d,e\}} = \succsim_{M \cup \{e\}}$. $\qquad\square$

The Result is part of Theorem 1, and was stated only for expository purposes. We therefore prove only Theorem 1.

Proof of Theorem 1:
The strategy of the proof is as follows. The notion of case equivalence allows us to reduce the discussion to vectors of non-negative integers. We define the set of *types* of cases to be the $\sim$-equivalence classes: $\mathbb{T} = \mathbb{C} / \sim$. Assume, for simplicity, that there are finitely many types and finitely many eventualities. Rather than referring to sets of specific cases (memories M), we focus on vectors of non-negative integers. Such a vector $I : \mathbb{T} \to \mathbb{Z}_+$ represents many equivalent memories by counting how many cases of each type are in each of these memories. Thus, instead of dealing with subsets of the set $\mathbb{C}$, most of the discussion will be conducted in the space $\mathbb{Z}_+^{\mathbb{T}}$. Next, using the combination axiom, we extend the family rankings $\{\succsim_I\}$ from $I \in \mathbb{Z}_+^{\mathbb{T}}$ to $I \in \mathbb{Q}_+^{\mathbb{T}}$.

Focusing on two eventualities, x and y, we divide the vectors $I \in \mathbb{Q}_+^{\mathbb{T}}$ to those that render x more likely than y, and to those that induce the opposite ranking. Completeness and combination are the key axioms that allow us to invoke a separating hyperplane theorem. With the aid of the Archimedean axiom, one can prove that the separating hyperplane precisely characterizes the memories for which x is (strongly or weakly) more likely than y.

If one has only two eventualities, the proof is basically complete. Most of the work is in showing that the hyperplanes, which were obtained for each *pair* of eventualities, can be represented by a single matrix. More concretely, the separation theorem applied to a pair x, y yields a vector v^{xy}, unique up to multiplication by a positive constant, such that x is at least as likely as y given memory I iff $v^{xy} \cdot I \geq 0$. One now wishes to find a vector v^x for each eventuality x such that v^{xy} is a positive multiple of $(v^x - v^y)$ (simultaneously for all x, y).

This can be done if and only if there is a selection of vectors $\{v^{xy}\}_{x,y}$ (where each is given only up to a multiplicative constant) such that $v^{xz} = v^{xy} + v^{yz}$ for every triple x, y, z. It turns out that, due to transitivity, this can be done for every triple x, y, z *separately*. The diversity axiom guarantees that this can also be done for sets of four eventualities, and the proof proceeds by induction.

The final two steps of the proof deal with extensions to infinitely many types and to infinitely many eventualities.

We finally turn to the formal proof. Let $\mathbb{T} = \mathbb{C}/\sim$ be the set of *types* of cases.[11] We prove the theorem in three steps. First we assume that there are finitely many types, that is, that $|\mathbb{T}| < \infty$. In this case the proof relies on an auxiliary result that is of interest in its own right. Since the proof of this theorem applies to an infinite set of eventualities X, we do not restrict the cardinality of X in this case. Step 2 proceeds to deal with the case in which $|\mathbb{T}|$ is unrestricted, but X is finite. Lastly, Step 3 deals with the general case in which both $|X|$ and $|\mathbb{T}|$ are unrestricted.

In all three steps, memories in $\mathbb{M}$ are represented by vectors of non-negative integers, counting how many cases of each type appear in memory. Formally, for every $T \subset \mathbb{T}$ define $\mathbb{J}_T = \mathbb{Z}_+^T = \{I \mid I : T \to \mathbb{Z}_+\}$ where $\mathbb{Z}_+$ stands for the non-negative integers. $I \in \mathbb{J}_T$ is interpreted as a counter vector, where $I(t)$ counts how many cases of type t appear in the memory represented by I. For $I \in \mathbb{J}_T$, if $\{t \mid I(t) > 0\}$ is finite, define $\succsim_I \subset X \times X$ as follows. Choose $M \in \mathbb{M}$ such that $M \subset \cup_{t \in T} t$ (recall that $t \subset \mathbb{C}$ is an equivalence class of cases) and $I(t) = \#(M \cap t)$ for all $t \in T$, and define $\succsim_I = \succsim_M$. Such a set M exists since, by the richness assumption, $|t| \geq \aleph_0$ for all $t \in \mathbb{T}$. For this reason, such a set M is not unique. However, if both $M_1, M_2 \in \mathbb{M}$

---

[11] $\mathbb{C}/\sim$ is the set of equivalence classes of $\sim$.

satisfy these properties, then $M_1 \sim M_2$ and $\succsim_{M_1} = \succsim_{M_2}$. Hence $\succsim_I$ is well-defined.

Moreover, this definition implies the following property, which will prove useful in the sequel: if $I \in \mathbb{J}_T$ and $I' \in \mathbb{J}_{T'}$ where $T \subset T'$, $I'(t) = I(t)$ for $t \in T$ and $I'(t) = 0$ for $t \in T' \setminus T$, then $\succsim_I = \succsim_{I'}$. Another obvious observation, to be used later, is that for every $M \in \mathbb{M}$ there exist a finite $T \subset \mathbb{T}$ and $I \in \mathbb{J}_T$ such that $M \subset \cup_{t \in T} t$ and $I(t) = \#(M \cap t)$ for all $t \in T$.

Step 1: The case $|\mathbb{T}| < \infty$.

Denote the set of all counter vectors by $\mathbb{J} = \mathbb{J}_T = Z_+^T$. For $I \in \mathbb{J}$, define $\succsim_I \subset X \times X$ as above. We now re-state the main theorem for this case, in the language of counter vectors. In the following, algebraic operations on $\mathbb{J}$ are performed pointwise.

A1* Order: For every $I \in \mathbb{J}$, $\succsim_I$ is complete and transitive on X.

A2* Combination: For every $I, J \in \mathbb{J}$ and every $x, y \in X$, if $x \succsim_I y$ $(x \succ_I y)$ and $x \succsim_J y$, then $x \succsim_{I+J} y$ $(x \succ_{I+J} y)$.

A3* Archimedean Axiom: For every $I, J \in \mathbb{J}$ and every $x, y \in X$, if $x \succ_I y$, then there exists $l \in N$ such that $x \succ_{lI+J} y$.

Observe that in the presence of Axiom 2, Axiom 3 also implies that for every $I, J \in \mathbb{J}$ and every $x, y \in X$, if $x \succ_I y$, then there exists $l \in N$ such that for all $k \geq l$, $x \succ_{kI+J} y$.

A4* Diversity: For every list (x, y, z, w) of distinct elements of X there exists $I \in \mathbb{J}$ such that $x \succ_I y \succ_I z \succ_I w$. If $|X| < 4$, then for any strict ordering of the elements of X there exists $I \in \mathbb{J}$ such that $\succ_I$ is that ordering.

Theorem 2: *Let there be given X, $\mathbb{T}$, and $\{\succsim_I\}_{I \in \mathbb{J}}$ as above. Then the following two statements are equivalent:*

(i) *$\{\succsim_I\}_{I \in \mathbb{J}}$ satisfy A1*–A4*;*
(ii) *There is a diversified matrix $v : X \times \mathbb{T} \to \mathbb{R}$ such that:*

$$
\begin{cases}
\text{for every } I \in \mathbb{J} \text{ and every } x, y \in X, \\
x \succsim_I y \quad \text{iff} \quad \sum_{t \in T} I(t)v(x, t) \geq \sum_{t \in T} I(t)v(y, t),
\end{cases}
\tag{3}
$$

Furthermore, in this case the matrix v is unique in the following sense: v and u both satisfy (3) iff there are a scalar $\lambda > 0$ and a matrix $\beta : X \times \mathbb{T} \to \mathbb{R}$ with identical rows (i.e., with constant columns) such that $u = \lambda v + \beta$.

Theorem 2 is reminiscent of the main result in Gilboa and Schmeidler (1997). In that work, cases are assumed to involve numerical payoffs, and algebraic and topological axioms are formulated in the payoff space. Here, by contrast, cases are not assumed to have any structure, and the algebraic and topological structures are given by the number of repetitions. This fact introduces two main difficulties. First, the space of "contexts" for which preferences are defined is not a Euclidean space, but only integer points thereof. This requires some care with the application of separation theorems. Second, repetitions can only be non-negative. This fact introduces several complications, and, in particular, changes the algebraic implication of the diversity condition.

Before proceeding with the proof, we find it useful to present a condition that is equivalent to diversification of a matrix. We will use it both for the matrix $v : X \times \mathbb{T} \to \mathbb{R}$ of Theorem 2 and the matrix $v : X \times \mathbb{C} \to \mathbb{R}$ of Theorem 1. We therefore state it for an abstract set of columns:

Auxiliary Proposition: *Let Υ be a set. Assume first $|X| \geq 4$. A matrix $v : X \times \Upsilon \to \mathbb{R}$ is diversified iff for every list (x, y, z, w) of distinct elements of X, the convex hull of differences of the row-vectors $(v(x, \cdot) - v(y, \cdot))$, $(v(y, \cdot) - v(z, \cdot))$, and $(v(z, \cdot) - v(w, \cdot))$ does not intersect $\mathbb{R}_-^{\Upsilon}$. Similar equivalence holds for the case $|X| < 4$.*

Proof. We prove the lemma for the case $|X| \geq 4$. The proof for $|X| < 4$ is similar. Assume first that a matrix v is diversified. Assume that the conclusion does not hold. Hence, there are distinct $x, y, z, w \in X$ and $\alpha, \beta, \gamma \geq 0$ with $\alpha + \beta + \gamma = 1$ such that

$$\alpha(v(x, \cdot) - v(y, \cdot)) + \beta(v(y, \cdot) - v(z, \cdot)) + \gamma(v(z, \cdot) - v(w, \cdot)) \leq 0.$$

If $\alpha > 0$, then

$$v(x, \cdot) \leq \frac{\alpha - \beta}{\alpha} v(y, \cdot) + \frac{\beta - \gamma}{\alpha} v(z, \cdot) + \frac{\gamma}{\alpha} v(w, \cdot)$$

which means that $v(x, \cdot)$ is dominated by an affine combination of $\{v(y, \cdot), v(z, \cdot), v(w, \cdot)\}$, in contradiction to the fact that v is diversified. If $\alpha = 0$, then, by a similar argument, if $\beta > 0$, then $v(y, \cdot)$ is dominated by an affine combination of $\{v(z, \cdot), v(w, \cdot)\}$. Finally, if $\alpha = \beta = 0$, then $v(z, \cdot)$ is dominated by $v(w, \cdot)$.

For the converse direction, assume that the convex hull of $\{(v(x, \cdot) - v(y, \cdot)), (v(y, \cdot) - v(z, \cdot)), (v(z, \cdot) - v(w, \cdot))\}$ (over all lists (x, y, z, w) of distinct elements in X) does not intersect $\mathbb{R}^Y_-$ but that, contrary to diversity of v, there are distinct $x, y, z, w \in X$ and $\lambda, \mu, \theta \in \mathbb{R}$ with $\lambda + \mu + \theta = 1$ such that

$$v(x, \cdot) \leq \lambda v(y, \cdot) + \mu v(z, \cdot) + \theta v(w, \cdot). \tag{3}$$

Since $\lambda + \mu + \theta = 1$, at least one of λ, μ, θ is non-negative. Assume, w.l.o.g., that $\theta \geq 0$. Hence $\lambda + \mu = 1 - \theta \leq 1$. This means that at least one of λ, μ cannot exceed 1. Assume, w.l.o.g., that $\lambda \leq 1$. Inequality (4) can be written as $v(x, \cdot) - \lambda v(y, \cdot) - \mu v(z, \cdot) - \theta v(w, \cdot) \leq 0$ or, equivalently, as

$$(v(x, \cdot) - v(y, \cdot)) + (1 - \lambda)(v(y, \cdot) - v(z, \cdot))$$
$$+ (1 - \lambda - \mu)(v(z, \cdot) - v(w, \cdot)) \leq 0.$$

Since $1 - \lambda \geq 0$ and $1 - \lambda - \mu = \theta \geq 0$, dividing by the sum of the coefficients yields a contradiction to the convex hull condition. □

Proof of Theorem 2:
We present the proof for the case $|X| \geq 4$. The proofs for the cases $|X| = 2$ and $|X| = 3$ will be described as by-products along the way.

We start by proving that (i) implies (ii). We first note that the following homogeneity property holds:

Claim 1: *For every $I \in \mathbb{Z}^T_+$ and every $k \in \mathbb{N}$, $\succsim_I = \succsim_{kI}$.*

Proof. Follows from consecutive application of the combination axiom. □

In view of this claim, we extend the definition of $\succsim_I$ to functions I whose values are non-negative rationals. Given $I \in \mathbb{Q}^T_+$, let $k \in \mathbb{N}$ be such that $kI \in \mathbb{Z}_+^{\mathrm{T}}$ and define $\succsim_I = \succsim_{kI}$. $\succsim_I$ is well-defined in view of Claim 1. By the definition and Claim 1 we also have:

Claim 2:   (***Homogeneity***) *For every* $I \in \mathbb{Q}_+^{\mathrm{T}}$ *and every* $q \in \mathbb{Q}$, $q > 0 : \succsim_{qI} = \succsim_I$.

Claim 2, A1*, and A2* imply:

Claim 3:   (*The* ***order axiom***) *For every* $I \in \mathbb{Q}_+^{\mathrm{T}}$, $\succsim_I$ *is complete and transitive on* X, *and (the* ***combination axiom***) *for every* $I, J \in \mathbb{Q}_+^{\mathrm{T}}$ *and every* $x, y \in X$ *and* $p, q \in \mathbb{Q}$, $p, q > 0$: *if* $x \succsim_I y$ ($x \succ_I y$) *and* $x \succsim_J y$, *then* $x \succsim_{pI+qJ} y$ ($x \succ_{pI+qJ} y$).

Two special cases of the combination axiom are of interest: (i) $p = q = 1$, and (ii) $p + q = 1$. Claims 2 and 3, and the Archimedean axiom, A3*, imply the following version of the axiom for the $\mathbb{Q}_+^{\mathrm{T}}$ case:

Claim 4:   (*The* ***Archimedean axiom***) *For every* $I, J \in \mathbb{Q}_+^{\mathrm{T}}$ *and every* $x, y \in X$, *if* $x \succ_I y$, *then there exists* $r \in [0, 1) \cap \mathbb{Q}$ *such that* $x \succ_{rI+(1-r)J} y$.

It is easy to conclude from Claim 3 and 4 that for every $I, J \in \mathbb{Q}_+^{\mathrm{T}}$ and every $x, y \in X$, if $x \succ_I y$, then there exists $r \in [0, 1) \cap \mathbb{Q}$ such that $x \succ_{pI+(1-p)J} y$ for every $p \in (r, 1) \cap \mathbb{Q}$.

The following notation will be convenient for stating the first lemma. For every $x, y \in X$ let $A^{xy} \equiv \{I \in \mathbb{Q}_+^{\mathrm{T}} \mid x \succ_I y\}$ and $B^{xy} \equiv \{I \in \mathbb{Q}_+^{\mathrm{T}} \mid x \succsim_I y\}$.

Observe that by definition and A1*: $A^{xy} \subset B^{xy}$, $B^{xy} \cap A^{yx} = \emptyset$, and $B^{xy} \cup A^{yx} = \mathbb{Q}_+^{\mathrm{T}}$. The first main step in the proof of the theorem is:

Lemma 1:   *For every distinct* $x, y \in X$ *there is a vector* $v^{xy} \in \mathbb{R}^{\mathrm{T}}$ *such that,*

(i) $B^{xy} = \{I \in \mathbb{Q}_+^{\mathrm{T}} \mid v^{xy} \cdot I \geq 0\}$;

(ii) $A^{xy} = \{I \in \mathbb{Q}_+^{\mathrm{T}} \mid v^{xy} \cdot I > 0\}$;

(iii) $B^{yx} = \{I \in \mathbb{Q}_+^{\mathrm{T}} \mid v^{xy} \cdot I \leq 0\}$;

(iv) $A^{yx} = \{I \in \mathbb{Q}_+^{\mathrm{T}} \mid v^{xy} \cdot I < 0\}$;

(v) *Neither* $v^{xy} \leq 0$ *nor* $v^{xy} \geq 0$;

(vi) $-v^{xy} = v^{yx}$.

Moreover, the vector v^{xy} satisfying (i)–(iv), *is unique up to multiplication by a positive number.*

The lemma states that we can associate with every pair of distinct eventualities $x, y \in X$ a separating hyperplane defined by $v^{xy} \cdot \xi = 0$ ($\xi \in \mathbb{R}^{\mathbb{T}}$), such that $x \succsim_I y$ iff I is in the half space defined by $v^{xy} \cdot I \geq 0$. Observe that if there are only two alternatives, Lemma 1 completes the proof of sufficiency: for instance, one may set $v^x = v^{xy}$ and $v^y = 0$. It then follows that $x \succsim_I y$ iff $v^{xy} \cdot I \geq 0$, i.e., iff $v^x \cdot I \geq v^y \cdot I$. More generally, we will show in the following lemmata that one can find a vector v^x for every alternative x, such that, for every $x, y \in X$, v^{xy} is a positive multiple of $(v^x - v^y)$.

Before starting the proof we introduce additional notation: let $\widehat{B}^{xy}$ and $\widehat{A}^{xy}$ denote the convex hulls (in $\mathbb{R}^{\mathbb{T}}$) of B^{xy} and A^{xy}, respectively. For a subset B of $\mathbb{R}^{\mathbb{T}}$ let $int(B)$ denote the set of interior points of B.

Proof of Lemma 1:
We break the proof into several claims.

Claim 5: *For every distinct $x, y \in X$, $A^{xy} \cap int(\widehat{A}^{xy}) \neq \emptyset$.*

Proof. By the diversity axiom $A^{xy} \neq \emptyset$ for all $x, y \in X, x \neq y$. Let $I \in A^{xy} \cap \mathbb{Z}_+^{\mathbb{T}}$ and let $J \in \mathbb{Z}_+^{\mathbb{T}}$ with $J(t) > 1$ for all $t \in \mathbb{T}$. By the Archimedean axiom there is an $l \in \mathbb{N}$ such that $K = lI + J \in A^{xy}$. Let $(\xi_j)_{j=1}^{2^{|\mathbb{T}|}}$ be the $2^{|\mathbb{T}|}$ distinct vectors in $\mathbb{R}^{\mathbb{T}}$ with coordinates 1 and -1. For $j, (j = 1, \ldots, 2^{|\mathbb{T}|})$, define $\eta_j = K + \xi_j$. Obviously, $\eta_j \in \mathbb{Q}_+^{\mathbb{T}}$ for all j. By Claim 4 there is an $r_j \in [0, 1) \cap \mathbb{Q}$ such that $\varsigma_j = r_j K + (1 - r_j)\eta_j \in A^{xy}$ (for all j). Clearly, the convex hull of $\{ \varsigma_j \mid j = 1, \ldots, 2^{|\mathbb{T}|}\}$, which is included in $\widehat{A}^{xy}$, contains an open neighborhood of K. □

Claim 6: *For every distinct $x, y \in X$, $\widehat{B}^{yx} \cap int(\widehat{A}^{xy}) = \emptyset$.*

Proof. Suppose, by way of negation, that for some $\xi \in int(\widehat{A}^{xy})$ there are $(\eta_i)_{i=1}^k$ and $(\lambda_i)_{i=1}^k$, $k \in \mathbb{N}$ such that for all i, $\eta_i \in B^{yx}$, $\lambda_i \in [0, 1]$, $\sum_{i=1}^k \lambda_i = 1$, and $\xi = \sum_{i=1}^k \lambda_i \eta_i$. Since $\xi \in int(\widehat{A}^{xy})$, there is a ball of radius $\varepsilon > 0$ around ξ included in $\widehat{A}^{xy}$. Let $\delta = \varepsilon/(2\sum_{i=1}^k \|\eta_i\|)$ and for each i let $q_i \in \mathbb{Q} \cap [0, 1]$ such that $|q_i - \lambda_i| < \delta$, and $\sum_{i=1}^k q_i = 1$. Hence, $\eta = \sum_{i=1}^k q_i \eta_i \in \mathbb{Q}_+^{\mathbb{T}}$ and $\|\eta - \xi\| < \varepsilon$, which, in turn, implies $\eta \in \widehat{A}^{xy} \cap \mathbb{Q}_+^{\mathbb{T}}$. Since for all i : $\eta_i \in B^{yx}$, consecutive application of the combination axiom (Claim 3) yields $\eta = \Sigma_{i=1}^{k} q_i \eta_i \in B^{yx}$. On the other hand, η is a convex combination of points in $A^{xy} \subset \mathbb{Q}_+^{\mathsf{T}}$ and thus it has a representation with rational coefficients (because the rationals are an algebraic field). Applying Claims 3 consecutively as above, we conclude that $\eta \in A^{xy}$ — a contradiction.   □

The main step in the proof of Lemma 1:
The last two claims imply that (for all $x, y \in X, x \neq y$) $\widehat{B}^{xy}$ and $\widehat{A}^{yx}$ satisfy the conditions of a separating hyperplane theorem. (Namely, these are convex sets, where the interior of one of them is non-empty and does not intersect the other set.) So there is a vector $v^{xy} \neq 0$ and a number c so that

$$v^{xy} \cdot I \geq c \quad \text{for every } I \in \widehat{B}^{xy}$$
$$v^{xy} \cdot I \leq c \quad \text{for every } I \in \widehat{A}^{yx}.$$

Moreover,

$$v^{xy} \cdot I > c \quad \text{for every } I \in int(\widehat{B}^{xy})$$
$$v^{xy} \cdot I < c \quad \text{for every } I \in int(\widehat{A}^{yx}).$$

By homogeneity (Claim 2), $c = 0$. **Parts (i)–(iv) of the lemma** are restated as a claim and proved below.

Claim 7:   *For all $x, y \in X, x \neq y$: $B^{xy} = \{I \in \mathbb{Q}_+^{\mathsf{T}} \mid v^{xy} \cdot I \geq 0\}$; $A^{xy} = \{I \in \mathbb{Q}_+^{\mathsf{T}} \mid v^{xy} \cdot I > 0\}$; $B^{yx} = \{I \in \mathbb{Q}_+^{\mathsf{T}} \mid v^{xy} \cdot I \leq 0\}$; and $A^{yx} = \{I \in \mathbb{Q}_+^{\mathsf{T}} \mid v^{xy} \cdot I < 0\}$.*

Proof.   (a) $B^{xy} \subset \{I \in \mathbb{Q}_+^{\mathsf{T}} \mid v^{xy} \cdot I \geq 0\}$ follows from the separation result and the fact that $z = 0$.

   (b) $A^{xy} \subset \{I \in \mathbb{Q}_+^{\mathsf{T}} \mid v^{xy} \cdot I > 0\}$: assume that $x \succ_I y$, and, by way of negation, $v^{xy} \cdot I \leq 0$. Choose a $J \in A^{yx} \cap int(\widehat{A}^{yx})$. Such a J exists by Claim 5. Since $z = 0$, J satisfies $v^{xy} \cdot J < 0$. By Claim 4 there exists $r \in [0, 1)$ such that $rI + (1 - r)J \in A^{xy} \subset B^{xy}$. By (a), $v^{xy} \cdot (rI + (1 - r)J) \geq 0$. But $v^{xy} \cdot I \leq 0$ and $v^{xy} \cdot J < 0$, a contradiction. Therefore, $A^{xy} \subset \{I \in \mathbb{Q}_+^{\mathsf{T}} \mid v^{xy} \cdot I > 0\}$.

   (c) $A^{yx} \subset \{I \in \mathbb{Q}_+^{\mathsf{T}} \mid v^{xy} \cdot I < 0\}$: assume that $y \succ_I x$ and, by way of negation, $v^{xy} \cdot I \geq 0$. By Claim 5 there is a $J \in A^{xy}$ with $J \in int(\widehat{A}^{xy}) \subset int(\widehat{B}^{xy})$. The inclusion $J \in int(\widehat{B}^{xy})$ implies $v^{xy} \cdot J > 0$. Using the

Archimedean axiom, there is an $r \in [0, 1)$ such that $rI + (1-r)J \in A^{yx}$. The separation theorem implies that $v^{xy} \cdot (rI + (1-r)J) \leq 0$, which is impossible if $v^{xy} \cdot I \geq 0$ and $v^{xy} \cdot J > 0$. This contradiction proves that $A^{yx} \subset \{I \in \mathbb{Q}_+^{\mathrm{T}} \mid v^{xy} \cdot I < 0\}$.

(d) $B^{yx} \subset \{I \in \mathbb{Q}_+^{\mathrm{T}} \mid v^{xy} \cdot I \leq 0\}$: assume that $y \succsim_I x$, and, by way of negation, $v^{xy} \cdot I > 0$. Let J satisfy $y \succ_J x$. By (c), $v^{xy} \cdot J < 0$. Define $r = (v^{xy} \cdot I)/(- v^{xy} \cdot J) > 0$. By homogeneity (Claim 2), $y \succ_{rJ} x$. By Claim 3, $I + rJ \in A^{yx}$. Hence, by (c), $v^{xy} \cdot (I + rJ) < 0$. However, direct computation yields $v^{xy} \cdot (I + rJ) = v^{xy} \cdot I + rv^{xy} \cdot J = 0$, a contradiction. It follows that $B^{yx} \subset \{I \in \mathbb{Q}_+^{\mathrm{T}} \mid v^{xy} \cdot I \leq 0\}$.

(e) $B^{xy} \supset \{I \in \mathbb{Q}_+^{\mathrm{T}} \mid v^{xy} \cdot I \geq 0\}$: follows from completeness and (c).

(f) $A^{xy} \supset \{I \in \mathbb{Q}_+^{\mathrm{T}} \mid v^{xy} \cdot I > 0\}$: follows from completeness and (d).

(g) $A^{yx} \supset \{I \in \mathbb{Q}_+^{\mathrm{T}} \mid v^{xy} \cdot I < 0\}$: follows from completeness and (a).

(h) $B^{yx} \supset \{I \in \mathbb{Q}_+^{\mathrm{T}} \mid v^{xy} \cdot I \leq 0\}$: follows from completeness and (b). □

Completion of the proof of the Lemma.

Part (v) of the Lemma, i.e., $v^{xy} \notin \mathbb{R}_+^{\mathrm{T}} \cup \mathbb{R}_-^{\mathrm{T}}$ for $x \neq y$, follows from the facts that $A^{xy} \neq \varnothing$ and $A^{yx} \neq \varnothing$. Before proving part (vi), we prove **uniqueness**.

Assume that both v^{xy} and u^{xy} satisfy (i)–(iv). In this case, $u^{xy} \cdot \xi \leq 0$ implies $v^{xy} \cdot \xi \leq 0$ for all $\xi \in \mathbb{R}_+^{\mathrm{T}}$. (Otherwise, there exists $I \in \mathbb{Q}_+^{\mathrm{T}}$ with $u^{xy} \cdot I \leq 0$ but $v^{xy} \cdot I > 0$, contradicting the fact that both v^{xy} and u^{xy} satisfy (i)-(iv).) Similarly, $u^{xy} \cdot \xi \geq 0$ implies $v^{xy} \cdot \xi \geq 0$. Applying the same argument for v^{xy} and u^{xy}, we conclude that $\{\xi \in \mathbb{R}_+^{\mathrm{T}} \mid v^{xy} \cdot \xi = 0\} = \{\xi \in \mathbb{R}_+^{\mathrm{T}} \mid u^{xy} \cdot \xi = 0\}$. Moreover, since $int(\widehat{A}^{xy}) \neq \varnothing$ and $int(\widehat{A}^{yx}) \neq \varnothing$, it follows that $\{\xi \in \mathbb{R}_+^{\mathrm{T}} \mid v^{xy} \cdot \xi = 0\} \cap int(\mathbb{R}_+^{\mathrm{T}}) \neq \varnothing$. This implies that $\{\xi \in \mathbb{R}^{\mathrm{T}} \mid v^{xy} \cdot \xi = 0\} = \{\xi \in \mathbb{R}^{\mathrm{T}} \mid u^{xy} \cdot \xi = 0\}$, i.e., that v^{xy} and u^{xy} have the same null set and are therefore a multiple of each other. That is, there exists α such that $u^{xy} = \alpha v^{xy}$. Since both satisfy (i)-(iv), $\alpha > 0$.

Finally, we prove **part (vi)**. Observe that both v^{xy} and $-v^{yx}$ satisfy (i)-(iv) (stated for the ordered pair (x, y)). By the uniqueness result,

$-v^{xy} = \alpha v^{yx}$ for some positive number α. At this stage we redefine the vectors $\{v^{xy}\}_{x,y \in X}$ from the separation result as follows: for every unordered pair $\{x, y\} \subset X$ one of the two ordered pairs, say (y, x), is arbitrary chosen and then v^{xy} is rescaled such that $v^{xy} = -v^{yx}$. (If X is uncountable the axiom of choice has to be used.) $\square$

Lemma 2:   *For every three distinct eventualities, $x, y, z \in X$, and the corresponding vectors v^{xy}, v^{yz}, v^{xz} from Lemma 1, there are unique $\alpha, \beta > 0$ such that:*

$$\alpha v^{xy} + \beta v^{yz} = v^{xz}.$$

The key argument in the proof of Lemma 2 is that, if v^{xz} is not a linear combination of v^{xy} and v^{yz}, one may find a vector I for which $\succ_I$ is cyclical.

If there are only three alternatives $x, y, z \in X$, Lemma 2 allows us to complete the proof as follows: choose an arbitrary vector v^{xz} that separates between x and z. Then choose the multiples of v^{xy} and of v^{yz} defined by the lemma. Proceed to define $v^x = v^{xz}$, $v^y = \beta v^{yz}$, and $v^z = 0$. By construction, $(v^x - v^z)$ is (equal and therefore) proportional to v^{xz}, hence $x \succsim_I z$ iff $v^x \cdot I \geq v^z \cdot I$. Also, $(v^y - v^z)$ is proportional to v^{yz} and it follows that $y \succsim_I z$ iff $v^y \cdot I \geq v^z \cdot I$. The point is, however, that, by Lemma 2, we obtain the same result for the last pair: $(v^x - v^y) = (v^{xz} - \beta v^{yz}) = \alpha v^{xy}$ and $x \succsim_I y$ iff $v^x \cdot I \geq v^y \cdot I$ follows.

Proof of Lemma 2:
First note that for every three distinct eventualities, $x, y, z \in X$, if v^{xy} and v^{yz} are colinear, then for all I either $x \succ_I y \Leftrightarrow y \succ_I z$ or $x \succ_I y \Leftrightarrow z \succ_I y$. Both implications contradict diversity. Therefore any two vectors in $\{v^{xy}, v^{yz}, v^{xz}\}$ are linearly independent. This immediately implies the uniqueness claim of the lemma. Next we introduce

Claim 8:   *For every distinct $x, y, z \in X$, and every $\lambda, \mu \in \mathbb{R}$, if $\lambda v^{xy} + \mu v^{yz} \leq 0$, then $\lambda = \mu = 0$.*

Proof.   Observe that Lemma 1(v) implies that if one of the numbers λ, and μ is zero, so is the other. Next, suppose, per absurdum, that $\lambda \mu \neq 0$, and consider $\lambda v^{xy} \leq \mu v^{zy}$. If, say, $\lambda, \mu > 0$, then $v^{xy} \cdot I \geq 0$ necessitates $v^{zy} \cdot I \geq 0$. Hence there is no I for which $x \succ_I y \succ_I z$, in contradiction to the diversity axiom. Similarly, $\lambda > 0 > \mu$ precludes $x \succ_I z \succ_I y$; $\mu > 0 > \lambda$ precludes $y \succ_I x \succ_I z$; and $\lambda, \mu < 0$ implies that for no $I \in \mathbb{Q}_+^{\mathbb{T}}$ is it the case that $z \succ_I y \succ_I x$. Hence the diversity axioms holds only if $\lambda = \mu = 0$. □

We now turn to the main part of the proof. Suppose that v^{xy}, v^{yz}, and v^{zx} are column vectors and consider the $|\mathbb{T}| \times 3$ matrix (v^{xy}, v^{yz}, v^{zx}) as a 2-person 0-sum game. If its value is positive, then there is an $\xi \in \Delta(\mathbb{T})$ such that $v^{xy} \cdot \xi > 0$, $v^{yz} \cdot \xi > 0$, and $v^{zx} \cdot \xi > 0$. Hence there is an $I \in \mathbb{Q}_+^{\mathbb{T}} \cap \Delta(\mathbb{T})$ that satisfies the same inequalities. This, in turn, implies that $x \succ_I y$, $y \succ_I z$, and $z \succ_I x$ — a contradiction.

Therefore the value of the game is zero or negative. In this case there are $\lambda, \mu, \zeta \geq 0$, such that $\lambda v^{xy} + \mu v^{yz} + \zeta v^{zx} \leq 0$ and $\lambda + \mu + \zeta = 1$. The claim above implies that if one of the numbers λ, μ and ζ is zero, so are the other two. Thus $\lambda, \mu, \zeta > 0$. We therefore conclude that there are $\alpha = \lambda/\zeta > 0$ and $\beta = \mu/\zeta > 0$ such that

$$\alpha v^{xy} + \beta v^{yz} \leq v^{xz} \tag{4}$$

Applying the same reasoning to the triple z, y, and x, we conclude that there are $\gamma, \delta > 0$ such that

$$\gamma v^{zy} + \delta v^{yx} \leq v^{zx}. \tag{5}$$

Summation yields

$$(\alpha - \delta)v^{xy} + (\beta - \gamma)v^{yz} \leq 0. \tag{6}$$

Claim 8 applied to inequality (7) implies $\alpha = \delta$ and $\beta = \gamma$. Hence inequality (6) may be rewritten as $\alpha v^{xy} + \beta v^{yz} \leq v^{xz}$, which together with (5) yields the desired representation.

Lemma 2 shows that, if there are more than three alternatives, the likelihood ranking of every triple of alternatives can be represented as in the theorem. The question that remains is whether these separate representations (for different triples) can be "patched" together in a consistent way.

Lemma 3: *There are vectors $\{v^{xy}\}_{x,y \in X, x \neq y}$, as in Lemma 1, such that for any three distinct acts, $x, y, z \in X$, the Jacobi identity $v^{xy} + v^{yz} = v^{xz}$ holds.*

Proof.   The proof is by induction, which is transfinite if X is uncountably infinite. The main idea of the proof is the following. Assume that one has rescaled the vectors v^{xy} for all alternatives x, y in some subset of acts $A \subset X$, and one now wishes to add another act to this subset, $w \notin A$. Choose $x \in A$ and consider the vectors v^{xw}, v^{yw} for $x, y \in A$. By Lemma 2, there are unique positive coefficients α, β such that $v^{xy} = \alpha v^{xw} + \beta v^{wy}$. One would like to show that the coefficient $\alpha = \alpha_y$ does not depend on the choice of $y \in A$. We will show that, if α_y did depend on y, one would find that there are $x, y, z \in A$ such that the vectors v^{xw}, v^{yw}, v^{zw} are linearly dependent, and this would contradict the diversity axiom.   □

Claim 9:   *Let $A \subset X, |A| \geq 3, w \in X \backslash A$. Suppose that there are vectors $\{v^{xy}\}_{x,y \in A, x \neq y}$, as in Lemma 1, and for any three distinct acts, $x, y, z \in X$, $v^{xy} + v^{yz} = v^{xz}$ holds. Then there are vectors $\{v^{xy}\}_{x,y \in A \cup \{w\}, x \neq y}$, as in Lemma 1, and for any three distinct acts, $x, y, z \in X$, $v^{xy} + v^{yz} = v^{xz}$ holds.*

Proof.   Choose distinct $x, y, z \in A$. Let $\hat{v}^{xw}, \hat{v}^{yw}$, and $\hat{v}^{zw}$ be the vectors provided by Lemma 1 when applied to the pairs $(x, w), (y, w)$, and (z, w), respectively. Consider the triple $\{x, y, w\}$. By Lemma 2 there are unique coefficients $\lambda(\{x, w\}, y), \lambda(\{y, w\}, x) > 0$ such that

$$v^{xy} = \lambda(\{x, w\}, y)\hat{v}^{xw} + \lambda(\{y, w\}, x)\hat{v}^{wy} \tag{7}$$

Applying the same reasoning to the triple $\{x, z, w\}$, we find that there are unique coefficients $\lambda(\{x, w\}, z), \lambda(\{z, w\}, x) > 0$ such that

$$v^{xz} = \lambda(\{x, w\}, z)\hat{v}^{xw} + \lambda(\{z, w\}, x)\hat{v}^{wz}.$$

or

$$v^{zx} = \lambda(\{x, w\}, z)\hat{v}^{wx} + \lambda(\{z, w\}, x)\hat{v}^{zw}. \tag{8}$$

We wish to show that $\lambda(\{x, w\}, y) = \lambda(\{x, w\}, z)$. To see this, we consider also the triple $\{y, z, w\}$ and conclude that there are unique coefficients $\lambda(\{y, w\}, z), \lambda(\{z, w\}, y) > 0$ such that

$$v^{yz} = \lambda(\{y, w\}, z)\hat{v}^{yw} + \lambda(\{z, w\}, y)\hat{v}^{wz}. \tag{9}$$

Since $x, y, z \in A$, we have

$$v^{xy} + v^{yz} + v^{zx} = 0$$

and it follows that the summation of the right-hand sides of (8), (9), and (10) also vanishes:

$$[\lambda(\{x, w\}, y) - \lambda(\{x, w\}, z)]\hat{v}^{xw} + [\lambda(\{y, w\}, z) - \lambda(\{y, w\}, x)]\hat{v}^{yw}$$
$$+ [\lambda(\{z, w\}, x) - \lambda(\{z, w\}, y)]\hat{v}^{zw} = 0.$$

If some of the coefficients above are not zero, the vectors $\{\hat{v}^{xw}, \hat{v}^{yw}, \hat{v}^{zw}\}$ are linearly dependent, and this contradicts the diversity axiom. For instance, if $\hat{v}^{xw}$ is a non-negative linear combination of $\hat{v}^{yw}$ and $\hat{v}^{zw}$, for no I will it be the case that $y \succ_I z \succ_I w \succ_I x$.

We therefore obtain $\lambda(\{x, w\}, y) = \lambda(\{x, w\}, z)$ for every $y, z \in A \setminus \{x\}$. Hence for every $x \in A$ there exists a unique $\lambda(\{x, w\}) > 0$ such that, for every distinct $x, y \in A$ $v^{xy} = \lambda(\{x, w\})\hat{v}^{xw} + \lambda(\{y, w\})\hat{v}^{wy}$. Defining $v^{xw} = \lambda(\{x, w\})\hat{v}^{xw}$ completes the proof of the claim. $\qquad\square$

To complete the proof of the lemma, we apply the claim consecutively. In case X is not countable, the induction is transfinite (and assumes that X can be well ordered). $\qquad\square$

Note that Lemma 3, unlike Lemma 2, guarantees the possibility to rescale *simultaneously* all the v^{xy}-s from Lemma 1 such that the Jacobi identity will hold on X.

We now complete the proof that (i) implies (ii). Choose an arbitrary act, say, g in X. Define $v^g = 0$, and for any other alternative, x, define $v^x = v^{xg}$, where the v^{xg}-s are from Lemma 3.

Given $I \in \mathbb{Q}_+^{\mathrm{T}}$ and $x, y \in X$ we have:

$$x \succsim_I y \Leftrightarrow v^{xy} \cdot I \geq 0 \Leftrightarrow (v^{xg} + v^{gy}) \cdot I \geq 0$$

$$\Leftrightarrow (v^{xg} - v^{yg}) \cdot I \geq 0 \Leftrightarrow v^x \cdot I - v^y \cdot I \geq 0 \Leftrightarrow v^x \cdot I \geq v^y \cdot I$$

The first implication follows from Lemma 1(i), the second from the Jacobi identity of Lemma 3, the third from Lemma 1(vi), and the fourth from the definition of the v^x-s. Hence, (3) of the theorem has been proved.

It remains to be shown that the vectors defined above are such that $conv(\{v^x - v^y, v^y - v^z, v^z - v^w\}) \cap \mathbb{R}^{\mathbb{T}}_- = \emptyset$. Indeed, in Lemma 1(v) we have shown that $v^x - v^y \notin \mathbb{R}^{\mathbb{T}}_-$. To see this one only uses the diversity axiom for the pair $\{x, y\}$. Lemma 2 has shown, among other things, that a non-zero linear combination of $v^x - v^y$ and $v^y - v^z$ cannot be in $\mathbb{R}^{\mathbb{T}}_-$, using the diversity axiom for triples. Linear independence of all three vectors was established in Lemma 3. However, the full implication of the diversity condition will be clarified by the following lemma. Being a complete characterization, we will also use it in proving the converse implication, namely, that part (ii) of the theorem implies part (i). The proof of the lemma below depends on Lemma 1. It therefore holds under the assumptions that for any distinct $x, y \in X$ there is an I such that $x \succ_I y$.

Lemma 4: *For every list* (x, y, z, w) *of distinct elements of* X, *there exists* $I \in \mathbb{J}$ *such that*

$$x \succ_I y \succ_I z \succ_I w \quad iff \ conv(\{v^{xy}, v^{yz}, v^{zw}\}) \cap \mathbb{R}^{\mathbb{T}}_- = \emptyset \,.$$

Proof. There exists $I \in \mathbb{J}$ such that $x \succ_I y \succ_I z \succ_I w$ iff there exists $I \in \mathbb{J}$ such that $v^{xy} \cdot I, v^{yz} \cdot I, v^{zw} \cdot I > 0$. This is true iff there exists a probability vector $p \in \Delta(\mathbb{T})$ such that $v^{xy} \cdot p, v^{yz} \cdot p, v^{zw} \cdot p > 0$.

Suppose that v^{xy}, v^{yz}, and v^{zw} are column vectors and consider the $|\mathbb{T}| \times 3$ matrix (v^{xy}, v^{yz}, v^{zw}) as a 2-person 0-sum game. The argument above implies that there exists $I \in \mathbb{J}$ such that $x \succ_I y \succ_I z \succ_I w$ iff the maximin in this game is positive. This is equivalent to the minimax being positive, which means that for every mixed strategy of player 2 there exists $t \in \mathbb{T}$ that guarantees player 1 a positive payoff. In other words, there exists $I \in \mathbb{J}$ such that $x \succ_I y \succ_I z \succ_I w$ iff for every convex combination of $\{v^{xy}, v^{yz}, v^{zw}\}$ at least one entry is positive, i.e., $conv(\{v^{xy}, v^{yz}, v^{zw}\}) \cap \mathbb{R}^{\mathbb{T}}_- = \emptyset$. $\qquad\square$

This completes the proof that (i) implies (ii). $\qquad\square$

Part 2: (ii) implies (i)

It is straightforward to verify that if $\{\succsim_I\}_{i\in\mathbb{Q}_+^{\mathbb{T}}}$ are representable by $\{v^x\}_{x\in X}$ as in (3), they have to satisfy Axioms 1–3. To show that Axiom 4 holds, we quote Lemma 4 of the previous part. $\qquad\square$

Part 3: Uniqueness

It is obvious that if $u^x = \alpha v^x + \beta$ for some scalar $\alpha > 0$, a vector $\beta \in \mathbb{R}^{\mathbb{T}}$, and all $x \in X$, then part (ii) of the theorem holds with the matrix u replacing v.

Suppose that $\{v^x\}_{x\in X}$ and $\{u^x\}_{x\in X}$ both satisfy (3), and we wish to show that there are a scalar $\alpha > 0$ and a vector $\beta \in \mathbb{R}^{\mathbb{T}}$ such that for all $x \in X$, $u^x = \alpha v^x + \beta$. Recall that, for $x \neq y$, $v^x \neq \lambda v^y$ and $u^x \neq \lambda u^y$ for all $0 \neq \lambda \in \mathbb{R}$ by A4.

Choose $x \neq g$ ($x, g \in X$, g satisfies $v^g = 0$). From the uniqueness part of Lemma 1 there exists a unique $\alpha > 0$ such that $(u^x - u^g) = \alpha(v^x - v^g) = \alpha v^x$. Define $\beta = u^g$.

We now wish to show that, for any $y \in X$, $u^y = \alpha v^y + \beta$. It holds for $y = g$ and $y = x$, hence assume that $x \neq y \neq g$. Again, from the uniqueness part of Lemma 1 there are unique $\gamma, \delta > 0$ such that $(u^y - u^x) = \gamma(v^y - v^x)$ and $(u^g - u^y) = \delta(v^g - v^y)$. Summing up these two with $(u^x - u^g) = \alpha(v^x - v^g)$, we get

$$0 = \alpha(v^x - v^g) + \gamma(v^y - v^x) + \delta(v^g - v^y) = \alpha v^x + \gamma(v^y - v^x) - \delta v^y.$$

Thus $(\alpha - \gamma)v^x + (\gamma - \delta)v^y = 0$.

Since $v^x \neq v^g = 0$, $v^y \neq v^g = 0$, and $v^x \neq \lambda v^y$ if $0 \neq \lambda \in \mathbb{R}$, we get $\alpha = \gamma = \delta$. Plugging $\alpha = \gamma$ into $(u^y - u^x) = \gamma(v^y - v^x)$ proves that $u^y = \alpha v^y + \beta$. $\qquad\square$

This completes the proof of Theorem 2. $\qquad\square\!\square$

We now turn to complete the proof of Step 1. First we prove that (i) implies (ii). Assume that $\{\succsim_M\}_M$ satisfy A1–A4. It follows that $\{\succsim_I\}_I$ satisfy A1*–A4*. Therefore, there is a representation of $\{\succsim_I\}_I$ by a matrix $v : X \times \mathbb{T} \to \mathbb{R}$ as in (3) of Theorem 2. We abuse notation and extend v to specific cases. Formally, we define $v : X \times \mathbb{C} \to \mathbb{R}$ as follows. For $x \in X$ and $c \in \mathbb{C}$, define $v(x, c) = v(x, t)$ for

$t \in \mathbb{T} \equiv \mathbb{C}/\sim$ such that $c \in t$. With this definition, (2) of Theorem 1 holds. Obviously, $c \sim d$ implies $v(\cdot, c) = v(\cdot, d)$. The converse also holds: if $v(\cdot, c) = v(\cdot, d)$, (2) implies that $c \sim d$. Finally, observe that, for every distinct four eventualities $x, y, z, w \in X$, the vectors $v(x, \cdot), v(y, \cdot), v(z, \cdot), v(w, \cdot) \in \mathbb{R}^C$ are obtained from the corresponding vectors in $\mathbb{R}^\mathbb{T}$ by replication of columns. Since $v : X \times \mathbb{T} \to \mathbb{R}$ is diversified, we also get that $v : X \times \mathbb{C} \to \mathbb{R}$ is diversified.

We now turn to prove that (ii) implies (i). Assume that a diversified matrix $v : X \times \mathbb{C} \to \mathbb{R}$, respecting case equivalence, is given. One may then define $v : X \times \mathbb{T} \to \mathbb{R}$ by $v(x, t) = v(x, c)$ for $t \in \mathbb{T} = \mathbb{C}/\sim$ such that $c \in t$, which is unambiguous because $v(\cdot, c) = v(\cdot, d)$ whenever $c \sim d$. Obviously, (3) of Theorem 2 follows from (2) of Theorem 1, and $v : X \times \mathbb{T} \to \mathbb{R}$ is diversified as well. Defining $\{\succcurlyeq_I\}_I$ by the matrix $v : X \times \mathbb{T} \to \mathbb{R}$ and (3), we find that $\{\succcurlyeq_I\}_I$ satisfy A1*–A4*. Also, $\succcurlyeq_M = \succcurlyeq_{I_M}$ for every $M \in \mathbb{M}$. Hence $\{\succcurlyeq_M\}_M$ satisfy A1–A4.

To see that uniqueness holds, assume that $v, u : X \times \mathbb{C} \to \mathbb{R}$ both satisfy (2) of Theorem 1, and respect case equivalence. Define $v, u : X \times \mathbb{T} \to \mathbb{R}$ as above. The uniqueness result in Theorem 2 yields the desired result. $\qquad\qquad\square$

Step 2: The case of arbitrary $|\mathbb{T}|$ and finite $|X|$.

We first prove that (i) implies (ii). Observe that a representation as in (ii) is guaranteed for every finite $T \subset \mathbb{T}$, provided that T is rich enough to satisfy the diversity axiom A4. We therefore restrict attention to such sets T, and show that the representations obtained for each of them can be "patched" together.

For every ordered list $(x, y, z, w) \in X$, choose $M \in \mathbb{M}$ such that $x \succ_M y \succ_M z \succ_M w$. Such an M exists by A4. Let M_0 be the union of all sets M so obtained. Since X is finite, so is M_0, i.e., $M_0 \in \mathbb{M}$. Let T_0 be the set of types (equivalence classes) of cases in M_0. Choose $g \in X$. Apply Theorem 2 to obtain a representation of $\{\succcurlyeq_I\}_{I \in \mathbb{J}_{T_0}}$ by $v_{T_0} : X \times T_0$ and (3) for all $I \in \mathbb{J}_{T_0} \equiv \mathbb{Z}_+^{T_0}$, such that $v_{T_0}(g, \cdot) = 0$. For every finite $T \subset \mathbb{T}$ such that $T_0 \subset T$, apply Theorem 2 again to obtain a representation of $\{\succcurlyeq_I\}_{I \in \mathbb{J}_T}$ by $v_T : X \times T$ and (3) for all $I \in \mathbb{J}_T \equiv \mathbb{Z}_+^T$, such that $v_T(g, \cdot) = 0$ and such that v_T extends v_{T_0}. v_T is uniquely defined by these conditions. Moreover, if $T \subset T_1 \cap T_2$, $T_0 \subset T$, and T_1

and T_2 are finite, then the restriction of v_{T_1} and of v_{T_2} to T coincide. The union of $\{v_T\}_{|T|<\infty}$ defines $v : X \times \mathbb{T} \to \mathbb{R}$ satisfying (3) for all $I \in \mathbb{J}_T$ for some finite $T \subset \mathbb{T}$. Defining v on $X \times \mathbb{C}$ as above yields a function that satisfies (2) of Theorem 1 and that respects case equivalence.

We now turn to prove that (ii) implies (i). Given a representation via a matrix $v : X \times \mathbb{C} \to \mathbb{R}$ as in (2), it follows that $\{\succcurlyeq_M\}_M$ satisfy A1 and A2. A3 also holds since v respects case equivalence. It remains to show that the above, for a diversified v, imply A4. Assume not. Then there are distinct $(x, y, z, w) \in X$ such that for no finite memory M is it the case that $x \succ_M y \succ_M z \succ_M w$. We wish to show that this condition contradicts the fact that v is diversified.

By diversification of v we know that

$$conv\{(v(x, \cdot) - v(y, \cdot)), (v(y, \cdot) - v(z, \cdot)), (v(z, \cdot) - v(w, \cdot))\} \cap \mathbb{R}^{\mathbb{C}}_- = \varnothing.$$

This implies that, for every vector (α, β, γ) in the two-dimensional simplex Δ^2, it is not the case that

$$\alpha(v(x, \cdot) - v(y, \cdot)) + \beta(v(y, \cdot) - v(z, \cdot)) + \gamma(v(z, \cdot) - v(w, \cdot)) \le 0.$$

In other words, for every $(\alpha, \beta, \gamma) \in \Delta^2$ there exists a case $c \in \mathbb{C}$ such that

$$\alpha(v(x, c) - v(y, c)) + \beta(v(y, c) - v(z, c)) + \gamma(v(z, c) - v(w, c)) > 0.$$

Thus

$$\{(\alpha, \beta, \gamma) \in \Delta^2 | \alpha(v(x, c) - v(y, c)) + \beta(v(y, c) - v(z, c))$$
$$+ \gamma(v(z, c) - v(w, c)) > 0\}_{c \in \mathbb{C}}$$

is an open cover of Δ^2 in the relative topology. But Δ^2 is compact in this topology. Hence it has an open sub-cover. But this implies that there is a finite memory $M \in \mathbb{M}$ such that, restricting v to $X \times M$,

$$conv\{(v(x, \cdot) - v(y, \cdot)), (v(y, \cdot) - v(z, \cdot)), (v(z, \cdot) - v(w, \cdot))\} \cap \mathbb{R}^M_- = \varnothing.$$

Let T be the set of types of cases appearing in M. Define $v :$ $X \times T \rightarrow \mathbb{R}$ as above. It also follows that

$$conv\{(v(x,\cdot) - v(y,\cdot)), (v(y,\cdot) - v(z,\cdot)), (v(z,\cdot) $$
$$- v(w,\cdot))\} \cap \mathbb{R}^T_- = \varnothing.$$

By Theorem 2 this implies that there exists $I \in \mathbb{J}_T$ for which $x \succ_I$ $y \succ_I z \succ_I w$. Let M' be a set of cases such that $I(t) = \#(M' \cap t)$, and $M' \subset \cup_{t \in T} t$. It follows that $x \succ_{M'} y \succ_{M'} z \succ_{M'} w$, a contradiction.

Finally, uniqueness follows from the uniqueness result in Step 1. $\square$

Step 3: The case of infinite $X, \mathbb{T}$.

We first prove that (i) implies (ii). Choose $e, f, g, h \in X$. For $A_0 = \{e, f, g, h\}$ there exists a diversified function $v_{A_0} : A_0 \times \mathbb{C} \rightarrow \mathbb{R}$ satisfying (2) and respecting case equivalence, as well as $v_{A_0}(e, \cdot) = 0$. Moreover, all such functions differ only by a multiplicative positive constant. Fix such a function $\widehat{v}_{A_0}$. For every finite set $A \subset X$ such that $A_0 \subset A$, there exists a diversified function $v_A : A \times \mathbb{C} \rightarrow \mathbb{R}$ satisfying (2) and respecting case equivalence. Moreover, there exists a unique v_A that extends $\widehat{v}_{A_0}$. Let us denote it by $\widehat{v}_A$. We now define $v : X \times \mathbb{C} \rightarrow \mathbb{R}$. Given $x \in X$, let A be a finite set such that $A_0 \cup \{x\} \subset A$. Define $v(x, \cdot) = \widehat{v}_A(x, \cdot)$. This definition is unambiguous, since, for every two finite sets A_1 and A_2 such that $A_0 \cup \{x\} \subset A_1, A_2$, we have $\widehat{v}_{A_1}(x, \cdot) = \widehat{v}_{A_1 \cup A_2}(x, \cdot) = \widehat{v}_{A_2}(x, \cdot)$. To see that v satisfies (2), choose $x, y \in X$ and consider $A = A_0 \cup \{x, y\}$. Since $v(x, \cdot) = \widehat{v}_A(x, \cdot)$, $v(y, \cdot) = \widehat{v}_A(y, \cdot)$ and $\widehat{v}_A$ satisfies (2) on A, v satisfies (2) on X. Next consider respecting case equivalence, namely, that $v(\cdot, c) = v(\cdot, d)$ iff $c \sim d$. The "if" part follows from the fact that, if $c \sim d$, then for every finite A, $\widehat{v}_A(\cdot, c) = \widehat{v}_A(\cdot, d)$. As for the "only if" part, it follows from the representation by (2) as in Step 1. Finally, to see that v is diversified, let there be given x, y, z, w and choose $A = A_0 \cup \{x, y, z, w\}$. Since $\widehat{v}_A$ is diversified, the desired conclusion follows.

The fact that (ii) implies (i) follows from the corresponding proof in Step 2, because each of the axioms A1–A4 involves only finitely many eventualities. Finally, uniqueness is proven as in Step 1. $\square$

Proof of Proposition — Insufficiency of A1–3: To see that without the diversity axiom representability is not guaranteed, let $X = [0,1]^2$ and let $\succsim_L$ be the lexicographic order on X.[12] Define, for every non-empty $M \in \mathbb{M}$, $\succsim_M = \succsim_L$, and $\succsim_\varnothing = X \times X$. It is easy to see that $\{\succsim_M\}_{M \in \mathbb{M}}$ satisfy A1-3. However, there cannot be a representation as in (2) since for any non-empty M, $\succsim_M$ is not representable by a real-valued function. $\qquad\square$

Proof of Remark: Consider an example in which $\{\succsim_M\}_M$ rank eventualities by relative frequencies, with a tie-breaking rule that is reflected by small additions to the value of v. These small additions, however, vary from case to case and their sum converges. Specifically, let $X = \{1, 2, 3, 4\}$. Define $\mathbb{T} = \{1, 2, 3, 4\}$. $\mathbb{T}$ will indeed end up to be the set of types of cases, as will become clear once we define $\{\succsim_M\}_M$. For the time being we will abuse the term and will refer to elements of $\mathbb{T}$ as "types". Let the set of cases be $\mathbb{C} \equiv T \times N$. We now turn to define $v : X \times \mathbb{C} \to \mathbb{R}$. For $x \in X$, $t \in \mathbb{T}$, and $i \in \mathbb{N}$, if $x \neq t$, $v(x, (t, i)) = 0$. Otherwise (i.e., if $x = t$), if $x \in \{1, 2, 3\}$, then $v(x, (t, i)) = 1$. Finally, $v(4, (4, i)) = 1 + \frac{1}{2^i}$ for $i \in \mathbb{N}$. Define $\{\succsim_M\}_M$ by v via (2).

We claim that two cases $(t, i), (s, j) \in \mathbb{T} \times \mathbb{N}$ are equivalent $((t, i) \sim (s, j))$ iff $t = s$. It is easy to see that if $t \neq s$, then (t, i) and (s, j) are not equivalent. (For instance, $t \succ_{\{(t,i)\}} s$ but $s \succ_{\{(s,j)\}} t$.) Moreover, if $t = s \in \{1, 2, 3\}$, then $v(\cdot, (t, i)) = v(\cdot, (s, j))$. By (2), $(t, i) \sim (s, j)$. It remains to show that, for all $i, j \in \mathbb{N}$, $(4, i) \sim (4, j)$ despite the fact that $v(\cdot, (4, i)) \neq v(\cdot, (4, j))$.

Observe, first, that $\{\succsim_M\}_M$ agree with relative frequency rankings. Specifically, consider a memory $M \in \mathbb{M}$. Let $I_M \in \mathbb{Z}_+^4$ be defined by $I_M(t) = \#\{i \in \mathbb{N} \mid (t, i) \in M\}$ for $t \in \{1, 2, 3, 4\}$. For any $s, t \in \{1, 2, 3, 4\}$, if $I_M(t) > I_M(s)$, it follows that $t \succ_M s$. Also, if $I_M(t) = I_M(s)$ and $s, t < 4$, then $t \approx_M s$. Finally, if, for $t \in \{1, 2, 3\}$, $I_M(t) = I_M(4)$, then $4 \succ_M t$.

Let there be given $M \in \mathbb{M}$ such that $(4, i), (4, j) \notin M$. The memories $M \cup \{(4, i)\}$ and $M \cup \{(4, j)\}$ agree on relative frequencies of the

---

[12]A1-3 do not suffice for the existence of a representation as in (2) even if X is finite. See Gilboa and Schmeidler (1997) for an example with $|X| = 4$, which can be easily adapted to the present set-up.

types, that is, $I_{M \cup \{(4,i)\}} = I_{M \cup \{(4,j)\}}$. Hence $\succsim_{M \cup \{(4,i)\}} = \succsim_{M \cup \{(4,j)\}}$ and $(4, i) \sim (4, j)$ follows.

Thus v satisfies (2) but does not respect case equivalence.[13] $\square$

References

Akaike, H. (1954), "An Approximation to the Density Function", *Annals of the Institute of Statistical Mathematics*, **6**: 127–132.

Cover, T. and P. Hart (1967), "Nearest Neighbor Pattern Classification", IEEE *Transactions on Information Theory*, **13**: 21–27.

de Finetti, B. (1937), "La Prevision: Ses Lois Logiques, Ses Sources Subjectives", *Annales de l'Institute Henri Poincare*, **7**: 1–68.

de Groot, M. H. (1975), *Probability and Statistics*, Reading, MA: Addison-Wesley Publishing Co.

Devroye, L., L. Gyorfi, and G. Lugosi (1996), *A Probabilistic Theory of Pattern Recognition*, New York: Springer-Verlag.

Fix, E. and J. Hodges (1951), "Discriminatory Analysis. Nonparametric Discrimination: Consistency Properties". Technical Report 4, Project Number 21-49-004, USAF School of Aviation Medicine, Randolph Field, TX.

— (1952), "Discriminatory Analysis: Small Sample Performance". Technical Report 21-49-004, USAF School of Aviation Medicine, Randolph Field, TX.

Forsyth, R. and R. Rada (1986), *Machine Learning: Applications in Expert Systems and Information Retrieval*, New-York: John Wiley and Sons.

Gilboa, I. and D. Schmeidler (1995), "Case-Based Decision Theory", *Quarterly Journal of Economics*, **110**: 605–639.

— (1997), "Act Similarity in Case-Based Decision Theory", *Economic Theory*, **9**: 47–61.

— (2002), "A Cognitive Foundation of Probability", *Mathematics of Operations Research*, **27**: 68–81.

— (2001), *A Theory of Case-Based Decisions*, Cambridge: Cambridge University Press.

---

[13] Observe that the relations $\{\succsim_M\}_M$ satisfy A1 and A2 (as they do whenever they are defined by some v via (2)), as well as A4, but not A3. Indeed, such an example cannot be generated if A3 holds as well. Specifically, one can prove the following result: if $\{\succsim_M\}_M$ are defined by v via (2), and satisfy A3 and A4, then $v(x, c) - v(y, c) = v(x, d) - v(y, d)$ whenever $c \sim d$. If, for instance, $v(e, \cdot) \equiv 0$ for some $e \in X$, then v respects case equivalence.

Gilboa, I., D. Schmeidler and P. P. Wakker (1999), "Utility in Case-Based Decision Theory", Foerder Institute for Economic Research Working Paper No. 31–99. Published in *Journal of Economic Theory*, **105** (2002): 483–502.

Hacking, I. (1975), *The Emergence of Probability*, Cambridge: Cambridge University Press.

Hume, D. (1748), *Enquiry into the Human Understanding*, Oxford: Clarendon Press.

Myerson, R. B. (1995), "Axiomatic Derivation of Scoring Rules Without the Ordering Assumption", *Social Choice and Welfare*, **12**: 59–74.

von Neumann, J. and O. Morgenstern (1944), *Theory of Games and Economic Behavior*, Princeton, NJ: Princeton University Press.

Parzen, E. (1962), "On the Estimation of a Probability Density Function and the Mode", *Annals of Mathematical Statistics*, **33**: 1065–1076.

Ramsey, F. P. (1931), "Truth and Probability", *The Foundation of Mathematics and Other Logical Essays*, New York: Harcourt, Brace and Co.

Riesbeck, C. K. and R. C. Schank (1989), *Inside Case-Based Reasoning*, Hillsdale, NJ, Lawrence Erlbaum Associates, Inc.

Rosenblatt, M. (1956), "Remarks on Some Nonparametric Estimates of a Density Function", *Annals of Mathematical Statistics*, **27**: 832–837.

Royall, R. (1966), *A Class of Nonparametric Estimators of a Smooth Regression Function*, Ph.D. Thesis, Stanford University, Stanford, CA.

Savage, L. J. (1954), *The Foundations of Statistics*, New York: John Wiley and Sons.

Schank, R. C. (1986), *Explanation Patterns: Understanding Mechanically and Creatively*, Hillsdale, NJ: Lawrence Erlbaum Associates.

Scott, D. W. (1992), *Multivariate Density Estimation: Theory, Practice, and Visualization*, New York: John Wiley and Sons.

Silverman, B. W. (1986), *Density Estimation for Statistics and Data Analysis*, London and New York: Chapman and Hall.

Stone, C. (1977), "Consistent Nonparametric Regression", *Annals of Statistics*, **5**: 689–705.

Young, H. P. (1975), "Social Choice Scoring Functions", *SIAM Journal of Applied Mathematics*, **28**: 824–838.

Chapter 5

A Derivation of Expected Utility Maximization in the Context of a Game*

Itzhak Gilboa[†] and David Schmeidler[‡]

Reprinted with Permission from *Games and Economic Behavior*, *44* (2003): 184–194.

A decision maker faces a decision problem, or a game against nature. For each probability distribution over the state of the world (nature's strategies), she has a weak order over her acts (pure strategies). We formulate conditions on these weak orders guaranteeing that they can be jointly represented by expected utility maximization with respect to an almost-unique state-dependent utility, that is, a matrix assigning real numbers to act-state pairs. As opposed to a utility function that is derived in another context, the utility matrix derived in the game will incorporate all psychological or sociological determinants of well-being that result from the very fact that the outcomes are obtained in a given game.

JEL classification numbers: C70, D80, D81

5.1. Introduction

5.1.1. *Motivation*

Do players maximize expected utility when playing a game? The experimental outcomes involving ultimatum and dictator games might

---

*We thank an associate editor and two anonymous referees for comments and references. This work was partially supported by ISF Grant No. 790/00.
†Tel-Aviv University and Cowles Foundation, Yale University. igilboa@post.tau.ac.il
‡Tel-Aviv University and The Ohio State University. schmeid@tau.ac.il

seem to suggest that they do not. (See Guth and Tietz (1990) and Roth (1992) for surveys.) For instance, a player who moves second in an ultimatum game, and rejects an offer of a positive amount of money, evidently does not maximize her monetary payoff. Similarly, a dictator in a dictator game, who chooses to leave some money to her dummy opponent, fails to maximize her payoff under conditions of certainty, let alone her expected payoff under conditions of uncertainty.

Some authors argue that these experimental results constitute a violation of game theoretic predictions. Indeed, if one insists that the utility function be defined over monetary payoffs alone, such a conclusion appears unavoidable. But many game theorists hold that the utility function need not be defined on monetary prizes alone. Indeed, an "outcome" should specify all the relevant features of the situation, including feelings of envy, guilt, preferences for fairness, and so forth. Moreover, recent developments in economic theory call for explicit modeling of such determinants of utility. (See, for instance, Frank (1988), Elster (1998), Rabin (1998), and Loewenstein (2000).) Further, if one adopts a purely behavioral approach, one has no choice but to incorporate into the utility function all psychological and sociological effects on well-being. The very fact that, say, a dictator prefers taking less money to taking more money implies that the utility of the former exceeds that of the latter. As long as players do not violate the axioms of von-Neumann and Morgenstern (vNM, 1944), they can be described as if they are maximizing the expected value of an appropriately chosen utility function. From this viewpoint, the experimental results of dictator and ultimatum games might challenge the implicit assumption that monetary payoff is the sole determinant of utility, but not the assumption of expected utility maximization itself.

We find that this argument is essentially correct: the debate aroused by dictator and ultimatum games is about determinants of the utility function, not about expected utility theory (EUT). Yet, we do not believe that vNM's axiomatic derivation of EUT is a very compelling argument in this context. vNM's result assumes a preference relation over lotteries with given probabilities, and derives a utility function over outcomes, such that the maximization of its expectation represents preferences over lotteries. vNM then assumed that, when players evaluate mixed strategies in a game, they use the same utility function for the calculation of expected payoff, and attempt to maximize this expectation. Thus, the vNM derivation implicitly assumes that the utility function that one obtains in the context of a single person decision problem will apply to the context of a game.

This assumption seems implausible precisely in the context of games such as ultimatum and dictator, where utility is heavily dependent on inter-personal comparisons and interactions. For instance, if player two considers an outcome of 10% of the pie, she cannot ignore the fact that player one (the devider) is about to pocket 9 times as much. Similarly, player one (the dictator) in a dictator game cannot be assumed to treat the outcome "I get $90" as equivalent to "I chose to take $90 and to leave $10 to my opponent." Preferences over fairness distinguish the former from the latter. Moreover, the very fact that the dictator has *chosen* a particular division of the money implies that she might experience guilt even if she has no preference for fairness *per se*. Finally, suppose that player two in an ultimatum game chooses to reject an offer not because it is unfair, but because she finds it insulting. That is, she does not envy player one, but she finds that he should be punished for his greed. In this case, she distinguishes between "I get $10, player one gets $90, and this was decided by Nature" and "I get $10, player one gets $90, and this was decided by Player one." Such distinctions are precisely about the difference between a single-player decision problem and a game. If one were to measure a player's utility over such outcomes in a laboratory, one would have to generate outcomes that simulate all the interactive effects of a game. That is, one would have to measure utility *in the context of the game itself.*

Similar issues arise when a single player is concerned. Consider, for instance, the effect of regret. It has long been argued that regret may color the way individuals evaluate outcomes. (See, for instance, Luce and Raiffa (1957), Loomes and Sugden (1982), and Gul (1991).) Thus, the utility function of a certain outcome, when measured in isolation, may not reflect the way this outcome is perceived in a game. "Getting $10" is not the same as "Getting $10 when I could have gotten $20." In order to measure the relevant utility of the latter, one would have to simulate the entire choice situation, that is, to measure utility in the context of the game.

In order to defend the expected utility paradigm in face of experimental evidence as well as of the theoretical considerations mentioned above, it does not suffice to show that it *can* explain the data with an appropriate definition of the utility function. One needs to show that this new definition also relies on sound axiomatic foundations. That is, one needs an axiomatic derivation of EUT that would parallel that of vNM, but will only use preferences in the game itself as data.

5.1.2. *The present contribution*

In this paper we axiomatize expected utility maximization in the context of a game.[1] We assume that every player can rank his pure strategies, given *any* distribution over the pure strategies of the other players. Such a distribution is interpreted as the player's subjective beliefs. In a two-person game, these beliefs might also coincide with the other player's mixed strategy. Equivalently, one may consider a single person decision problem under uncertainty (a "game against nature"), where, for each vector of probabilities over the state of nature (representing the decision maker's beliefs), the decision maker has a weak order over the possible acts. The set of acts may be finite or infinite, and it is not assumed to have any algebraic, topological, or other structure. The set of states of nature may also be finite or infinite.

Pairs of acts and states (or combinations of pure strategies) can be thought of as defining outcomes. In the formal model we do not assume any specification of physical outcomes. When such specifications exist, they may be part of the definition of outcomes, but they need not summarize all the utility-relevant information pertaining to the outcomes. In particular, one may not infer dominance relations between strategies from physical descriptions of outcomes.

When a player compares two acts, given a distribution over the states of nature, she may be viewed as comparing lotteries, namely, distributions over outcomes. Observe, however, that we do not assume that the player can compare any pair of lotteries defined over the

---

[1] See Hammond (1997) for a different approach, employing classes of games.

possible outcomes. First, we do not consider lotteries that assign positive probability to outcomes that result from different acts. Second, we do not assume that the player can compare any two such lotteries. Rather, we assume that the player can compare only lotteries that induce the same marginal distribution over states. In particular, the data assumed in our results will not include a comparison of a certain outcome (i.e., a degenerate lottery) to a non-degenerate lottery.

We assume that the rankings over the acts (the player's pure strategies) satisfy two axioms that relate preferences given different beliefs (different mixed strategies of the opponent): first, we assume *convexity*: if act *a* is preferred to *b* given probability *p*, as well as given probability *q*, then the same preference will be observed for any convex combination of *p* and *q*. Second, we assume *continuity*: if *a* is strictly preferred to *b* given belief *p*, the same preference should hold in a neighborhood of *p*. Finally, we also need an axiom of *diversity*, requiring that any four pure acts can be ranked, in any given strict order, for at least one belief vector *p*. (See the following section for a more precise formulation of the axioms.) These axioms imply that there is a utility matrix, such that, for every belief (opponent's mixed strategy) *p*, the decision maker (player) ranks her acts (pure strategies) according to their expected utility, computed for the relevant *p*.

We do not claim that the expected utility paradigm is broad enough to encompass all types of psychological or social payoffs. Indeed, there are modes of behavior that would violate expected utility maximization for *any* utility function, and would therefore also violate our axioms. Our goal is not to argue for the universality of the expected utility paradigm, but to contribute to the precise delineation of its scope of applicability.

Since in our model the utility we derive is defined over act-state pairs, it is a state-dependent utility function. (See Dreze (1961) and Karni, Schmeidler, and Vind (1983).) This also implies that there is some freedom in the choice of the utility function: one may add a separate constant to each column in the matrix without changing the expected utility rankings. Indeed, our utility matrix is unique up to such shifts, and up to multiplication of the entire matrix by a positive number.

The diversity axiom implies that the matrix we obtain satisfies a certain condition, which we dub "diversification": no row in the matrix is dominated by an affine combination of (up to) three other rows in it. In particular, it does not allow domination relations between pure strategies. However, in the absence of the diversity assumption, the other axioms do not imply the existence of the numerical representation we seek.

Further discussion of our axioms and the result is deferred to Section 3. We now turn to the formal statement of the model and result.

5.2. Result

The result presented in this section is reminiscent of the main results in Gilboa and Schmeidler (1997, 1999). All these results derive a representation of a family of weak orders by a matrix of real numbers, as follows. The objects to be ranked corresponds to rows in the matrix. A "context", which induces a weak order over these objects, is defined by a function, attaching a real number to each column. Given such a context, the ranking corresponding to it is represented by the inner products of the context with each of the rows in the matrix. While our new results are similar in spirit to those in previous papers, some differences exist. In particular, the extension to an infinite state space is new.

There are several reasons for which one may be interested in an infinite space. First, there are many situations in which Nature or other players actually have infinitely many strategies. Second, even when other players have finitely many pure strategies, one need have preferences that are linear in the probabilities defining their mixed strategies. In such situations, we may introduce mixed strategies explicitly as columns, and still assume that the player's preferences are linear in her own beliefs over these columns. However, such a construction naturally leads to an infinite matrix. Finally, a player may have psychological payoffs depending on the intentions, beliefs, and more generally, type of other players. These may also call for infinitely many columns in the matrix.[2]

---

[2]We thank the associate editor for this last point.

Assume that a decision maker is facing a decision problem with a non-empty set of acts A and a measurable space of states of the world (Ω, Σ), where Σ is a σ-algebra of subsets of Ω. Further, assume that Σ includes all singletons. Let $\mathbf{B}(\Omega, \Sigma)$ be the space of bounded Σ-measurable real-valued functions on Ω. Recall that $\mathbf{ba}(\Omega, \Sigma)$, the space of finitely additive bounded measures on Σ, is the dual of $\mathbf{B}(\Omega, \Sigma)$. Let $\mathbb{P}$ denote the subset of $\mathbf{ba}(\Omega, \Sigma)$ consisting of finitely additive probability measures on Σ. Assume that, for every probability measure $p \in \mathbb{P}$, the decision maker has a binary preference relation $\succsim_p$ over A. As usual, we denote by $\succ_p$ and $\sim_p$ the asymmetric and symmetric parts of $\succsim_p$, respectively. The axioms on $\{\succsim_p\}_{p \in \mathbb{P}}$ are:

A1 Ranking: For every $p \in \mathbb{P}$, $\succsim_p$ is complete and transitive on A.

A2 Combination: For every $p, q \in \mathbb{P}$ and every $a, b \in A$, if $a \succsim_p b$ ($a \succ_p b$) and $a \succsim_q b$, then $a \succsim_{\alpha p + (1-\alpha)q} b$ ($a \succ_{\alpha p + (1-\alpha)q} b$) for every $\alpha \in (0, 1)$.

A3 Continuity: For every $a, b \in A$ the set $\{p \in \mathbb{P} \mid a \succ_p b\}$ is open in the relative weak* topology.

A4 Diversity: For every list (a, b, c, d) of distinct elements of A there exists $p \in \mathbb{P}$ such that $a \succ_p b \succ_p c \succ_p d$. If $|A| < 4$, then for any strict ordering of the elements of A there exists $p \in \mathbb{P}$ such that $\succ_p$ is that ordering.

Axioms 1 and 3 are rather standard. Axiom 2 states that the set of beliefs, for which act a is preferred over act b, is convex. It is rather natural if payoffs, including psychological and sociological ones, do not depend on probability mixtures. By contrast, it should be expected to fail if probability mixing itself generates affective reactions, or if it defines social characteristics. The main justification for Axiom 4 is mathematical (see below). Observe that it precludes cases in which one act dominates another.

We need the following definition: a matrix of real numbers is called *diversified* if no row in it is dominated by an affine combination of three (or less) other rows in it. Formally:

Definition: A matrix $u : A \times \Omega \to \mathbb{R}$, where $|A| \geq 4$, is *diversified* if there are no distinct four elements $a, b, c, d \in A$ and $\lambda, \mu, \theta \in \mathbb{R}$ with $\lambda + \mu + \theta = 1$ such that $u(a, \cdot) \leq \lambda u(b, \cdot) + \mu u(c, \cdot) + \theta u(d, \cdot)$.

If $|A| < 4$, u is diversified if no row in u is dominated by an affine combination of the others.

We can now state

Theorem 1:   *The following two statements are equivalent:*

(i) *$\{\succsim_p\}_{p \in \mathbb{P}}$ satisfy $A1, A2, A3, A4$;*
(ii) *There exists a diversified matrix $u : A \times \Omega \to \mathbb{R}$ such that $u(a, \cdot) \in$*
     $\mathbf{B}(\Omega, \Sigma)$ for all $a \in A$ and
                 for every $p \in \mathbb{P}$ and every $a, b \in A$,
     $(**)$
$$a \succsim_p b \quad \text{iff} \quad \int_\Omega u(a, \cdot) dp \geq \int_\Omega u(b, \cdot) dp ,$$

Furthermore, *if* (i) *(equivalently,* (ii)*) holds, the matrix $u(\cdot, \cdot)$ is unique in the following sense: a matrix $w : A \times \Omega \to \mathbb{R}$ with $w(a, \cdot) \in \mathbf{B}(\Omega, \Sigma)$ for all $a \in A$ satisfies $(**)$ iff there are a scalar $\lambda > 0$ and a function $v \in \mathbf{B}(\Omega, \Sigma)$ such that $w(a, \cdot) = \lambda u(a, \cdot) + v$ for all $a \in A$.*

The proof of this theorem is given in an appendix.

We do not know of a set of axioms that are necessary and sufficient for a representation as in $(*)$ by a matrix u that need not be diversified. We do know that dropping A4 will not do. (The counter-examples in Gilboa and Schmeidler (1997, 1999) can be easily adapted to our case.) It will be clear from the proof that weaker versions of A4 suffice for a representation as in $(*)$. Ashkenazi and Lehrer (2001) also offer a condition that is weaker than A4, and that also suffices for a similar representation. The diversity axiom is stated here in its simplest and most elegant form, rather than in its mathematically weakest form.

5.3. Discussion

Fishburn (1976) and Fishburn and Roberts (1978) provide derivations of expected utility maximization in the context of a game. In these papers, a player is assumed to have preferences over lotteries that are generated by her own mixed strategies and by mixed strategies of the opponents. These results do not suffice for our purposes for two reasons. First, they assume that all lotteries, obtained by independent mixed strategies, can be compared. But each player in a game can only choose her own strategies. Thus, to make such preferences observable one would, again, have to resort to experimental settings that are external to the game. Second, a player's preference over her own mixed strategies has been criticized as shaky data. It is not clear when a player's actual choices (of pure strategies) reflect preferences over mixed strategies. Moreover, it has been argued that players never actually play mixed strategies. (See Rubinstein (2000).) We do not assume preferences over a player's own mixed strategies,[3] and interpret a mixed strategy of a player merely as the beliefs of other players regarding her (pure strategy) choice. (See Aumann and Brandenburger (1995).)

The question of observability

We assume that a player can rank pure strategies given *any* belief over the opponents' strategies. There are situations in which this assumption is rather plausible. For instance, suppose that a player is anonymously matched with other players and is given statistical data regarding past plays of the game by other players from the same population. In an experimental set-up, one may induce any probability vector p as the player's beliefs. Observe that this assumption is compatible with the player's belief in rationalizability: as dictator and ultimatum games indicate, knowledge of the physical outcomes of the game does not imply knowledge of dominance relations for other players.

By contrast, consider a situation in which a player is matched with another player, whom she knows. There might be beliefs p that the player would not entertain regarding her opponent. For instance, if Mary is happily married to John, and the two are matched to play an ultimatum game, Mary might be convinced that John, as player 1, will never make an ungenerous offer. In this case, we will never know what she would do if she did believe that John is ungenerous. In particular, we will not be able to tell whether Mary is nice to John because he is generous in his dealings with her, or because she will like him even if he treats her badly.

To address this difficulty, we point out that our result can be extended. Our theorem holds also when beliefs are restricted to a

---

[3] Naturally, we also do not derive representation of preferences of a player over her own mixed strategies.

convex subset of probability measures. In this case the continuity axiom implies that the set of probability measures has a nonempty interior relative to $\mathbb{P}$. Further, one may strengthen the result by weakening the continuity axiom so that it applies only relative to the convex set of probability measures. Observe, however, that requiring the diversity axiom on a subset of $\mathbb{P}$ is more restrictive than requiring it on $\mathbb{P}$ in its entirety. Moreover, if one insists that preferences can only be elicited given the players' actual beliefs, our approach does not apply.[4]

One may wonder how can an experimenter observe a binary relation, whereas, in reality, only a choice out of a given set is made by the player.[5] One way to experimentally observe an entire binary relation is the following. A player is told that not all acts need be available, and asked to provide a ranking, such that the best available act according to this ranking will eventually be played. If the additional complication in the experimental procedure does not introduce new affective reactions, the player will reveal a meaningful binary relation.

Roles of axiomatizations

Axiomatic derivations can be useful also when the presumed data are not directly or easily observable. First, one may use an axiomatic system such as ours for normative purposes. In this case one need not verify that the axioms are satisfied by a given player, or even observe the player's preferences at all. Rather, a player who finds our axioms compelling will be glad to know that they are consistent and, furthermore, that the only way to satisfy them is by maximization of expected utility, for an appropriately defined utility function. The player may then map her utility function, incorporating psychological and social factors, and use it for decision making.

Second, the rhetorical use of axioms can also convince one that expected utility theory is a useful descriptive tool even when it cannot be directly tested. Specifically, some researchers cast doubts on the usefulness of the expected utility paradigm in the presence of psychological and social factors. Assume that such a researcher finds

---

[4]Hammond (1997) addresses this problem. His approach, however, requires choices in hypothetical games, and not just in the game that is actually being played.

[5]We thank an anonymous referee for raising this point.

that these factors do not lead to violations of our axioms. In this case, she may conclude that maximization of expected utility can still be a valid description of behavior even when such factors are present.

Third, an axiomatic derivation delineates the scope of refutability of a theory. Thus, should one suspect that any mode of behavior can be justified as expected utility maximization for a carefully chosen utility function, axioms such as ours show that this is not the case. More generally, our axioms show precisely what is meant by expected utility maximization when the utility function may incorporate various game-specific determinants of utility.

Finally, one may use axioms to define and measure a utility function. The definition of the utility function serves a theoretical purpose, and characterizes the degree to which this function is unique. The measurement of utility is done by elicitation of observable preference data. This process can be greatly simplified using the structure provided by our result.

Appendix: Proof

Proof of Theorem 1:
Theorem 1 is reminiscent of the main results in Gilboa and Schmeidler (1997, 1999; see also 2001). Although the spaces discussed are different, some steps in the proof are practically identical. We therefore provide here only a sketch of steps that appear elsewhere in detail.

We present the proof for the case $|A| \geq 4$. The proofs for the cases $|A| = 2$ and $|A| = 3$ will be described as by-products along the way. For $u \in \mathbf{B}(\Omega, \Sigma)$ and $p \in \mathbb{P}$, let $u \cdot p$ denote $\int_\Omega u dp$.

The following notation will be convenient for stating the first lemma. For every $a, b \in A$ let

$$\Upsilon^{ab} \equiv \{p \in \mathbb{P} \mid a \succ_p b\} \quad \text{and}$$

$$W^{ab} \equiv \{p \in \mathbb{P} \mid a \succsim_p b\}.$$

Observe that by definition and A1: $\Upsilon^{ab} \subset W^{ab}$, $W^{ab} \cap \Upsilon^{ba} = \emptyset$, and $W^{ab} \cup \Upsilon^{ba} = \mathbb{P}$. The first main step in the proof of the theorem is:

Lemma 1:   *There exists $\{u^{ab}\}_{a,b \in A, a \neq b} \subset \mathbf{B}(\Omega, \Sigma)$ such that, for every distinct $a, b \in A$:*

(i) $W^{ab} = \{p \in \mathbb{P} \mid u^{ab} \cdot p \geq 0\}$;
(ii) $\Upsilon^{ab} = \{p \in \mathbb{P} \mid u^{ab} \cdot p > 0\}$;
(iii) $W^{ba} = \{p \in \mathbb{P} \mid u^{ab} \cdot p \leq 0\}$;
(iv) $\Upsilon^{ba} = \{p \in \mathbb{P} \mid u^{ab} \cdot p < 0\}$;
(v) *Neither* $u^{ab} \leq 0$ *nor* $u^{ab} \geq 0$;
(vi) $-u^{ab} = u^{ba}$.

Moreover, $\{u^{ab}\}_{a,b \in A, a \neq b}$ is unique in the following sense: for every distinct $a, b \in A$, u^{ab} and u^{ba} may be multiplied by an arbitrary positive constant $\lambda_{\{a,b\}} > 0$.

The lemma states that we can associate with every pair of distinct acts $a, b \in A$ a separating hyperplane defined by $u^{ab} \cdot p = 0$ $(p \in \mathbb{P})$, such that $a \succsim_p b$ iff p is on a given side of the plane (i.e., iff $u^{ab} \cdot p \geq 0$). Observe that if there are only two acts, Lemma 1 completes the proof of sufficiency: for instance, one may set $u^a = u^{ab}$ and $u^b = 0$. More generally, we will show in the following lemmata that one can find a function u^a for every act a, such that, for every $a, b \in A$, u^{ab} is a positive multiple of $(u^a - u^b)$.

For a subset B of $\mathbb{P}$ let $int(B)$ denote the set of interior points of B (relative to $\mathbb{P}$).

Proof of Lemma 1:

The combination axiom implies that the sets Υ^{ba} and W^{ab} are convex. The continuity axiom implies that the sets Υ^{ba} are open (in the relative topology). This, in turn, implies that the sets W^{ab} are closed and therefore compact in the weak* topology. This allows the use of a (weak) separating hyperplane theorem between two disjoint and convex sets, one of which is compact: the convex hull of W^{ab} and the origin (i.e., $\{\alpha p \mid \alpha \in [0,1], p \in W^{ab}\}$) on the one hand, and $\{\alpha p \mid \alpha \in (0,1], p \in \Upsilon^{ba}\}$ on the other. That is, we obtain a non-zero function $u^{ab} \in \mathbf{B}(\Omega, \Sigma)$ such that $u^{ab} \cdot p \geq 0$ for all $p \in W^{ab}$ and $u^{ab} \cdot p \leq 0$ for all $p \in \Upsilon^{ba}$. Further, we argue that u^{ab} does not vanish on $W^{ab} \cup \Upsilon^{ba} = \mathbf{ba}_+^1(\Omega, \Sigma) = \mathbb{P}$. If it did, then it would also vanish on $\mathbf{ba}_+(\Omega, \Sigma)$, and therefore also on $\mathbf{ba}_-(\Omega, \Sigma)$. But in this case it would vanish on all of $\mathbf{ba}(\Omega, \Sigma)$, in view of Jordan's decomposition theorem, in contradiction to the fact that u^{ab} is non-zero.

We argue that for some $p \in \Upsilon^{ba}$, $u^{ab} \cdot p < 0$. If not, $u^{ab} \cdot p = 0$ for all $p \in \Upsilon^{ba}$. Since u^{ab} does not vanish on $W^{ab} \cup \Upsilon^{ba}$, there has to exist a $q \in W^{ab}$ with $u^{ab} \cdot q > 0$. But then for all $\varepsilon > 0$, $u^{ab} \cdot (\varepsilon q + (1-\varepsilon)p) > 0$, while $\varepsilon q + (1-\varepsilon)p \in \Upsilon^{ba}$ for small enough ε by the continuity axiom. Next, we argue that for all $q \in \Upsilon^{ab}$ we have $u^{ab} \cdot q > 0$. Indeed, if $u^{ab} \cdot q = 0$ for $q \in \Upsilon^{ab}$, $u^{ab} \cdot (\varepsilon p + (1-\varepsilon)q) < 0$ for all $\varepsilon > 0$. By a similar argument, $u^{ab} \cdot p < 0$ for all $p \in \Upsilon^{ba}$.

Thus $\Upsilon^{ba} \subset \{p \mid u^{ab} \cdot p < 0\}$. Since we also have $W^{ab} \subset \{p \mid u^{ab} \cdot p \geq 0\}$, $\Upsilon^{ba} \supset \{p \mid u^{ab} \cdot p < 0\}$. That is, $\Upsilon^{ba} = \{p \mid u^{ab} \cdot p < 0\}$ and $W^{ab} = \{p \mid u^{ab} \cdot p \geq 0\}$. We have also shown that $\Upsilon^{ab} \subset \{p \mid u^{ab} \cdot p > 0\}$. To show the converse inclusion, assume that $u^{ab} \cdot p > 0$ but $a \sim_p b$. Choose $q \in \Upsilon^{ba}$. By the combination axiom, $\alpha p + (1-\alpha)q \in \Upsilon^{ba}$ for all $\alpha \in (0,1)$. But for α close enough to 1 we have

$u^{ab} \cdot (\alpha p + (1-\alpha)q) > 0$, a contradiction. Hence $\Upsilon^{ab} = \{p \mid u^{ab} \cdot p > 0\}$ and $W^{ba} = \{p \mid u^{ab} \cdot p \leq 0\}$.

Observe that u^{ab} can be neither non-positive nor non-negative due to the diversity axiom (applied to the pair a, b).

We now turn to prove uniqueness. Assume that $u^{ab}, v^{ab} \in \mathbf{B}(\Omega, \Sigma)$ both satisfy conditions (i)–(v) of Lemma 1. Consider a two-person zero-sum game with a payoff matrix $(u^{ab}, -v^{ab})$. Specifically, (i) the set of pure strategies of player 1 (the row player) is Ω; (ii) the set of pure strategies of player 2 (the column player) is $\{L, R\}$; and (iii) if player 1 chooses $\omega \in \Omega$, and player 2 chooses L, the payoff to player 1 will be $u^{ab}(\omega)$, whereas if player 2 chooses R, the payoff to player 1 will be $-v^{ab}(\omega)$. Since both u^{ab}, v^{ab} satisfy conditions (i)–(iv), there is no $p \in \mathbb{P}$ for which $u^{ab} \cdot p > 0$, $-v^{ab} \cdot p > 0$. Hence the maximin in this game is non-positive. Therefore, so is the minimax. It follows that there exists a mixed strategy of player 2 that guarantees a non-positive payoff against any pure strategy of player 1. In other words, there are $\alpha, \beta \geq 0$ with $\alpha + \beta = 1$ such that $\alpha u^{ab}(\omega) \leq \beta v^{ab}(\omega)$ for all $\omega \in \Omega$. Moreover, by condition (v) $\alpha, \beta > 0$. Hence for $\gamma = \beta/\alpha > 0$, $u^{ab} \leq \gamma v^{ab}$. Applying the same argument to the game $(-u^{ab}, v^{ab})$, we find that there exists $\delta > 0$ such that $u^{ab} \geq \delta v^{ab}$. Therefore, $\gamma v^{ab} \geq u^{ab} \geq \delta v^{ab}$ for $\gamma, \delta > 0$. In view of part (v), there exists $\omega \in \Omega$ with $v^{ab}(\omega) > 0$, implying $\gamma \geq \delta$. By the same token there exists $\omega' \in \Omega$ with $v^{ab}(\omega') < 0$, implying $\gamma \leq \delta$. Hence $\gamma = \delta$ and $u^{ab} = \gamma v^{ab}$.

Finally, we prove part (vi). Observe that both u^{ab} and $-u^{ba}$ satisfy (i)–(iv) (stated for the ordered pair (a, b)). By the uniqueness result, $-u^{ab} = \alpha u^{ba}$ for some positive number α. At this stage we redefine the functions $\{u^{ab}\}_{a,b \in A}$ from the separation result as follows: for every unordered pair $\{a, b\} \subset A$ one of the two ordered pairs, say (b, a), is arbitrarily chosen and then u^{ab} is rescaled such that $u^{ab} = -u^{ba}$. (If A is of an uncountable power, the axiom of choice has to be used.) □

Lemma 2: *For every three distinct acts, $f, g, h \in A$, and the corresponding vectors u^{fg}, u^{gh}, u^{fh} from Lemma 1, there are unique $\alpha, \beta > 0$ such that:*

$$\alpha u^{fg} + \beta u^{gh} = u^{fh}.$$

The key argument in the proof of Lemma 2 is that, if u^{fh} is not a linear combination of u^{fg} and u^{gh}, one may find $p \in \mathbb{P}$ for which $\succ_p$ is cyclical. The detailed proof follows closely that of the corresponding lemma in Gilboa and Schmeidler (1999).

If there are only three acts $f, g, h \in A$, Lemma 2 allows us to complete the proof as follows: choose a function $u^{fh} \in \mathbf{B}(\Omega, \Sigma)$ that separates between f and h. Then choose the multiples of u^{fg} and of u^{gh} defined by the lemma, and proceed to define $u^f = u^{fh}, u^g = \beta u^{gh}$, and $u^h = 0$.

If there are more than three acts, Lemma 2 shows that the ranking of every triple of acts can be represented as in the theorem. The question that remains is whether these separate representations (for different triples) can be "patched" together in a consistent way.

Lemma 3: *There are functions $\{u^{ab}\}_{a,b \in A, a \neq b} \subset \mathbf{B}(\Omega, \Sigma)$, as in Lemma 1, such that, for any three distinct acts, $f, g, h \in A$, the Jacobi identity $u^{fg} + u^{gh} = u^{fh}$ holds.*

The proof is by induction, which is transfinite if A is uncountably infinite. The main idea of the proof is the following. Assume that one has rescaled the functions u^{ab} for all acts a, b in some subset of acts $X \subset A$, and one now wishes to add another act to this subset, $d \notin X$. Choose $a \in X$ and consider the functions u^{ad}, u^{bd} for $b \in X$. By Lemma 2, there are unique positive coefficients α, β such that $u^{ab} = \alpha u^{ad} + \beta u^{db}$. One would like to show that the coefficient α does not depend on the choice of $b \in X$. Indeed, if it did, one would find that there are $a, b, c \in X$ such that the vectors u^{ad}, u^{bd}, u^{cd} are linearly dependent, and this contradicts the diversity axiom. Again, details are to be found in Gilboa and Schmeidler (1997, 1999).

Note that Lemma 3, unlike Lemma 2, guarantees the possibility to rescale *simultaneously* all the u^{ab}-s from Lemma 1 such that the Jacobi identity will hold on A.

We now complete the proof that (i) implies (ii). Choose an arbitrary act, say, e in A. Define $u^e = 0$, and for any other act, a, define $u^a = u^{ae}$, where the u^{ae}-s are from Lemma 3.

Given $p \in \mathbb{P}$ and $a, b \in A$ we have:

$$a \succsim_p b \Leftrightarrow u^{ab} \cdot p \geq 0 \Leftrightarrow (u^{ae} + u^{eb}) \cdot p \geq 0 \Leftrightarrow$$

$$(u^{ae} - u^{be}) \cdot p \geq 0 \Leftrightarrow u^a \cdot p - u^b \cdot p \geq 0 \Leftrightarrow u^a \cdot p \geq u^b \cdot p$$

Defining $u(a, \cdot) = u^a(\cdot)$, (∗∗) of the theorem has been proved.

It remains to be shown that the functions defined above form a diversified matrix. This follows from the following results (adapted from Gilboa and Schmeidler (1999, revised version — 2001)):

Proposition 2:   *Let Υ be a set. Assume first $|A| \geq 4$. A matrix $u : X \times \Upsilon \to R$ is diversified iff for every list (a, b, c, d) of distinct elements of A, the convex hull of differences of the row-vectors $(u(a, \cdot) - u(b, \cdot)), (u(b, \cdot) - u(c, \cdot))$, and $(u(c, \cdot) - u(d, \cdot))$ does not intersect $\mathbb{R}_-^\Upsilon$. Similar equivalence holds for the case $|A| < 4$.*

Lemma 4:   *For every list (a, b, c, d) of distinct elements of A,*

$$conv(\{u^{ab}, u^{bc}, u^{cd}\}) \cap \mathbb{R}_-^\Omega = \emptyset.$$

iff there exists $p \in \mathbb{P}$ such that $a \succ_p b \succ_p c \succ_p d$.

This completes the proof that (i) implies (ii).   □

Part 2: (ii) implies (i)
It is straightforward to verify that if $\{\succsim_p\}_{i \in \mathbb{P}}$ are representable by $\{u(a, \cdot)\}_{a \in A} \subset \mathbf{B}(\Omega, \Sigma)$ as in (∗), they have to satisfy Axioms 1–3. To show that Axiom 4 holds, we quote Lemma 4 and Proposition 2 of the previous part.   □

Part 3: Uniqueness
Similar to the proof in Gilboa and Schmeidler (1997, 1999).   □□

References

Ashkenazi, G. and E. Lehrer (2001), "Well-Being Indices", mimeo.

Aumann, R. J. and A. Brandenburger (1995), "Epistemic Conditions for Nash Equilibrium", *Econometrica*, **63**: 1161–1180.

Dreze, J. H. (1961), "Le foundements logique de l'utilite cardinale et de la probabilite subjective", La Decision. Paris: CRNS. Translated and reprinted as "Decision Theory with Moral Hazard and State-Dependent

Preferences", in Dreze, J. H. (1987), *Essays on Economic Decisions under Uncertainty*, Cambridge: Cambridge University Press.

Elster, J. (1998), "Emotions and Economic Theory", *Journal of Economic Literature*, **36**: 47–74.

Fishburn, P. C. (1976), "Axioms for Expected Utility in *n*-Person Games", *International Journal of Game Theory*, **5**: 137–149.

Fishburn, P. C. and F. S. Roberts (1978), "Mixture Axioms in Linear and Multilinear Utility Theories", *Theory and Decision*, **9**: 161–171.

Frank, R. H. (1988), *Passions Within Reason: The Strategic Role of the Emotions*, WW Norton.

Gilboa, I. and D. Schmeidler (1997), "Act Similarity in Case-Based Decision Theory", *Economic Theory*, **9**: 47–61.

Gilboa, I. and D. Schmeidler (1999), "Inductive Inference: An Axiomatic Approach", forthcoming, *Econometrica*, **71** (2003): 1–26. Reprinted as Chapter 4 in this volume.

Gilboa, I. and D. Schmeidler (2001), *A Theory of Case-Based Decisions*, Cambridge: Cambridge University Press.

Gul, F. (1991), "A Theory of Disappointment Aversion", *Econometrica*, **59**: 667–686.

Guth, W. and R. Tietz (1990), "Ultimatum Bargaining Behavior: A Survey and Comparison of Experimental Results", *Journal of Economic Behavior and Organization*, **11**: 417–449.

Hammond, P. J. "Consequentialism and Bayesian Rationality in Normal Form Games", in W. Leinfellner and E. Kohler (eds.), Game Theory, Experience, Rationality. Foundations of Social Sciences, Economics and Ethics. In Honor of John C. Harsanyi. (Vienna Circle Institute Yearbook 5) (Kluwer Academic Publishers, 1997).

Karni, E., D. Schmeidler and K. Vind (1983), "On State Dependent Preferences and Subjective Probabilities", *Econometrica*, **51**: 1021–1031.

Loewenstein, G. (2000), "Emotions in economic theory and economic behavior", *American Economic Review: Papers and Proceedings*, **90**: 426–432.

Loomes, G. and R. Sugden (1982), "Regret Theory: An Alternative Theory of Rational Choice Under Uncertainty", *Economic Journal*, **92**: 805–824.

Luce, R. D. and H. Raiffa (1957), "Games and Decisions". New York: Wiley.

Rabin, M. (1998), "Psychology and Economics", *Journal of Economic Literature*, **36**: 11–46.

Roth, A. E. (1992), "Bargaining Experiments", in J. Kagel and A. E. Roth (eds.), *Handbook of Experimental Economics*, Princeton: Princeton University Press.

Rubinstein, A. (2000), *Economics and Language*, Cambridge: Cambridge University Press.

von Neumann, J. and O. Morgenstern (1944), *Theory of Games and Economic Behavior*, Princeton, NJ: Princeton University Press.

Chapter 6

Subjective Distributions*

Itzhak Gilboa[†] and David Schmeidler[‡]

Reprinted from *Theory and Decision*, 56 (2004): 345–357.

A decision maker has to choose one of several random variables, with uncertaintly known distributions. As a Bayesian she behaves as if she knew the distributions. In this paper we suggest an axiomatic derivtion of these (subjective) distributions, which is much more economical than the derivations by de Finetti or Savage. They derive the whole joint distribution of all the available random variables.

6.1. Introduction

This paper provides an axiomatic derivation of subjective probabilities in the context of expected value maximization without a state space. As opposed to the usual approaches (Ramsey (1931), de Finetti (1937), and Savage (1954)), our model does not assume that the decision maker can or need estimate probabilities beyond those that are directly used for expected value maximization.

In many decision problems under uncertainty acts are not given as functions from states of nature to outcomes. For instance, suppose that the decision maker has to select one of several investments options (portfolios) of equal present value. For each portfolio she would like to know the distribution of its values at, say, a year later. To present our

---

*We wish to thank Edi Karni for many helpful comments. This work was partially supported by ISF Grant No. 790/00.
[†]Tel-Aviv University. Fellow of the Cowles Foundation, Yale University. igilboa@post.tau.ac.il
[‡]Tel-Aviv University and The Ohio State University. schmeid@post.tau.ac.il

approach we turn to the stylized example of m urns containing colored balls. A ball will be drawn at random from each urn, and the outcome will be determined by the color of the ball drawn from the urn chosen by the decision maker. There are n possible colors, but the distributions of colors in the urns are unknown. The standard approach calls for the definition of states of nature as functions from acts to outcomes. Thus, there are n^m states in this problem, and $n^m - 1$ parameters for the decision maker to assess. However, expected value (or expected utility) maximization will only make use of $m(n-1)$ numbers, namely, the distribution of colors in each urn separately. If we view the color of the ball being drawn from each urn as a random variable, only the marginal distributions of these random variables are needed. In the standard approach, by contrast, the decision maker is asked to assess the entire joint distribution.

An obvious question arises: can one provide an axiomatic derivation of subjective probabilities (coupled with expected value maximization) that would *not* resort to a probability on an entire algebra of events (defining a large state space), but would make do with the probability values that are of actual use?

The purpose of this paper is to provide such an axiomatization. We assume as given a set of acts (such as the urns), and a set of physical consequences (such as the color of the balls drawn out of the urns). To each physical consequence we can attach a real-valued payoff, to be thought of as a monetary prize, or a utility index. Given such a payoff assignment, we assume that the decision maker can express preferences over the set of acts. We provide conditions under which there exist, for each act, a (subjective) distribution (over physical consequences), such that, for each payoff assignment, the decision maker ranks acts according to their expected payoff with respect to these distributions. In other words: each act is identified with a probability vector over the set of physical consequences. An entry in such a vector is interpreted as the subjective probability that, should this act be chosen, the corresponding consequence would result. We show that for every payoff assignment, the corresponding preference relation over acts agrees with maximization of expected value with respect to the given payoff assignment and the subjective distributions we derive.

Treating each available act as a random variable, taking values in the set of physical consequences, our result identifies a subjective distribution for each such variable. We do not identify their joint distribution, as do de Finetti and Savage. Rather, we only deal with the marginal distributions, which are those that will be used in choice between actually available acts. It also follows that our model does not address the question of (subjective) independence of or correlation between these random variables.

Our result serves several purposes. First, it relates the notion of subjective probability to preferences in a new way. As such, it might make this notion meaningful even when Savage questionnaires are too complex and counter-intuitive to be considered observable data. Second, it outlines the elicitation of subjective marginal distributions of a set of random variables, without resorting to the elicitation of the joint distribution thereof. Lastly, it can be used to judge the expected payoff paradigm, both descriptively and normatively, in situations where the de Finetti-Savage approach requires too detailed preference data. In particular, it shows that the expected payoff paradigm might be cognitively plausible in situations in which its known axiomatic derivations are not. That is, it is possible that decision makers generate subjective marginal distributions and choose a random variable that maximizes the corresponding subjective expected utility, without having Bayesian beliefs over the entire algebra of events generated by the random variables in question. In the language of preferences, decision makers may only have well-defined preferences between acts that are actually available to them, but not between hypothetical acts that can be defined by the former.

The next section presents the model and results. Proofs are relegated to an appendix. We defer further discussion to Section 3.

6.2. Model and Results

Assume that a decision maker is facing a decision problem with a finite and non-empty set of *acts*, A. Each act will result in one (and only one) physical *consequence* from the set $N = \{1, \ldots, n\}$ for $n \geq 1$. A *context* is a real-valued function on N. The space of all contexts, $\mathbb{R}^N$, is identified with $\mathbb{R}^n$, endowed with the natural topology and the standard algebraic operations. Given a context $x \in \mathbb{R}^n$, $\succsim_x \subset A \times A$ is a binary relation over acts.

The interpretation is as follows. The physical consequences are abstract, and they do not determine the decision maker's utility. Rather, it is the context x which associates a utility value to each possible consequence. Put differently, the set of contexts is the set of possible utility functions on the abstract set of consequences N. As in the introduction, a consequence might be, for instance, "a ball drawn from urn 1 is red". We do not assume that different acts have disjoint sets of possible consequences. Rather, every physical consequence might, a-priori, result from any act. It is assumed that we can observe the decision maker's preferences over acts given *any* utility function.

We now formulate axioms on $\{\succsim_x\}_{x \in \mathbb{R}^n}$:

A1 Order: For every $x \in \mathbb{R}^n$, $\succsim_x$ is complete and transitive on A.

A2 Additivity: For every $x, y \in \mathbb{R}^n$ and every $a, b \in A$, if $a \succsim_x b$ and $a \succ_y b$, then $a \succ_{x+y} b$.

A3 Continuity: For every $a, b \in A$ the sets $\{x \mid a \succ_x b\}$ and $\{x \mid b \succ_x a\}$ are open.

A4 Diversity: For every list (a, b, c, d) of distinct elements of A there exists $x \in \mathbb{R}^n$ such that $a \succ_x b \succ_x c \succ_x d$. If $|A| < 4$, then for any strict ordering of the elements of A there exists $x \in \mathbb{R}^n$ such that $\succ_x$ is that ordering.

A5 Neutrality: For every constant $c \in \mathbb{R}^n$ (i.e., $c_i = c_j$ for all $i, j \in N$), and every $a, b \in A$, $a \sim_c b$.

Axiom 1 is standard. Axiom 2 is the most crucial axiom, as it guarantees that the set of contexts (utility functions) for which act a is preferred to act b is convex. Axiom 3 states that this set is also open. The diversity axiom (A4) rules out certain preferences. For instance, it does not allow one lottery to be always preferred to another. Finally, A5 is a weak consequentialism axiom. It holds whenever the decision maker cares only about the final utility derived from consequences: if this utility function happens to be constant, no act should be preferred to any other.

The statement of the theorem requires two additional definitions. A function $P : N \rightarrow [0, 1]$ with $\sum_{i \in N} P(i) = 1$ is called a *lottery*. Algebraic operations on lotteries are performed pointwise. In particular, the α-mixture of lotteries P and Q, $\alpha P + (1 - \alpha)Q$ is also a lottery. A collections of lotteries $\{P_a\}_{a \in A}$ is called *4-independent* if every four (or fewer) lotteries in it are linearly independent.[1]

Theorem 1: *The following two statements are equivalent*:

(i) $\{\succsim_x\}_{x \in \mathbb{R}^n}$ *satisfy* $A1 - A5$;
(ii) *There is a collection of* 4-*independent lotteries* $P = \{P_a\}_{a \in A}$ *such that*:

$$\text{for every } x \in \mathbb{R}^n \text{ and every } a, b \in A,$$

$(*)$

$$a \succsim_x b \quad \text{iff} \quad \sum_{i \leq n} P_a(i)x(i) \geq \sum_{i \leq n} P_b(i)x(i) ,$$

To what extent are the lotteries $\{P_a\}_{a \in A}$ unique? Clearly, if $\{P_a\}_{a \in A}$ satisfy $(*)$, then, for any lottery R and any $\alpha \in (0, 1]$, the collection $\{\alpha P_a + (1 - \alpha)R\}_{a \in A}$ also satisfies $(*)$.[2] Observe that in $\{P_a\}_{a \in A}$ differences between lotteries are more pronounced than in $\{\alpha P_a + (1 - \alpha)R\}_{a \in A}$. This gives rise to the following definition.

For two collections of lotteries, $P = \{P_a\}_{a \in A}$ and $Q = \{Q_a\}_{a \in A}$, we say that P is *more extreme than* Q if there exists a lottery R and $\alpha \in (0, 1)$ such that $\alpha P_a + (1 - \alpha)R = Q_a$ for all $a \in A$. We can now state the uniqueness result.

Proposition 2: *There exists a unique collection of* 4-*independent lotteries* P *that satisfies* $(*)$ *and that is more extreme than any other collection* Q *that satisfies* $(*)$.

We do not know of a set of axioms that are necessary and sufficient for a representation as in $(*)$ by a collection of lotteries that need not be 4-independent. We do know that dropping A4 will not do.[3]

---

[1] That is, if $|A| < 4$, $\{P_a\}_{a \in A}$ is 4-independent if it is linearly independent.
[2] This is tantamount to saying that the indepence axiom of vNM (1944) is necessary for expected utility maximization.
[3] In Gilboa and Schmeidler (1997) we provide two counter-examples that show that A1–A3, coupled with a weaker neutrality axiom, do not imply a representation as in $(*)$. The first counter-example uses a finite set A and satisfies the stronger version of A5 used here, and therefore applies to our case as well.

It is clear from the proof of the main result in Gilboa and Schmeidler (1997) that weaker versions of A4 suffice for a representation as in (∗). Ashkenazi and Lehrer (2001) also offer a condition that is weaker than A4, and that also suffices for such a representation. The diversity axiom is stated here in its simplest and most elegant form, rather than in its mathematically weakest form.

6.3. Discussion

Our derivation of expected utility with subjective lotteries assumes that legitimate data are only preferences between acts that are actually available, but that these can be observed for any assignment of utility values to consequences. One obvious drawback of this axiomatization is that, like that of de Finetti (1937), it assumes linearity (or convexity) in payoffs. This can be interpreted in several ways. First, one may assume that payoffs are monetary, and that the decision maker is risk neutral. This makes the data easily observable, but the axiom becomes rather implausible since decision makers are typically not assumed to be risk neutral, especially when large sums of money are involved.

Second, numerical payoffs can be viewed as utility values. That is, one may assume that there is an independent measurement of a cardinal utility function, and that contexts x refer to values of this utility function. No loss of generality is involved here, since every expected utility maximizer will satisfy this axiom in this formulation. But the allegedly observable data are then supposed to be choices given such assignments of utility values, and it is not entirely clear where one can get such a utility function to begin with.

A third interpretation involves small amounts of money, and it relies on the fact that an expected utility maximizer (with a differentiable utility function) behaves like a risk neutral decision maker when small amounts of money are involved. To be more concrete, assume that there are preference orders $\{\succsim'_x\}_{x \in \mathbb{R}^n}$ on the set of act A for any monetary payoff function $x \in \mathbb{R}^n$. For a vector $x \in \mathbb{R}^n$ define a new relation, $\succsim_x$ as follows: $a \succsim_x b$ iff there exists $\widehat{\varepsilon} > 0$ such that, for all $\varepsilon \in (0, \widehat{\varepsilon})$, $a \succsim'_{\varepsilon x} b$. Obviously, $a \succsim_x b$ iff $a \succsim_{\lambda x} b$ for every $x \in \mathbb{R}^n$

and every $\lambda > 0$. One can now assume that $\{\succsim_x\}_{x \in \mathbb{R}^n}$ satisfy A1–A5. In this interpretation, the completeness axioms becomes less obvious. However, the additivity axiom (A2) is more palatable. Naturally, the preferences $\{\succsim_x\}_{x \in \mathbb{R}^n}$, namely, preferences given arbitrarily small monetary prizes, suffice for the derivation of subjective lotteries. But this approach suggests that elicitation of probabilities be made using prizes about which the decision maker does not really care. It is not clear that one would like to base the derivation of subjective probabilities entirely on preferences between asymptotically small prizes.

Rather than adopting one of these interpretation, one can also attempt to derive the utility function in conjunction with the subjective lotteries. Building on the techniques of Wakker (1989), one may re-state the additivity axiom so that it correspond to addition of utilities, rather than of prizes. Indeed, in Gilboa, Schmeidler, and Wakker (1999) we follow a similar tack in adapting the original result of Gilboa and Schmeidler (1997), which assumed a given utility function, to a result that derives the utility function as well. In fact, one may start with the first axiomatization in Gilboa, Schmeidler, and Wakker (1999), strengthen axiom A5 as above, and continue to obtain a derivation of expected utility with subjective lotteries and a general utility function.

Our approach assumes that there are physical consequences, to which various payoffs can be assigned. For instance, the physical consequences could be colors of balls drawn from urns, whereas the monetary payoffs attached to them are arbitrary. There are applications in which this arbitrariness is unwarranted. For instance, if the physical consequence of death cannot be ascribed arbitrary utility value.

Yet, there are many economic applications in which one may define a physical consequence that does not uniquely define the decision maker's well-being. For instance, the physical consequences could be the success or failure of certain new technologies, while the payoffs attached to them are defined by market conditions. Alternatively, physical consequences might correspond to prices of stocks, whereas the decision maker's payoff is defined by a various derivatives she holds.

In situations such as this, it might be easier for a decision maker to imagine various payoffs attached to physical consequences, than to imagine conceivable acts defined on the entire state space in which actual acts are embedded.

Observe that, if one starts with acts and consequences and uses them to define states of nature, all preference questions in our model have corresponding questions in de Finetti's model. Whereas we assume that physical consequences might yield arbitrary utility values, de Finetti would use states that define the actual utility (through the consequence) and then attach to them hypothetical utility. By contrast, the de Finetti questionnaire would include many hypothetical questions that are not needed in our model.

It might be insightful to embed our model in a de Finetti-Savage model. Let there be given m acts, each of which may result in one of n physical consequences. Given an act a and a payoffs vector $x \in \mathbb{R}^n$, the payoff x_i will result if a is chosen by the decision maker and consequence i is chosen by nature. The corresponding de Finetti model has n^m states. An act a and a payoff vector x define a de Finetti act $y \in \mathbb{R}^{n^m}$ as follows: for a state of nature j, where the choice of act a results in consequence i, $y_j = x_i$. Act b with the same context $x \in \mathbb{R}^n$ is represented by $z \in \mathbb{R}^{n^m}$ where, for a state of nature k in which act b results in consequence l, $z_k = x_l$. The de Finetti acts y and z in $\mathbb{R}^{n^m}$ assume the same coordinate values, which are those of $x \in \mathbb{R}^n$, but not at the same coordinates. Conversely, taking the same act a with another payoff vector $x' \in \mathbb{R}^n$ would yield a de Finetti act $y' \in \mathbb{R}^{n^m}$ that is obtained from y by replacing each x_i by x_i'. If we fix an act a and consider all the vectors $y \in \mathbb{R}^{n^m}$ that correspond to a and to some $x \in \mathbb{R}^n$, we obtain a subspace of $\mathbb{R}^{n^m}$ whose dimension is n. Ranging over all acts a, the entire sets of de Finetti acts that result in the union of m such subspaces. This is clearly a much smaller set that $\mathbb{R}^{n^m}$. Moreover, in our axiomatization an act a in the context $x \in \mathbb{R}^n$ is compared only with the other $m - 1$ acts in the same context x. By contrast, in a de Finetti-Savage axiomatization, each vector $y \in \mathbb{R}^{n^m}$ is compared with all other vectors in $\mathbb{R}^{n^m}$.

Appendix: Proof

Proof of Theorem 1:

We first quote the main result in Gilboa and Schmeidler (1997). In this theorem, A1–A4 are identical to ours. A5 is weaker:

A5* Weak Neutrality: For every $a, b \in A$, $a \sim_0 b$.
where 0 denotes the origin in $\mathbb{R}^n$. The theorem states

Theorem 3 (Gilboa-Schmeidler, 1997): *The following two statements are equivalent if $|A| \geq 4$:*

(i) *$\{\succsim_x\}_{x \in \mathbb{R}^n}$ satisfy $A1-A4, A5^*$;*
(ii) *There is a collection $\{v^a\}_{a \in A} \subset \mathbb{R}^n$ such that:*
   for every $x \in \mathbb{R}^n$ and every $a, b \in A$,
   (**)
$$a \succsim_x b \quad \text{iff} \quad \sum_{i \leq n} v^a_i x_i \geq \sum_{i \leq n} v^b_i x_i ,$$
   and, for every distinct $a, b, c, d \in A$, the vectors $\{v^a - v^b, v^b - v^c, v^c - v^d\}$ are linearly independent.
   Furthermore, $\{v^a\}_{a \in A}$ are unique in the following sense: $\{u^a\}_{a \in A}$ also satisfy (**) iff there exists $\alpha > 0$ and $\beta \in \mathbb{R}^n$ such that $u^a = \alpha v^a + \beta$ for all $a \in A$.

Clearly, the representation in (**) is basically identical to that in (*), with $P_a(i) = v^a_i$. As in Theorem 1, Theorem 3 restricts the type of collections of vectors $\{v^a\}_{a \in A}$ that may arise. We now show that, if v^a is a lottery on $\{1, \ldots, n\}$, this restriction is equivalent to that of Theorem 1.

Proposition 4: *Assume first $|A| \geq 4$. A collection of lotteries $\{v^a\}_{a \in A}$ is 4-independent iff for every list (a, b, c, d) of distinct elements of A, the vectors $\{v^a - v^b, v^b - v^c, v^c - v^d\}$ are linearly independent. Similar equivalence holds for the case $|A| < 4$.*

Proof. Assume that $|A| \geq 4$. Let a collection of lotteries $\{v^a\}_{a \in A}$ be given, and assume that it is 4-independent. Consider distinct a, b, c, d. If the vectors $\{v^a - v^b, v^b - v^c, v^c - v^d\}$ are dependent, so are $\{v^a, v^b, v^c, v^d\}$, contrary to our assumption.

For the converse direction, assume that the vectors $\{(v^a - v^b),$ $(v^b - v^c), (v^c - v^d)\}$ are independent (for all quadruples of distinct a, b, c, d) but that, contrary to 4-independence of $\{v^a\}_{a \in A}$, there are distinct $a, b, c, d \in A$ and $\lambda, \mu, \theta \in \mathbb{R}$ such that

$$v^a = \lambda v^b + \mu v^c + \theta v^d.$$

Summation yields $\sum_{i=1}^n v^a = \lambda \sum_{i=1}^n v^b + \mu \sum_{i=1}^n v^c + \theta \sum_{i=1}^n v^d$. Since v^a, v^b, v^c, v^d are lotteries, we obtain $\lambda + \mu + \theta = 1$. This implies that

$$\lambda(v^a - v^b) + \mu(v^a - v^c) + \theta(v^a - v^d) = 0$$

or

$$(\lambda + \mu + \theta)(v^a - v^b) + (\mu + \theta)(v^b - v^c) + \theta(v^c - v^d) = 0.$$

Not all coefficients are zero, (again, since $\lambda + \mu + \theta = 1$), hence $\{(v^a - v^b), (v^b - v^c), (v^c - v^d)\}$ are dependent, a contradiction.   □

Next we mention that Theorem 3 is easily proved also for the case $|A| < 4$. (This is a by-product of the proof in Gilboa-Schmeidler (1997).)

We now turn to the proof of Theorem 1. We first prove sufficiency of the axioms. Assume, then, that A1–A5 hold. Thus A1–A4, A5* hold. Hence, by Theorem 3, there is a representation by $\{v^a\}_{a \in A}$ as in (**) such that, for every distinct a, b, c, d, the vectors $\{v^a - v^b, v^b - v^c, v^c - v^d\}$ are linearly independent. Further, applying A5 for $c = (1, \ldots, 1)$, one obtains that $\sum_{i \in N} v_i^a$ is independent of a.

Since A is finite, one may define $\beta_i = -\min_{a \in A} v_i^a$ and set $u_i^a = v_i^a + \beta_i$. Thus, $\{u_i^a\}_{a \in A}$ also satisfy (**), $\sum_{i \in N} u_i^a$ is independent of a, and $\min_{a \in A} v_i^a = 0$ for all $i \in N$. Let $c = \sum_{i \in N} u_i^a$ for some (hence, all) $a \in A$. If $c = 0$, $u_i^a = 0$ for all $a \in A$ and $i \in N$. Hence $c > 0$. Define $\alpha = c^{-1}$ and $P_a(i) = \alpha u_i^a$. Thus $P = \{P_a\}_{a \in A}$ are lotteries that satisfy (*). By Proposition 4, P is 4-independent.

To see the converse, assume that $P = \{P_a\}_{a \in A}$ is given, such that (*) holds and P is 4-independent. Define $v_i^a = P_a(i)$ and use Theorem 3 to show that A1–A4 hold. Then observe that, since $\sum_{i \in N} P_a(i) = 1$ for all a, (*) implies that A5 holds as well.

This completes the proof of Theorem 1. □□

Proof of Proposition 2:

Consider the collection of lotteries P defined in the proof of Theorem 1. We claim that it is more extreme than any other collection of lotteries Q satisfying $(*)$. Let Q be such a collection, with $Q \neq P$. By the uniqueness part of Theorem 3, there are $\alpha > 0$ and, for each $i \in N$, $\beta_i \in \mathbb{R}$, such that

$$Q_a(i) = \alpha P_a(i) + \beta_i$$

for every $a \in A$ and $i \in N$.

By construction of P, $\min_{a \in A} P_a(i) = 0$ for every $i \in N$. It follows that $\beta_i \geq 0$ (otherwise, we get $Q_a(i) < 0$ for some $a \in A$ and $i \in N$). Observe that, for all a,

$$1 = \sum_{i \in N} Q_a(i) = \alpha \sum_{i \in N} P_a(i) + \sum_{i \in N} \beta_i = \alpha + \sum_{i \in N} \beta_i$$

If $\beta_i = 0$ for all $i \in N$, $\alpha = 1$ and $P = Q$. Since Q differs from P, there exists $\beta_i > 0$. Hence $\alpha < 1$. Define $R(i) = \beta_i/(1 - \alpha) \geq 0$. Observe that $\sum_{i \in N} R(i) = \sum_{i \in N} \beta_i/(1 - \alpha) = 1$. Hence R is a lottery such that

$$Q_a(i) = \alpha P_a(i) + \beta_i = \alpha P_a(i) + (1 - \alpha)R(i)$$

for every $a \in A$ and $i \in N$. This proves that P is more extreme than Q. □□

References

Allais, M. (1953), "Le Comportement de L'Homme Rationel devant le Risque: critique des Postulates et Axioms de l'Ecole Americaine", *Econometrica*, **21**: 503–546.

Ashkenazi, G. and E. Lehrer (2001), "Well-Being Indices", mimeo.

de Finetti, B. (1937), "La Prevision: Ses Lois Logiques, Ses Sources Subjectives", *Annales de l'Institute Henri Poincare*, 7: 1–68.

Ellsberg, D. (1961), "Risk, Ambiguity and the Savage Axioms", *Quarterly Journal of Economics*, 75: 643–669.

Gilboa, I. and D. Schmeidler (1995), "Case-Based Decision Theory", *The Quarterly Journal of Economics*, **110**: 605–639.

Gilboa, I. and D. Schmeidler (1997), "Act Similarity in Case-Based Decision Theory", *Economic Theory*, **9**: 47–61.

Gilboa, I. and D. Schmeidler (1999), "Inductive Inference: An Axiomatic Approach", mimeo. Revised, 2001. Published in *Econometrica*, **71** (2003): 1–26. Reprinted as Chapter 4 in this volume.

Gilboa, I. and D. Schmeidler (2001), *A Theory of Case-Based Decisions*. Forthcoming, Cambridge University Press.

Gilboa, I., D. Schmeidler and P. Wakker (1999), "Utility in Case-Based Decision Theory", mimeo. Published in *Journal of Economic Theory*, **105** (2002): 483–502.

Kahneman, D. and A. Tversky (1979), "Prospect Theory: An Analysis of Decision Under Risk," *Econometrica*, **47**: 263–291.

Ramsey, F. P. (1931), "Truth and Probability", *The Foundation of Mathematics and Other Logical Essays*, New York: Harcourt, Brace and Co.

Savage, L. J. (1954), *The Foundations of Statistics*, New York: John Wiley and Sons.

von Neumann, J. and O. Morgenstern (1944), *Theory of Games and Economic Behavior*, Princeton, N.J.: Princeton University Press.

Wakker, P. P. (1989), *Additive Representations of Preferences: A New Foundation of Decision Analysis*, Kluwer Academic Publishers, Dordrecht.

Chapter 7

Probabilities as Similarity-Weighted Frequencies*

Antoine Billot[†], Itzhak Gilboa[‡], Dov Samet[§]
and David Schmeidler[¶]

Reprinted from *Econometrica*, 73 (2005): 1125–1136.

A decision maker is asked to express her beliefs by assigning probabilities to certain possible states. We focus on the relationship between
her database and her beliefs. We show that, if beliefs given a union
of two databases are a convex combination of beliefs given each of
the databases, the belief formation process follows a simple formula:
beliefs are a similarity-weighted average of the beliefs induced by each
past case.

7.1. Introduction

A physician administers a certain treatment to her patient. She is
asked to describe her prognosis by assigning probabilities to each

---

*We are grateful to Larry Epstein, David Levine, Offer Lieberman, and three anonymous referees
for their comments. Gilboa and Schmeidler gratefully acknowledge ISF grant no. 975/03,
Samet — ISF grant no. 891/04, Samet and Schmeidler — The Henry Crown Institute of Business
Research in Israel, for financial support.

[†]University de Paris II, IUF, and CERAS-ENPC. billot@u-paris2.fr
[‡]Tel-Aviv University and Yale University. igilboa@post.tau.ac.il
[§]Tel-Aviv University. samet@tauex.tau.ac.il
[¶]Tel-Aviv University and The Ohio State University. schmeid@post.tau.ac.il

of several possible outcomes $\Omega = \{1, \ldots, n\}$ of the treatment. The physician has a lot of data on past outcomes of the treatment, and she can readily quote the empirical frequencies of these outcomes. Yet, patients are not identical. They differ in age, gender, heart condition, and several other measurable variables that may affect the treatment outcome. Let us assume that these form a vector of real-valued variables $X = (X^1, \ldots, X^k)$ and that X was measured for all past cases. Thus, case j is a $(k + 1)$-tuple $(x_j, \omega_j) \in \mathbb{R}^k \times \Omega$ where, $x_j \in \mathbb{R}^k$ is the value of X observed in case j, and $\omega_j \in \Omega$ is the observed outcome of the treatment in case j. The new patient is defined by the values $x_t \in \mathbb{R}^k$ of X. How should these measurements affect the probability assessment of the physician?

It makes sense to restrict attention to those past cases that had the same X values as the one at hand, and compute relative frequencies only for these data. That is, to estimate the probability of state ω by its relative frequency in the sub-database consisting of all cases j for which $x_j = x_t$. However, large as the original database may be, the sub-database of patients whose X value is identical to x_t might be quite small or even empty. Therefore, we wish to have a procedure for assessments of probabilities over Ω that makes use of data with different X values, while taking differences in these values into account.

Assume that the physician can judge which past cases are more similar to the one at hand, and which are less similar. In evaluating the probability of a state, she may assign a higher weight to more similar cases. Formally, suppose that there exists a function $s : \mathbb{R}^k \times \mathbb{R}^k \to \mathbb{R}_{++}$, where $s(x_t, x_j)$ measures the degree to which, in the physician's judgment, a patient whose presenting conditions are given by $x_t \in \mathbb{R}^k$ is similar to another patient whose presenting conditions are $x_j \in \mathbb{R}^k$. Given a database of past cases $((x_j, \omega_j))_j$, we suggest to assign probabilities to the possible outcomes of treatment for a new patient with conditions x_t by the formula,

$$p_t = \frac{\sum_j s(x_t, x_j)\delta^j}{\sum_j s(x_t, x_j)} \in \Delta^{n-1} \qquad (7.1)$$

where $\delta^j \in \Delta^{n-1}$ is the unit vector assigning probability 1 to ω_j.

Observe that (unqualified) empirical frequencies (of states in Ω) constitute a special case of this formula, where the function s is constant. Another special case is given by $s(x_t, x_j) = 1_{\{x_t = x_j\}}$.[1] In this case, (7.1) boils down to the empirical frequencies (of states in Ω) in the sub-database defined by x_t. Thus, formula (7.1) may be viewed as offering a continuous spectrum between the unconditional empirical frequencies and conditional empirical frequencies given x_t.

In this paper we study the probability assignment problem axiomatically. We consider the relationship between various databases, modeled as sequences of cases, and the probabilities they induce. We impose two axioms on the probability assignment function. The first, *invariance*, states that the order of cases in the database is immaterial. This axiom is not very restrictive if the description of a case is informative enough, including, for instance, the time of occurrence of the case. The second axiom, *concatenation*, requires that, for every two databases, the probability induced by their concatenation is a convex combination of the probabilities induced by each of them separately. In behavioral terms, this axiom states that, if each of two databases induces a preference for one act over another, then the same preference will be induced by their concatenation. Under a minor additional condition, these two axioms are equivalent to the existence of a similarity function such that the assignment of probabilities is done as a similarity-weighted average of the probabilities induced by single cases. Two additional assumptions then yield the representation (7.1).

In our theorem, the function s is derived from presumably observable probability assignments given various possible databases. We interpret this function as a similarity function. Yet, it need not satisfy any particular properties, and may not even be symmetric. One may impose additional conditions, as in Billot, Gilboa, and Schmeidler (2004), under which there exists a norm $\mathbf{n}$ on $\mathbb{R}^k$ such that

$$s(x_t, x_j) = e^{-\mathbf{n}(x_t - x_j)}. \tag{7.2}$$

---

[1] We assumed that the function s is strictly positive. This simplifies the analysis as one need not deal with vanishing denominators. Yet, for the purposes of the present discussion it is useful to consider the more general case, allowing zero similarity values. This case is not axiomatized in this paper.

Such a function s satisfies symmetry and multiplicative transitivity (that is, $s(x, z) \geq s(x, y)s(y, z)$ for all x, y, z).[2]

The Bayesian approach calls for the assignment of a prior probability measure to a state space, and for the updating of this prior by Bayes's law given new information. Ramsey (1931), de Finetti (1937), Savage (1954), and Anscombe and Aumann (1963) provided compelling axiomatizations that justify the Bayesian approach from a normative viewpoint. But these axiomatizations do not help a predictor to form a prior if she does not already have one. In this context, our approach can be viewed as providing a belief-generation tool that may be an aid to a predictor who wishes to develop a Bayesian prior.

Such a predictor may be convinced by our axiomatization that, in certain situations, it might be desirable to generate beliefs according to formula (7.1). Yet, just as Bayesian axiomatizations do not serve to choose a prior, our axiomatization does not provide help in choosing the similarity function. Even if one adopts a certain functional form as in (7.2), the question still remains, which specific similarity function should we choose?

We believe that this question is, in the final analysis, an empirical one. Hence, the similarity function should be estimated from past data. Gilboa, Lieberman, and Schmeidler (2004) axiomatize formula (7.1) for the case $n = 2$ (not dealt with in this paper), and develop the statistical theory required for the estimation of the function s, assuming that such a function governs the data generating process. The present paper provides an axiomatization for the case $n > 2$. In certain situations, it allows to reduce the question of belief formation to the problem of similarity assessment. Developing the corresponding statistical theory is beyond the scope of this paper.

---

[2] Billot, Gilboa, and Schmeidler (2004) deal with a similarity-weighted average for a single real-valued variable, assuming that values of the same variables were observed in the past. Their axioms may be applied to any single component of the probability vector discussed here.

7.2. Model and Result

Let $\Omega = \{1, \ldots, n\}$ be a set of *states of nature*, $n \geq 3$.[3] Let C be a non-empty set of *cases*. C may be an abstract set of arbitrarily large cardinality. A *database* is a sequence of cases, $D \in C^r$ for $r \geq 1$. The set of all databases is denoted $C^* = \cup_{r \geq 1} C^r$. The concatenation of two databases, $D = (c_1, \ldots, c_r) \in C^r$ and $E = (c_1', \ldots, c_t') \in C^t$ is denoted by $D \circ E$ and it is defined by $D \circ E = (c_1, \ldots, c_r, c_1', \ldots, c_t') \in C^{r+t}$.

Observe that the same element of C may appear more than once in a given database. This structure implicitly assumes that additional observations of the same case do in fact add information. Indeed, when one estimates probabilities by relative frequencies, one subscribes to the same assumption.

For the statement of our main result we need not assume that C and Ω are a-priori related. We therefore impose no structure on C, simplifying notation and obtaining a more general result. Yet, the intended interpretation is as in the Introduction, namely, that C is a subset of $\mathbb{R}^k \times \Omega$. The prediction problem at hand, described above by $x_t \in \mathbb{R}^k$, is fixed throughout this discussion. We therefore suppress it from the notation when no confusion is likely to arise. As usual, $\Delta(\Omega)$ denotes the simplex of probability vectors over Ω.

For each $D \in C^*$, the predictor has a probabilistic belief $p(D) \in \Delta(\Omega)$ about the realization of $\omega \in \Omega$ in the problem under discussion.

For $r \geq 1$, let Π_r be the set of all permutations on $\{1, \ldots, r\}$, i.e., all bijections $\pi : \{1, \ldots, r\} \to \{1, \ldots, r\}$. For $D \in C^r$ and a permutation $\pi \in \Pi_r$, let πD be the permuted database, that is, $\pi D \in C^r$ is defined by $(\pi D)_i = D_{\pi(i)}$ for $i \leq r$.

We formulate the following axioms.

Invariance: For every $r \geq 1$, every $D \in C^r$, and every permutation $\pi \in \Pi_r$, $p(D) = p(\pi D)$.

---

[3] Our result only holds when the range of the probability assignment function is not contained in a line segment. The condition $n \geq 3$ is obviously a necessary but insufficient condition for this requirement to hold. We mention it here in order to highlight the fact that the case $n = 2$ is not covered by our result. See Gilboa, Lieberman, and Schmeidler (2004).

Concatenation: For every $D, E \in C^*$, $p(D \circ E) = \lambda p(D) + (1-\lambda)p(E)$ for some $\lambda \in (0, 1)$.

The Invariance axiom might appear rather restrictive, as it does not allow cases that appear later in D to have a greater impact on probability assessments than do cases that appear earlier. But this does not mean that cases that are chronologically more recent cannot have a greater weight than less recent ones. Indeed, should one include time as one of the variables in X, all permutations of a sequence of cases would contain the same information. In general, cases that are not judged to be exchangeable differ in values of some variables. Once these variables are brought forth, the Invariance axiom seems quite plausible.

The Concatenation axiom states that the beliefs induced by the concatenation of two databases cannot lie outside the interval connecting the beliefs induced by each database separately. If an expected payoff maximizer is faced with a decision problem where the states of nature are Ω, the Concatenation axiom could be re-stated as follows: for every two acts a and b, if a is (weakly) preferred to b given database D as well as given database E, then a is (weakly) preferred to b given the database $D \circ E$, and a strict preference given one of $\{D, E\}$ suffices for a strict preference given $D \circ E$.

We can now state our main result.

Theorem 1:   *Let there be given a function $p : C^* \to \Delta(\Omega)$. The following are equivalent:*

(i) *p satisfies the Invariance axiom, the Combination axiom, and not all $\{p(D)\}_{D \in C^*}$ are collinear;*

(ii) *There exists a function $\hat{p} : C \to \Delta(\Omega)$, where not all $\{\hat{p}(c)\}_{c \in C}$ are collinear, and a function $s : C \to \mathbb{R}_{++}$ such that, for every $r \geq 1$ and every $D = (c_1, \ldots, c_r) \in C^r$,*

$$p(D) = \frac{\sum_{j \leq r} s(c_j)\hat{p}(c_j)}{\sum_{j \leq r} s(c_j)}. \tag{*}$$

Moreover, in this case the function $\hat{p}$ is unique, and the function s is unique up to multiplication by a positive number.

This theorem may be extended to a general measurable state space Ω with no additional complications, because for every D only a finite number of measures are involved in the formula for $p(D)$.

Theorem 1 deals with an abstract set of cases C. Let us now assume, as in the Introduction, that a case c_j is a $(k + 1)$-tuple $(x_j, \omega_j) \in \mathbb{R}^k \times \Omega$, and that the function p is defined for every database D, and a given point $x_t \in \mathbb{R}^k$. The theorem then states that, under the non-collinearity condition, a function $p(D) = p(x_t, D)$ on C^* satisfies the Invariance and Concatenation axioms if and only if there are functions $s(c_j) = s(x_t, c_j)$ and $\hat{p}(c_j) = \hat{p}(x_t, c_j)$ on C such that ($*$) holds for $p(D) = p(x_t, D)$.

This application of formula ($*$) is more general than formula (7.1) in two ways: first, $\hat{p}(x_t, c_j)$ need not equal δ^j, namely, the unit vector assigning probability 1 to state ω_j. Second, $s(x_t, c_j)$ may depend on ω_j and not only on (x_t, x_j). To obtain the representation (7.1), one therefore needs two additional assumptions. First, assume that a state ω that has never been observed in the database is assigned probability zero. This guarantees that $\hat{p}(x_t, c_j) = \delta^j$. Second, assume that if the names of the states of nature are permuted in the entire database, then the resulting probability vector is accordingly permuted. This would guarantee the independence of $s(x_t, c_j)$ of ω_j.

Limitations

Formula (7.1) might be unreasonable when the entire database is very small. Specifically, if there is only one observation, resulting in state ω_i, p_t assigns probability 1 to ω_i for any x_t. This appears to be quite extreme. However, for large databases it may be acceptable to assign zero probability to a state that has never been observed. Moreover, a state that has never been observed may not be conceived of to begin with. That is, for many applications it seems natural to define Ω as the set of states that have been observed in the past. In this case, (7.1) assigns a positive probability to each state.

The intended application of formula (7.1) is for the assignment of probabilities given databases that are large, but that are not large enough to condition on every possible combination of values of $(X^1, \ldots, X^k)$. Indeed, one may assume that the function p is defined only on a restricted domain of large databases, such as $C_L^* = \cup_{n \geq L} C^n$ for a large $L \geq 1$. It is straightforward to extend our result to such restricted domains.

The Concatenation axiom that we use in this paper is very similar in spirit to the Combination axiom used in Gilboa and Schmeidler (2003). Much of the discussion of this axiom in that paper applies here as well. In particular, there are two important classes of examples wherein the Concatenation axiom does not seem plausible. The first includes situations where the similarity function is learnt from the data.[4] The second class of examples involves both inductive and deductive reasoning. For instance, if we try to learn the parameter of a coin, and then use this estimate to make predictions over several future tosses, the Concatenation axiom is likely to fail.

---

[4]The estimation procedure in Gilboa, Liebermen, and Schmeidler (2004) estimates the similarity function from the data, but assumes that these data were generated according to a *fixed* (though unknown) similarity function. However, when the data generating process itself involves an evolving similarity function, our formulae and estimation procedures are no longer valid.

Appendix: Proof

It is obvious that (ii) implies the Invariance axiom. Hence we may restrict attention to functions p that satisfy the Invariance axiom, and show that for such functions, (ii) is equivalent to the Concatenation axiom combined with the condition that not all $\{p(D)\}_{D \in C^*}$ are collinear.

In light of the Invariance axiom, a database $D \in C^*$ can be identified with a counter vector $I_D : C \to \mathbb{Z}_+$, where $I_D(c)$ is the number of times that c appears in D. Formally, for $D = (c_1, \ldots, c_r)$ let $I_D(c) = \#\{i \leq r \mid c_i = c\}$. The set of counter vectors obtained from all databases $D \in C^*$ is $\mathcal{I} = \{I : C \to \mathbb{Z}_+ \mid 0 < \sum_{j \in C} I(j) < \infty\}$. For $I \in \mathcal{I}$, define $p(I) = p(D)$ for a $D \in C^*$ such that $I = I_D$. It is straightforward that for each $I \in \mathcal{I}$ such a D exists, and that, due to the Invariance axiom, $p(D)$ is well-defined.

We now turn to state a version of our theorem for the counter vector set-up. Observe that the concatenation of two databases D and E corresponds to the pointwise addition of their counter vectors. Formally, $I_{D \circ E} = I_D + I_E$. The Concatenation axioms is therefore re-stated as the following.

Combination: For every $I, J \in \mathcal{I}$, $p(I + J) = \lambda p(I) + (1 - \lambda)p(J)$ for some $\lambda \in (0, 1)$.

Theorem 2: *Let there be given a function $p : \mathcal{I} \to \Delta(\Omega)$. The following are equivalent:*

(i) *p satisfies the Combination axiom, and not all $\{p(I)\}_{I \in \mathcal{I}}$ are collinear;*

(ii) *There are probability vectors $\{p^j\}_{j \in C} \subset \Delta(\Omega)$, not all collinear, and positive numbers $\{s_j\}_{j \in C}$ such that, for every I,*

$$p(I) = \frac{\sum_{j \in C} s_j I(j) p^j}{\sum_{j \in C} s_j I(j)}. \qquad (*)$$

Moreover, in this case the probabilities $\{p^j\}_{j \in C}$ are unique, and the weights $\{s_j\}_{j \in C}$ are unique up to multiplication by a positive number.

Observe that Theorems 1 and 2 are equivalent. We now turn to prove Theorem 2. It is straightforward to see that (ii) implies (i). Similarly, the uniqueness part of the theorem is easily verified. We therefore only prove that (i) implies (ii).

We start with the case of a finite C, say, $C = \{1, \ldots, m\}$.

Remark: For every $I \in \mathcal{I}$, $k \geq 1$, $p(kI) = p(I)$.

Proof.   Using the fact that $p(I + J) \in [p(I), p(J)]$ inductively.[1]   □

This Remark allows an extension of the domain of p to rational-coordinate vectors. Specifically, given $I \in \mathbb{Q}^C_+$, choose k such that $kI \in \mathbb{Z}^C_+$, and define $p(I)$ as identical to $p(kI)$. The Remark guarantees that the selection of k is immaterial. It follows that one may restrict attention to $p(I)$ only for $I \in \mathbb{Q}^C_+ \cap \Delta(C)$, that is, for rational points in the simplex of the case types. Restricted to this domain, p is a mapping from $\mathbb{Q}^C_+ \cap \Delta(C)$ into $\Delta(\Omega)$. We now state an auxiliary result that will complete the proof of (ii).[2]

Proposition 1:   *Assume that* $p : \mathbb{Q}^m_+ \cap \Delta^{m-1} \to \Delta^{n-1}$ *satisfies the following conditions: (i) for every* $q, q' \in \mathbb{Q}^m_+ \cap \Delta^{m-1}$, *and every rational* $\alpha \in (0, 1)$, $p(\alpha q + (1 - \alpha)q') = \lambda p(q) + (1 - \lambda)p(q')$ *for some* $\lambda \in (0, 1)$; *and (ii) not all* $\{p(q)\}_{q \in \mathbb{Q}^m_+ \cap \Delta^{m-1}}$ *are collinear. Then there are probability vectors* $\{p^j\}_{j \leq m} \subset \Delta^{n-1}$, *not all of which are collinear, and positive numbers* $\{s_j\}_{j \leq m}$ *such that, for every* $q \in \mathbb{Q}^m_+ \cap \Delta^{m-1}$,

$$p(q) = \frac{\sum_{j \leq m} s_j q_j p^j}{\sum_{j \leq m} s_j q_j}. \tag{•}$$

Proof.   For $j \leq m$, let q^j denote the j-unit vector in $\mathbb{R}^m$, i.e., the j-th extreme point of Δ^{m-1}. Obviously, one has to define $p^j = p(q^j)$. Observe that, since $p(\alpha q + (1 - \alpha)q')$ is a convex combination of $p(q)$ and $p(q')$, not all $\{p(q^j) = p^j\}_{j \leq m}$ are collinear.

---

[1]Throughout this paper, the interval defined by two vectors, p and q, is given by $[p, q] = \{\lambda p + (1 - \lambda)q \mid \lambda \in [0, 1]\}$.
[2]The following proposition is a manifestation of a general principle, stating that functions that map intervals onto intervals are projective mappings. Another manifestation of this principle in decision theory can be found in Chew (1983).

We have to show that there are positive numbers $\{s_j\}_{j \leq m}$ such that $(\bullet)$ holds for every $q \in \mathbb{Q}_+^m \cap \Delta^{m-1}$. $\square$

Step 1: $m = 3$.

Let $q^* = \frac{1}{3}(q^1 + q^2 + q^3)$. Choose positive numbers s_1, s_2, s_3 such that $(\bullet)$ holds for q^*. Observe that such s_1, s_2, s_3 exist and are unique up to multiplication by a positive number. Define $p_s(q) = \dfrac{\sum_{j \leq m} s_j q_j p^j}{\sum_{j \leq m} s_j q_j}$ for all $q \in \mathbb{Q}_+^3 \cap \Delta^2$. Denote $E = \{q \in \mathbb{Q}_+^3 \cap \Delta^2 | p_s(q) = p(q)\}$. We know that $\{q^1, q^2, q^3, q^*\} \subset E$, and we wish to show that $E = \mathbb{Q}_+^3 \cap \Delta^2$.

Step 1.1: Simplicial points are in E:

The first simplicial partition of $\mathbb{Q}_+^3 \cap \Delta^2$ is a partition to four triangles separated by the segments connecting $\{(\frac{1}{2}q^1 + \frac{1}{2}q^2), (\frac{1}{2}q^2 + \frac{1}{2}q^3), (\frac{1}{2}q^3 + \frac{1}{2}q^1)\}$. The second simplicial partition is obtained by similarly partitioning each of the four triangles to four smaller triangles, and the k-th simplicial partition is defined recursively. The simplicial points of the k-th simplicial partition are all the vertices of triangles of this partition.

We now state the following

Claim: If the vertices and the center of gravity of a simplicial triangle are in E, then so are the vertices and center of gravity of all of its four simplicial sub-triangles.

Proof. If four points that are not collinear, a, b, c, d, are in E, then the point defined by the intersection of the segments $[a, b]$ and $[c, d]$ is also in E. The proof is conducted by applying this fact inductively as suggested by Figure 7.1.

Explicitly, let $\{q_k^1, q_k^2, q_k^3\}$ be the vertices of a triangle in the k-th simplicial partition. Assume that $q_k^1, q_k^2, q_k^3, \frac{1}{3}(q_k^1 + q_k^2 + q_k^3) \in E$. We first show that $(\frac{1}{2}q_k^1 + \frac{1}{2}q_k^2), (\frac{1}{2}q_k^2 + \frac{1}{2}q_k^3), (\frac{1}{2}q_k^3 + \frac{1}{2}q_k^1) \in E$. Indeed, $(\frac{1}{2}q_k^1 + \frac{1}{2}q_k^2)$ is the intersection of the line connecting q_k^3 and $\frac{1}{3}(q_k^1 + q_k^2 + q_k^3)$, and the line connecting q_k^1 and q_k^2. Hence both $p(\frac{1}{2}q_k^1 + \frac{1}{2}q_k^2)$ and $p_s(\frac{1}{2}q_k^1 + \frac{1}{2}q_k^2)$ have to be the intersection of the line connecting $p(q_k^3) = p_s(q_k^3)$ and $p(\frac{1}{3}(q_k^1 + q_k^2 + q_k^3)) = p_s(\frac{1}{3}(q_k^1 + q_k^2 + q_k^3))$, and the line connecting $p(q_k^1) = p_s(q_k^1)$ and $p(q_k^2) = p_s(q_k^2)$. Since not all

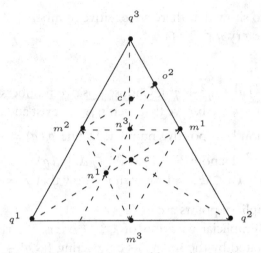

Fig. 7.1.   The vertices and center of gravity of four sub-triangles

The point m^1 is the intersection of the lines $q^2 q^3$ and $q^1 c$. The points m^2 and m^3 are similarly constructed. The point n^3 is the intersection on $m^1 m^2$ and $q^3 c$. The point n^1 is similarly constructed. The point o^2 is the intersection of $n^1 n^3$ and $q^2 q^3$. Finally, the center of gravity of $m^1 m^2 q^3$ is the intersection of $m^2 o^2$ and $q^3 m^3$ at c'.

$p(q)$ are collinear, this intersection is unique. Hence $(\frac{1}{2} q_k^1 + \frac{1}{2} q_k^2) \in E$. Similarly, we also have $(\frac{1}{2} q_k^2 + \frac{1}{2} q_k^3), (\frac{1}{2} q_k^3 + \frac{1}{2} q_k^1) \in E$.

Next consider the center of gravity of the four sub-triangles. For the triangle $conv\{(\frac{1}{2} q_k^1 + \frac{1}{2} q_k^2), (\frac{1}{2} q_k^2 + \frac{1}{2} q_k^3), (\frac{1}{2} q_k^3 + \frac{1}{2} q_k^1)\}$, the center of gravity is equal to that of $conv\{q_k^1, q_k^2, q_k^3\}$, which is already known to be in E. Next consider the center of gravity of one of the three sub-triangles that have a vertex is common with $conv\{q_k^1, q_k^2, q_k^3\}$. Assume, without loss of generality, that it is the triangle defined by $\{q_k^3, (\frac{1}{2} q_k^1 + \frac{1}{2} q_k^3), (\frac{1}{2} q_k^2 + \frac{1}{2} q_k^3)\}$. We first note that $\frac{1}{2}(\frac{1}{2} q_k^1 + \frac{1}{2} q_k^3) + \frac{1}{2}(\frac{1}{2} q_k^2 + \frac{1}{2} q_k^3)$ is in E because it is the intersection of $[q^3, (\frac{1}{2} q_k^1 + \frac{1}{2} q_k^2)]$ and $[(\frac{1}{2} q_k^2 + \frac{1}{2} q_k^3), (\frac{1}{2} q_k^3 + \frac{1}{2} q_k^1)]$. Similarly, $\frac{1}{2}(\frac{1}{2} q_k^1 + \frac{1}{2} q_k^3) + \frac{1}{2}(\frac{1}{2} q_k^1 + \frac{1}{2} q_k^2)$ is in E. The point $\frac{1}{2} q_k^3 + \frac{1}{2}(\frac{1}{2} q_k^2 + \frac{1}{2} q_k^3) = \frac{3}{4} q_k^3 + \frac{1}{4} q_k^2$ is on the line connecting $\frac{1}{2}(\frac{1}{2} q_k^1 + \frac{1}{2} q_k^3) + \frac{1}{2}(\frac{1}{2} q_k^2 + \frac{1}{2} q_k^3)$ and $\frac{1}{2}(\frac{1}{2} q_k^1 + \frac{1}{2} q_k^3) + \frac{1}{2}(\frac{1}{2} q_k^1 + \frac{1}{2} q_k^2)$ and on the line connecting q_k^2 and q_k^3. Hence $\frac{3}{4} q_k^3 + \frac{1}{4} q_k^2$ is in E. The center of gravity of the triangle $conv\{q_k^3, (\frac{1}{2} q_k^1 + \frac{1}{2} q_k^3), (\frac{1}{2} q_k^2 + \frac{1}{2} q_k^3)\}$ is the intersection of $[q^3, \frac{1}{2} q_k^1 + \frac{1}{2} q_k^2]$ and $[(\frac{1}{2} q_k^1 + \frac{1}{2} q_k^3), (\frac{3}{4} q_k^3 + \frac{1}{4} q_k^2)]$. Hence the center of gravity of the triangle $conv\{q_k^3, (\frac{1}{2}q_k^1 + \frac{1}{2}q_k^3), (\frac{1}{2}q_k^2 + \frac{1}{2}q_k^3)\}$ is in E. $\square$

Applying the claim inductively, we conclude that E contains all points that are vertices of simplicial sub-triangles of $conv\{q_k^1, q_k^2, q_k^3\}$. $\square$

Step 1.2: Completion:

Observe that, if $q \in \mathbb{Q}_+^3 \cap conv(q, q', q'')$, then $p(q) \in conv(p(q), p(q'), p(q''))$. Consider an arbitrary $q \in conv\{q^1, q^2, q^3\}$. Take a sequence of simplicial triangles, $conv\{q_k^1, q_k^2, q_k^3\}$, such that $q \in conv\{q_k^1, q_k^2, q_k^3\}$ and that $\lim_{k\to\infty} q_k^j = q$ for all $j = 1, 2, 3$. Since p_s is a continuous function, $\lim_{k\to\infty} p_s(q_k^j) = p_s(q)$ for all $j = 1, 2, 3$. Moreover, because both p and p_s satisfy the Combination axiom, it follows that $p(q), p_s(q) \in conv\{p(q_k^1) = p_s(q_k^1), p(q_k^2) = p_s(q_k^2), p(q_k^3) = p_s(q_k^3)\}$. This is possible only if $p(q) = p_s(q)$. Hence $q \in E$. Since the choice of q was arbitrary, $E = \mathbb{Q}_+^3 \cap \Delta^2$.

Step 2: $m > 3$.

Step 2.1: Defining s_j:

Consider a triple $j, k, l \le m$ such that $\{p^j, p^k, p^l\}$ are not collinear. Apply Step 1 to obtain a representation

$$p(q) = \sum_{v\in\{j,k,l\}} s_v^{\{j,k,l\}} q_v p^v(\{j,k,l\}) / \sum_{v\in\{j,k,l\}} s_v^{\{j,k,l\}} q_v$$

for all $q \in \mathbb{Q}_+^m \cap conv(\{q^j, q^k, q^l\})$. Moreover, for all $v \in \{j, k, l\}$, $p^v(\{j, k, l\}) = p(q^v) = p^v$, and the coefficients $\{s_v^{\{j,k,l\}}\}_{v\in\{j,k,l\}}$ are unique up to multiplication by a positive number.

Next consider all triples $j, k, l \le m$ such that $\{p^j, p^k, p^l\}$ are not collinear. We argue that, for given j, k, $s_j^{\{j,k,l\}} / s_k^{\{j,k,l\}}$ is independent of l. To see this, assume that l and l' are such that neither $\{p^j, p^k, p^l\}$ nor $\{p^j, p^k, p^{l'}\}$ are collinear. Restricting attention to rational combinations of q^j and q^k, one observes that $s_j^{\{j,k,l\}} / s_k^{\{j,k,l\}} = s_j^{\{j,k,l'\}} / s_k^{\{j,k,l'\}}$. Denote this ratio by γ_{jk}. Observe that it is defined for every distinct $j, k \le m$, because for every j, k there exists at least one l such that $\{p^j, p^k, p^l\}$ are not collinear. Further, note that if $\{p^j, p^k, p^l\}$ are not collinear, then $\gamma_{jk}\gamma_{kl}\gamma_{lj} = 1$.

Define $s_1 = 1$ and $s_j = \gamma_{j1}$ for $1 < j \leq m$. We wish to show that, for every triple $j, k, l \leq m$ such that $\{p^j, p^k, p^l\}$ are not collinear, $\{s_v^{\{j,k,l\}}\}_{v \in \{j,k,l\}}$ is proportional to $\{s_j, s_k, s_l\}$. Without loss of generality, it suffices to show that $s_j^{\{j,k,l\}} / s_k^{\{j,k,l\}} = s_j/s_k$, or that $\gamma_{jk} = s_j/s_k$. If $\{p^1, p^j, p^k\}$ are not collinear, then this equation follows from $\gamma_{1j}\gamma_{jk}\gamma_{k1} = 1$. If, however, $\{p^1, p^j, p^k\}$ are collinear, then $\{p^1, p^j, p^l\}$ and $\{p^1, p^k, p^l\}$ are not collinear. Hence $\gamma_{kl} = s_k/s_l$ and $\gamma_{lj} = s_l/s_j$. In this case, $\gamma_{jk} = 1/\gamma_{kl}\gamma_{lj} = s_j/s_k$.

Given $s = (s_j)_{j \leq m}$, define $p_s(q) = \frac{\sum_{j \leq m} s_j q_j p^j}{\sum_{j \leq m} s_j q_j}$. Thus, we wish to show that $p(q) = p_s(q)$ for all $q \in \mathbb{Q}_+^m \cap \Delta^{m-1}$.

Step 2.2: Completion:

We prove the following claim by induction on k, $3 \leq k \leq m$:

Claim: For every subset $K \subset \{1, \dots, m\}$ with $|K| = k$, if $\{p^j\}_{j \in K}$ are not collinear, then $p(q) = p_s(q)$ holds for every $q \in \Delta_K \equiv \mathbb{Q}_+^m \cap conv(\{q^j \mid j \in K\})$.

Proof.   The case $k = 3$ was proven in Step 1. We assume that the claim is correct for $k \geq 3$, and we prove it for $k + 1$. Let there be given $K \subset \{1, \dots, m\}$ with $|K| = k+1$, such that $\{p^j\}_{j \in K}$ are not collinear. Let $J = \{j \in K \mid \{p^l\}_{l \in K \setminus \{j\}}$ are not collinear$\}$. Observe that, for every $j \in J$, $p(q) = p_s(q)$ holds for every $q \in \Delta_{K \setminus \{j\}}$.

We argue that $|J| \geq k$. To see this, assume that there were two distinct elements j and k, in $K \setminus J$. Then all $\{p^l\}_{l \neq j}$ are collinear, as are all $\{p^l\}_{l \neq k}$. Since $|K| = k + 1 \geq 4$, there are at least two distinct elements in $K \setminus \{j, k\}$. Both p^j and p^k are collinear with $\{p^l\}_{l \neq j,k}$, and it follows that all $\{p^l\}_{l \in K}$ are collinear, a contradiction.

Consider a rational point $q \in \mathbb{Q}_+^m$ in the relative interior of $conv(\{q^l \mid l \in K\})$. Denote $q = \sum_{l \in K} \alpha_l q^l$ with $\alpha_l > 0$. For every $j \in J$, Let $q(j)$ be the point in $conv(\{q^l \mid l \in K \setminus \{j\}\})$ that is on the line connecting q^j and q, that is, $q(j) = \sum_{l \in K \setminus \{j\}} \frac{\alpha_l}{1 - \alpha_j} q^l$. Obviously, $p_s(q^j) = p(q^j) = p^j$. Moreover, since $j \in J$, one may apply the claim to $K \setminus \{j\}$, yielding $p_s(q(j)) = p(q(j))$. Since p_s satisfies the Combination axiom, it follows that both $p(q)$ and $p_s(q)$ are on the interval $[p_s(q^j), p_s(q(j))] = [p^j, p(q(j))]$.

Next we wish to show that, for at least two elements $j, k \in J$, the intervals $[p^j, p(q(j))]$ and $[p^k, p(q(k))]$ cannot lie on the same line. Assume not, that is, that all intervals $\{[p^j, p(q(j))]\}_{j \in J}$ lie on a line L. If $J = K$, this implies that all $\{p^j\}_{j \in K}$ are collinear, a contradiction. Assume, then, that there is an i such that $J = K \backslash \{i\}$. In this case, p^i is not on L. For $j \in J$, consider $q(j)$ as a convex combination of q^i and a point $q' \in conv(\{q^l \mid l \in K \backslash \{i, j\}\})$. By the Combination axiom, $p(q')$ is on the line L. Moreover, since $p^i \neq p(q')$, $p(q(j))$ is in the open interval $(p^i, p(q'))$, and therefore not on L. But this contradicts the assumption that all intervals $\{[p^j, p(q(j))]\}_{j \in J}$ lie on L.

It follows that there are distinct $j, k \in J$ for which the intervals $[p^j, p(q(j))]$ and $[p^k, p(q(k))]$ do not lie on the same line. Hence these intervals can intersect in at most one point. Since both $p(q)$ and $p_s(q)$ are on both intervals, $p(q) = p_s(q)$ follows.

We conclude that $p(q) = p_s(q)$ holds for every rational q in the relative interior of $conv(\{q^j \mid j \in K\})$, as well as for all rational points in $conv(\{q^l \mid l \in K \backslash \{j\}\})$ for $j \in J$. It is left to show that $p(q) = p_s(q)$ for rational points in $conv(\{q^l \mid l \in K \backslash \{i\}\})$ for $i \in K \backslash J$. Assume not. Then, for some $q \in \mathbb{Q}_+^m \cap conv(\{q^l \mid l \in K \backslash \{i\}\})$, $p(q) \neq p_s(q)$. But $p(q^i) = p_s(q^i) = p^i$. Hence the interval (q^i, q) is mapped by p into $(p^i, p(q))$ and by p_s — into $(p^i, p_s(q))$. Note that these two open intervals are disjoint. But for any $q' \in (q^i, q)$ we should have $p(q') = p_s(q')$, a contradiction. $\qquad\square$

It is left to complete the proof of the sufficiency of the Combination axiom in case C is infinite. For every $B \subset C$, let $\mathcal{I}_B$ be the set of databases $I \in \mathcal{I}$ such that $\sum_{j \notin B} I(j) = 0$. For every $j \in C$, define p^j by $p(I_j)$ where I_j is defined by $I_j(j) = 1$ and $I_j(k) = 0$ for $k \neq j$. For every finite $B \subset C$, for which not all $\{p^j\}_{j \in B}$ are collinear, there is a function s_B such that ($*$) holds for every $I \in \mathcal{I}_B$. Moreover, this function is unique up to multiplication by a positive number. Fix one such finite set C_0 and choose a function s_{C_0}. For every other finite $B \subset C$, for which not all $\{p^j\}_{j \in B}$ are collinear, consider $B' = C_0 \cup B$. Over B' there exists a unique $s_{B'}$ that satisfies ($*$) for all $I \in \mathcal{I}_{B'}$ and that extends s_{C_0}. Define s_B as the restriction of $s_{B'}$ to B. To see that this construction is well-defined, suppose that B_1 and B_2 are two such sets with a non-empty intersection. Consider $B = B_1 \cup B_2$. Since s_{B_1} and s_{B_2} are both restrictions of s_B, they are equal on $B_1 \cap B_2$.   □□

References

Anscombe, F. J. and R. J. Aumann (1963), "A Definition of Subjective Probability", *The Annals of Mathematics and Statistics*, **34**: 199–205.

Billot, A., I. Gilboa and D. Schmeidler (2004), "An Axiomatization of an Exponential Similarity Function", mimeo. Published in *Mathematical Social Sciences*, **55** (2008): 107–115. Reprinted as Chapter 10 of this volume.

Chew, S. H. (1983), "A Generalization of the Quasilinear Mean with Applications to the Measurement of Income Inequality and Decision Theory Resolving the Allais Paradox", *Econometrica*, **51**: 1065–1092.

de Finetti, B. (1937), "La Prevision: Ses Lois Logiques, Ses Sources Subjectives", *Annales de l'Institute Henri Poincare*, 7: 1–68.

Gilboa, I. and D. Schmeidler (2003), "Inductive Inference: An Axiomatic Approach", *Econometrica*, **71**: 1–26.

Gilboa, I., O. Lieberman and D. Schmeidler (2004), "Empirical Similarity", mimeo. Published in *Review of Economics and Statistics*, **88** (2006): 433–444. Reprinted as Chapter 9 of this volume.

Ramsey, F. P. (1931), "Truth and Probability", *The Foundation of Mathematics and Other Logical Essays*, New York: Harcourt, Brace and Co.

Savage, L. J. (1954), *The Foundations of Statistics*, New York: John Wiley and Sons.

Chapter 8

Fact-Free Learning*

Enriqueta Aragones[†], Itzhak Gilboa[‡], Andrew Postlewaite[§]
and David Schmeidler[¶]

Reprinted from *American Economic Review*, 95 (2005):
1355–1368.

People may be surprised by noticing certain regularities that hold in
existing knowledge they have had for some time. That is, they may
learn without getting new factual information. We argue that this
can be partly explained by computational complexity. We show that,
given a knowledge base, finding a small set of variables that obtain
a certain value of R^2 is computationally hard, in the sense that this

*Earlier versions of this paper circulated under the titles "¿From Cases to Rules: Induction
and Regression" and "Accuracy versus Simplicity: A Complex Trade-Off". We have benefited
greatly from comments and references by the editor, an anonymous referee, Yoav Benjamini,
Joe Halpern, Offer Lieberman, Bart Lipman, Yishay Mansour, Nimrod Megiddo, Dov Samet,
Petra Todd, and Ken Wolpin, as well as the participants of the SITE conference on Behavioral
Economics at Stanford, August, 2003 and the Cowles Foundation workshop on Complexity
in Economic Theory at Yale, September, 2003. Aragones acknowledges financial support
from the Spanish Ministry of Science and Technology, grant number SEC2003-01961, and
CREA-Barcelona Economics. Aragones, Gilboa, and Schmeidler gratefully acknowledge support
from the Polarization and Conflict Project CIT-2-CT-2004-506084 funded by the European
Commission-DG Research Sixth Framework Programme. Gilboa and Schmeidler gratefully
acknowledge support from the Israel Science Foundation (Grant Nos. 790/00 and 975/03).
Postlewaite gratefully acknowledges support from National Science Foundation grant SBR-
9810693.
[†]Institut d'Anàlisi Econòmica, C.S.I.C. enriqueta.aragones@uab.es
[‡]Tel-Aviv University and Cowles Foundation, Yale University. igilboa@tau.ac.il
[§]University of Pennsylvania; apostlew@econ.sas.upenn.edu
[¶]Tel-Aviv University and the Ohio State University. schmeid@tau.ac.il

term is used in computer science. We discuss some of the implications of this result and of fact-free learning in general.

JEL D83

Keywords: bounded rationality, complexity, learning, regression.

Fact-Free Learning

"The process of induction is the process of assuming the simplest law that can be made to harmonize with our experience." — *Wittgenstein (1922)*

8.1. Introduction

Understanding one's social environment requires accumulating information and finding regularities in that information. Many theoretical models of learning focus on learning new facts, on their integration in an existing knowledge base, and on the way they modify beliefs. Within the Bayesian framework the integration of new facts and the modification of beliefs is done mechanically according to Bayes's rule. However, much of human learning has to do with making observations and finding regularities that, in principle, could have been determined using existing knowledge, rather than with the acquisition of new facts.

Consider technological innovations. In many cases, the main idea of an innovation involves combining well-known facts. For instance, putting wheels at the bottom of a suitcase allows it to roll easily. This idea was quite original when it was first introduced. But, since it only selected and combined facts that everyone had already known, it appears obvious in hindsight. It takes originality to come up with such an idea, but no particular expertise is needed to judge its value. This phenomenon is so pervasive that it has been canonized in literature: Sherlock Holmes regularly explains how the combination of a variety of clues lead inexorably to a particular conclusion, following which Watson exclaims "Of course!"

To consider an even more extreme case, assume that an individual follows a mathematical proof of a theorem. In order to check the proof, one need not resort to the knowledge of facts. The knowledge that the agent acquires in the process has always been, in principle, available to her. Yet, mathematics has to be studied. In fact, it is an entire discipline based solely on fact-free learning.

In this paper we focus on a particular type of fact-free learning. We consider an agent who has access to a database, involving many variables and many observations. The agent attempts to find regularities in the database. We model this learning problem and explain the difficulty in solving it optimally.[1]

The immediate consequence of this difficulty is that individuals typically will not discover all the regularities in their knowledge base, and may overlook the most useful regularities. Two people with the same knowledge base may notice different regularities, and may consequently hold different views about a particular issue. One person may change the beliefs and actions of another without communicating new facts, but simply by pointing to a regularity overlooked by the other person. On the other hand, people may agree to disagree even if they have the same knowledge base and are communicating. We elaborate on these consequences in Section 8.4.

For illustration, consider the following example.

Ann: "Russia is a dangerous country."

Bob: "Nonsense."

Ann: "Don't you think that Russia might initiate a war against a Western country?"

Bob: "Not a chance."

Ann: "Well, I believe it very well might."

Bob: "Can you come up with examples of wars that erupted between two democratic countries?"

Ann: "I guess so. Let me see... How about England and the US in 1812?"

Bob: "OK, save colonial wars."

---

[1] Herbert Simon (1955) argued a half century ago for incorporation of "the physiological and psychological limitations" in models of decision making.

Ann: "Well, then, let's see. OK, maybe you have a point. Perhaps Russia is not so dangerous."

Bob seems to have managed to change Ann's views without providing Ann with any new factual information. Rather, he pointed out a regularity in Ann's knowledge base of which she had been unaware: democratic countries have seldom waged war on each other.[2]

It is likely that Ann failed to notice that the democratic peace phenomenon holds in her own knowledge base simply because it had not occurred to her to categorize wars by the type of regime of the countries involved. For most people, wars are categorized, or "indexed", by chronology and geography, but not by regime. Once the variable "type of regime" is introduced, Ann will be able to reorganize her knowledge base and observe the regularity she had failed to notice earlier.

Fact-free learning is not always due to the introduction of a new variable, or a categorization that the individual has not been aware of. Often, one may be aware of all variables involved, and yet fail to see a regularity that involves a *combination* of such variables. Consider an econometrician who wants to understand the determinants of the rate of economic growth. She has access to a large database of realized growth rates for particular economies that includes a plethora of variables describing these economies in detail.[3] Assume that the econometrician prefers fewer explanatory variables to more. Her main difficulty is to determine what set of variables to use in her regression. We can formalize her problem as determining whether there exists a set of k regressors that give a particular level of R^2. This is a well-defined problem that can be relegated to a computer software. However, testing all subsets of k regressors out of, say, m variables involves running $\binom{m}{k} = O(m^k)$ regressions. When m and k are of

---

[2] In the field of international relations this is referred to as the "democratic peace phenomenon". (See, e.g., Zeev Maoz and Bruce Russett (1993).)

[3] As an example of the variety of variables that may potentially be relevant, consider the following quote from a recent paper by Raphael La Porta, Florencio Lopez-de-Silanes, Andrei Shleifer, and Robert Vishny (1998) on the quality of government: "We find that countries that are poor, close to the equator, ethnolinguistically heterogeneous, use French or socialist laws, or have high proportions of Catholics or Muslims exhibit inferior government performance."

realistic magnitude, it is impractical to perform this exhaustive search. For instance, choosing the best set of $k = 13$ regressors out of $m = 100$ potentially relevant variables involves $\binom{100}{13} \approx 7 * 10^{15}$ regressions. On a computer that can perform 10 million regression analyses per second, this task would take more than twenty-two years.

Linear regression is a structured and relatively well-understood problem, and one may hope that, using clever algorithms that employ statistical analysis, the best set of k regressors can be found without actually testing all $\binom{m}{k}$ subsets. Our main result is that this is not the case. Formally, we prove that finding whether k regressors can obtain a pre-specified value of R^2, r, is, in the language of computer science, NP-Complete.[4] Moreover, we show that this problem is hard (NP-Complete) for *every* positive value of r. Thus our regression problem belongs to a large family of combinatorial problems for which no efficient (polynomial) algorithm is known. An implication of this result is that, even for moderate size data sets, it will generally be impossible to know the trade-off between increasing the number of regressors and increasing the explanatory power of those regressors.[5]

Our interest is not in the problem econometricians face, but in the problems encountered by nonspecialists attempting to understand their environment. That is, we wish to model the reasoning of standard economic agents, rather than of economists analyzing data. We contend, however, that a problem that is difficult to solve for a working economist will also be difficult for an economic agent. If an econometrician cannot be guaranteed to find the "best" set of regressors, many economic agents may also fail to identify important relationships in their personal knowledge base.[6]

Neither economic agents nor social scientists typically look for the best set of regressors without any guiding principle. Rather than engaging in data mining they espouse and develop various theories

---

[4]In Section 8.3 we explain the concept of NP-completeness and provide references to formal definitions.

[5]In particular, principle components analysis, which finds a set of orthogonal components, is not guaranteed to find the best combination of predictors (with unconstrained correlations).

[6]We discuss this further in Section 8.4 below.

that guide their search for regularities. Our econometrician will often have some idea about which variables may be conducive to growth. She therefore need not exhaust all subsets of k regressors in her quest for the "best" regression. Our model does not capture the development of and selection among causal theories, but even the set of variables potentially relevant to our econometrician's theory is typically large enough to raise computational difficulties. More importantly, if the econometrician wants to test her scientific paradigm, and if she wants to guarantee that she is not missing some important regularities that lie outside her paradigm, she cannot restrict attention to the regressors she has already focused on.

While computational complexity is not the only reason for which individuals may be surprised to discover regularities in their own knowledge bases, it is one of the reasons that knowledge of facts does not imply knowledge of all their implications. Hence computational complexity, alongside unawareness, makes fact-free learning a common phenomenon.

In the next section we lay out our model and discuss several notions of regularities and the criteria to choose among them. The difficulty of discovering satisfactory sets of regressors is proven in Section 8.3. In the last section we discuss the results, their implications and related literature.

8.2. Regularities in a Knowledge Base

An individual's knowledge base consists of her observations, past experiences, as well as observations that were related to her by others. We will assume that observations are represented as vectors of numbers. An entry in the vector might be the value of a certain numerical variable, or a measure of the degree to which the observation has a particular attribute. Thus, we model the information available to an individual as a knowledge base consisting of a matrix of numbers where rows correspond to observations (distinct pieces of information) and columns to attributes.

We show below a fraction of a conceivable knowledge base pertinent to the democratic peace example. The value in a given entry represents the degree to which the attribute (column) holds for the observation (row). (The numbers are illustrative only.)

Observation	$M1$	$M2$	$D1$	$D2$	T	W
WWII[7]	.7	1	1	0	0	1
Cuban missile crisis	1	1	1	0	1	0
1991 Gulf war	1	.3	1	0	1	1

M_i — how strong was country i?
D_i — was country i a democracy?
T — was it after 1945?
W — did war result?

The democratic peace regularity states that if, for any given observation, the attribute W assumes the value 1, then at least one of the attributes $\{D_1, D_2\}$ does not assume that value.[8]

This model is highly simplified in several respects. It assumes that the individual has access to a complete matrix of data, whereas in reality certain entries in the matrix may not be known or remembered. The model implicitly assumes also that all variables are observed with accuracy. More importantly, in our model we assume that observations are already encoded in a particular way that facilitates identifying regularities.[9]

---

[7] We refer here to England's declaration of war on Germany on September 3, 1939.

[8] More precisely, this is the contrapositive of the democratic peace regularity.

[9] For instance, in this matrix above country "1" is always the democratic one. But, when representing a real-life case by a row in the matrix, one may not know which country should be dubbed "1" and which — "2". This choice of encoding is immaterial in the democratic peace phenomenon, because this rule is symmetric with respect to the countries. If, however, we were to consider the rule "a democratic country would never attack another country", encoding would matter. If the encoding system keeps country "1" as a designator of a democratic country (as long as one of the countries involved is indeed a democracy), this rule would take the form "if $D1 = 1$ then $A1 = 0$", where Ai stands for "country i attacked". If, however, the encoding system does not retain this regularity, the same rule will not be as simple to formulate. In fact, it would require a formal relation between variables, allowing to state "For every i, if $Di = 1$ then $Ai = 0$". Since such relations are not part of our formal model, the model would give rise to different regularities depending on the encoding system. Indeed, finding the "appropriate" encoding is part of the problem of finding regularities in the database.

We will prove that despite all these simplifying assumptions, it is hard to find regularities in the knowledge base. Finding regularities in real knowledge bases, which are not so tidy, would be even more difficult.

The democratic peace phenomenon is an example of an *association rule*. Such a rule states that *if*, for any given observation, the values of certain attributes are within stipulated ranges, *then* the values of other attributes are within prespecified ranges. An association rule does not apply to the entire knowledge base: its scope is the set of observations that satisfy its antecedent. It follows that association rules differ from each other in their generality, or scope of applicability. Adding variables to the antecedent (weakly) decreases the scope of such a rule, but may increase its accuracy. For example, we may refine the democratic peace rule by excluding observations prior to the first world war. This will eliminate some exceptions to the rule (e.g., the War of 1812 and the Boer War) but will result in a less general rule.

A second type of regularity is a *functional rule*: a rule that points to a functional relationship between several "explanatory" variables (attributes) and another one (the "predicted" variable). A well-known example of such a rule is linear regression, with which we deal in the formal analysis. All functional rules on a given knowledge base have the same scope of applicability, or the same generality.

Both association rules and functional rules may be ranked according to accuracy and simplicity. Each criterion admits a variety of measures, depending on the specific model. In the case of linear regression, it is customary to measure accuracy by R^2 while simplicity is often associated with a low number of variables. Irrespective of the particular measures used, people generally prefer high accuracy and low complexity. The preference for accuracy is perhaps the most obvious: rules are supposed to describe the knowledge base, and accuracy is simply the degree to which they succeed in doing so.

The preference for simplicity is subtler. A standard econometric exercise is to use a data base consisting of a number of observations to derive a linear relationship between a variable of interest and other variables. The goal is to use the linear relationship to predict the variable of interest in similar situations in the future. A typical example would consist of a number of past instances in which women with breast cancer were given different treatments. Each observation would consist of the treatment, a number of diagnostic tests such as blood chemistry, location of the tumor, size of the tumor in X-rays, etc., and the degree to which the treatment was successful. These observations would be used to determine a linear relationship between the diagnostic tests and the degree of success for each treatment. The resulting relationship is then used to predict the success of future cases.

When faced with a problem such as this, a scientist need not automatically prefer fewer explanatory variables to more. The literature in statistics and machine learning provides criteria for "model selection", and in particular, for the inclusion of explanatory variables, in such a way as to avoid spurious correlations and "over-fitting". Our interest, however, is not in the way a scientist or an econometrician would use a data base to predict future outcomes, but rather in the way an ordinary person might find relationships in his or her personal knowledge base. We maintain that, other things being equal, people tend to have more faith in the robustness of relationships that use fewer variables than in those that use more. That is, we suggest that the preference for parsimony and simplicity, as measured by the number of variables employed, is a natural tendency of the human mind.

Individuals may prefer fewer explanatory variables because of availability of data. Having a rule that involves more variables implies that more variables need to be gathered and maintained in order to use it. Importantly, it also makes it less likely that all the variables needed for the application of the rule will indeed be available in a related problem.

When fewer variables are involved, people will find it easier to make up explanations for a regularity in the data. This may be another reason for the preference for fewer variables. Be that as it may, the (normative) claim that people should prefer simpler theories to more complex ones goes back to William of Occam, and the (descriptive) claim that this is how the human mind works can also be found in Wittgenstein (1922).

In this paper we assume that people generally prefer rules that are as accurate and as simple as possible. Of course, these properties present one with non-trivial trade-offs. In the next section we discuss functional rules for a given knowledge base. We will show that the feasible set in the accuracy-simplicity space cannot be easily computed. A similar result can be shown for association rules. We choose to focus on linear regression for two reasons. First, in economics it is a more common technique for uncovering rules. Second, our main result is less straightforward in the case of linear regression.

8.3. The Complexity of Linear Regression

In this section we study the trade-off between simplicity and accuracy of functional rules in the case of linear regression. While regression analysis is a basic tool of scientific research, we here view it as an idealized model of non-professional human reasoning.[10] For a given variable, one attempts to find those variables that predict the variable of interest. A common measure of amount of variation in the variable of interest that is explained by the predicting variables is the coefficient of determination, R^2. A reasonable measure of complexity is the number of explanatory variables one uses. The "adjusted R^2" is frequently used as a measure of the quality of a regression, trading off accuracy and simplicity. Adjusted R^2 essentially levies a multiplicative penalty for additional variables to offset the spurious increase in R^2 that results from an increase in the number of predicting variables. In recent years statisticians and econometricians mostly use additive penalty functions in model specification (choosing the predicting variables) for a regression problem.[11] The different penalties are associated with different criteria determining the trade-off between parsimony and precision. Each penalty function can be viewed as defining preferences over the number of included variables and R^2, reflecting the trade-off between simplicity and accuracy. Rather than choose a specific penalty function, we assume more generally that an individual can be ascribed a function $u : \mathbb{R}_+ \times [0,1] \to \mathbb{R}$ that represents her preferences for simplicity and accuracy, where $u(k, r)$ is her utility for a regression that attains $R^2 = r$ with k explanatory variables. Thus, if $u(\cdot, \cdot)$ is decreasing

---

[10] See Margaret Bray and Nathan Savin (1986), who used regression analysis to model the learning of economic agents.

[11] See, e.g., Trevor Hastie *et al.* (2001) for a discussion of model specification and penalty functions.

in its first argument and increasing in the second, a person who chooses a rule so as to maximize u may be viewed as though she prefers both simplicity and accuracy, and trades them off as described by u.

Our aim is to demonstrate that finding "good" rules is a difficult computational task. We use the concept of NP-Completeness from computer science to formalize the notion of difficulty of solving problems. A yes/no problem is NP if it is easy (can be performed in polynomial worst-case time complexity) to verify that a suggested solution is indeed a solution to it. If an NP problem is also NP-Complete, then, there is at present no known algorithm, whose (worst-case time) complexity is polynomial, that can solve it. However, NP-Completeness means more than that there is no such known algorithm. The non-existence of such an algorithm is not due to the fact that the problem is new or unstudied. For NP-Complete problems it is known that, if a polynomial algorithm were found for one of them, such an algorithm could be translated into polynomial algorithms for all other problems in NP. Thus, a problem that is NP-Complete is at least as hard as many problems that have been extensively studied for years and for which no polynomial algorithm has yet been found.

We emphasize again that the rules we discuss do not necessarily offer complete theories, identify causal relationships, provide predictions, or suggest courses of action. Rules are regularities that hold in a given knowledge base, and they may be purely coincidental. Rules may be associated with theories, but we do not purport to model the process of developing and choosing among theories.

Assume that we are trying to predict a variable Y given the explanatory variables $X = (X_1, \ldots, X_m)$. For a subset K of $\{X_1, \ldots, X_m\}$, let R_K^2 be the value of the coefficient of determination R^2 when we regress $(y_i)_{i \leq n}$ on $(x_{ij})_{i \leq n, j \in K}$. We assume that the data are given in their entirety, that is, that there are no missing values.

How does one select a set of explanatory variables? First consider the feasible set of rules, projected onto the accuracy-complexity space. For a set of explanatory variables K, let the degree of complexity be $k = |K|$ and a degree of accuracy $- r = R_K^2$. Consider the $k - r$ space and, for a given knowledge base $X = (X_1, \ldots, X_m)$ and a variable Y, denote by $F(X, Y)$ the set of pairs (k, r) for which there exists a rule with these parameters. Because the set $F(X)$ is only defined for integer values of k, and for certain values of r, it is more convenient to visualize its comprehensive closure defined by:

$$F'(X, Y) \equiv \{(k, r) \in \mathbb{R}_+ \times [0, 1] \mid \exists (k', r') \in F(X, Y), k \geq k', r \leq r'\}$$

The set $F'(X, Y)$ is schematically illustrated in Fig. 8.1. Note that it need not be convex.

The optimization problem that such a person with utility function $u(\cdot, \cdot)$ faces is depicted in Fig. 8.2.

This optimization problem is hard to solve, because one generally cannot know its feasible set. In fact, for every $r > 0$, given X, Y, k, determining whether $(k, r) \in F'(X, Y)$ is computationally hard:

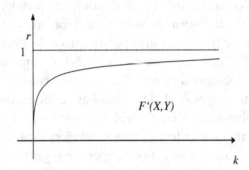

Fig. 8.1.

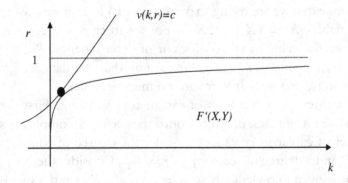

Fig. 8.2.

Theorem 1: *For every* $r \in (0,1]$, *the following problem is NP-Complete: Given explanatory variables* $X = (X_1, \ldots, X_m)$, *a variable* Y, *and an integer* $k \geq 1$, *is there a subset* K *of* $\{X_1, \ldots, X_m\}$ *such that* $|K| \leq k$ *and* $R_K^2 \geq r$?

Theorem 1 explains why people may be surprised to learn of simple regularities that exist in a knowledge base they have access to. A person who has access to the data should, in principle, be able to assess the veracity of all linear theories pertaining to these data. Yet, due to computational complexity, this capability remains theoretical. In practice one may often find that one has overlooked a simple linear regularity that, once pointed out, seems evident.

We show that, for any positive value of r, it is hard to determine whether a given k is in the r-cut of $F'(X, Y)$ when the input is (X, Y, k). By contrast, for a given k, computing the k-cut of $F'(X, Y)$ is a polynomial problem (when the input is (X, Y, r)), bounded by a polynomial of degree k. Recall, however, that k is bounded only by the number of columns in X. Moreover, even if k is small, a polynomial of degree k may assume large values if m is large. We conclude that, in general, finding the frontier of the set $F'(X, Y)$, as a function of X and Y, is a hard problem. The optimization problem depicted in Fig. 8.2 has a fuzzy feasible set, as described in Fig. 8.3.

A decision maker may choose a functional rule that maximizes $u(k, r)$ out of all the rules she is aware of, but the latter are likely to constitute only a subset of the set of rules defining the actual set

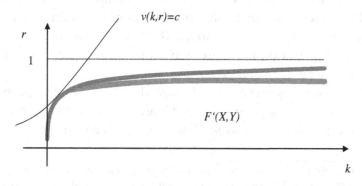

Fig. 8.3.

$F'(X, Y)$. Hence, many of the rules that people formulate are not necessarily the simplest (for a given degree of accuracy) or the most accurate (for a given degree of complexity).

We conclude this section with the observation that one may prove theorems similar to Theorem 1, which would make explicit reference to a certain function $u(k, r)$. The following is an example of such a theorem.

Theorem 2: *For every* $r \in (0, 1]$, *the following problem is NP-Complete: Given explanatory variables* $X = (X_1, \ldots, X_m)$ *and a variable* Y, *is there a subset* K *of* $\{X_1, \ldots, X_m\}$ *that obtains an adjusted* R^2 *of at least* r?

As is clear from the proof of Theorem 2, this result does not depend on the specific measure of the accuracy-simplicity trade-off, and similar results can be proven for a variety of functions $u(k, r)$.[12]

8.4. Discussion

8.4.1. *Approximation*

We posed a particular question — Does there exist a set of k explanatory variables for which the adjusted R^2 is at least r? — and showed that it is NP-complete. We argue that an implication of the result is that people will generally not know the regularities that exist in their knowledge base. But it is possible that, while it may be extremely difficult to get an *exact* answer to the question "What is the maximum R^2 possible with k variables?", it may be dramatically easier to obtain a very good *approximation* to such a question. If there are fast heuristics that do reasonably well on the regression problem, the scope of fact-free learning may be quite limited.

However, it is generally *not* the case that NP-Complete problems admit polynomial approximations. Consider, for instance, the NP-Complete problem *Minimum Exact Cover*, which can be described

---

[12]There are, however, functions v for which the result does *not* hold. For example, consider $v(k, r) = \min(r, 2 - k)$. This function obtains its maximum at $k = 1$ and it is therefore easy to maximize it.

as follows. Given a set S and a set of subsets of S, $\mathfrak{S}$, is there collection of pairwise disjoint subsets of S in $\mathfrak{S}$ whose union equals S? This is the yes/no problem we have used in the proof of Theorem 1.[13] To define the notion of approximation, one defines an optimization problem that corresponds to the yes/no problem. For instance, the Minimum Exact Cover problem can be viewed as corresponding to the following optimization problem: "Minimize the sum of the cardinalities of the sets in a collection that covers S"; if the solution is the cardinality of S, an exact cover has been identified.

How good an approximation can one get to the problem "Minimize the sum of the cardinalities of the sets that cover S" with an algorithm that is polynomial in the size of the problem? Suppose, for example, that one wanted an algorithm that had the property that, for all problems in this class, if the minimum possible sum for the problem were n, the algorithm would find a set of subsets with total cardinality λn for some $\lambda > 1$. (λ might be thought of as the accuracy of the approximation.) It is known that there does not exist such a polynomial algorithm, *no matter how large λ is*, unless $P = NP$ (Lund and Yannakakis (1994), Raz and Safra (1997)). In other words, finding an algorithm that assures *any* degree of reliability for large problems is as hard as solving NP-complete problems themselves.

We should emphasize that the difficulty in approximating the minimum exact cover problem doesn't assure that it is equally difficult to approximate our regression problem. An algorithm that provides a good approximation to one problem will not necessarily translate into a good approximation to other problems. While it is beyond the scope of this paper to determine how well one might approximate the regression problem analyzed above, we note that many (if not most) of the NP-Complete problems whose approximation have been studied turned out to be difficult to approximate.[14]

---

[13] That is, our proof consists of showing that any instance of the Minimum Exact Cover problem can be translated, via a polynomial algorithm, to an instance of the problem defined in Theorem 1, such that the answer to the latter is "yes" iff so is the answer to the former.

[14] See, for example, the descriptions of attainable approximations to NP-complete problems on the website "A Compendium of NP Optimization Problems" http://www.nada.kth.se/~viggo/problemlist/compendium.html.

8.4.2. *The relevance of NP-Completeness*

We maintain that a problem that is NP-Complete will be hard for economic agents to solve. Agents may obtain or learn the optimal solutions to particular instances of the general problem, especially if they are only interested in instances described by small inputs. But should economic agents encounter new instances of reasonable sizes on a regular basis, high computational complexity implies that it is unlikely that all, or most, agents in the economy would determine the optimal solutions in these instances.

In the case of fact-free learning, economic agents are called upon to find regularities in large knowledge bases. These regularities cannot be uncovered once and for all. The economic and political environment changes constantly and the lore of yesterday does not provide a blueprint for the decisions of tomorrow. It is therefore reasonable to model economic agents as problem solvers who constantly need to cope with new and large problems.

One can argue that NP-Completeness is a concept that relates to the way computers perform computations, and has little or no bearing on human reasoning. Indeed, there are problems such as natural language understanding or face recognition that toddlers perform better than do computers. But these are problems for which finding an appropriate mathematical model is a major part of the solution. By contrast, for well defined combinatorial problems such as those in the class NP, it is rarely the case that humans perform better than do computers. Our modest claim is that it is safe to assume that neither people nor computers can solve NP-Complete problems optimally.

One may question the use of complexity concepts that are defined by worst-case analysis. Indeed, why would we worry about an algorithm whose worst-case performance is exponential, if it is polynomial on average? Experience, however, indicates that NP-Complete problems do not tend to be efficiently solvable even in expectation, under any reasonable assumptions on the distribution of inputs.[15]

---

[15]See Christos Papadimitriou (1994) who makes this point, and emphasizes that the example of linear programming confirms this experience. Indeed, the Simplex algorithm has exponential worst-case time complexity but very good expected complexity. Linear programming, however,

We do not claim that the inability to solve NP-Complete problems is necessarily the most important cognitive limitation on people's ability to perform induction. As mentioned above, even polynomial problems can be difficult to solve when the knowledge base consists of many cases and many attributes. Moreover, it is often the case that looking for a general rule does not even cross someone's mind. Yet, the difficulty of performing induction shares with NP-Complete problems this central property: while it is hard to come up with a solution to such a problem, it is easy to verify whether a suggested solution is valid.

People need not be lazy or irrational to explain why they do not find all relevant rules. Rather, looking for simple regularities is a genuinely hard problem. There is nothing irrational about not being able to solve NP-Complete problems. Faced with the problem of selecting a set of explanatory variables, which is NP-Complete, people may use various heuristics to find prediction rules, but they cannot be sure, in general, that the rules they find are the simplest or most accurate ones.

8.4.3. *Implications*

Agreeing to disagree. Our model suggests two reasons for which people may have different beliefs, even if these beliefs are defined by rules that are derived from a shared knowledge base. First, two people may notice different regularities. Since finding the "best" regularities is a hard problem, we should not be surprised if one person failed to see a regularity that another came up with. Second, even if the individuals share the rules that they found, they may entertain different beliefs if they make different trade-offs between the accuracy and the simplicity of rules. Different people may well have different u functions, with some people more willing to sacrifice accuracy for simpler rules. If two individuals choose different levels of simplicity, they may also disagree on the relevance of a characteristic. In particular, a variable that is important when there are relatively few other variables in a regression may not be important if the number of variables considered increases.

---

is *not* an NP-Complete problem and there are now algorithms to solve linear programming problems with polynomial worst-case performance.

Thus, a particular attribute may play a large role in the rule one person uses but no role in the rule another employs.

Locally optimal rules. Our central point is that people use rules that are not fully optimal because of the complexity of the problem of finding fully optimal rules. When an individual uses a rule that is less than fully optimal, she may improve upon the rule by considering alternatives to it. A person faced with the regression problem may think of alternatives to her current "best" regression by adding or deleting variables from her current included set, or by replacing variables in the included set with others. While we do not formally model this search and revision process, one can imagine two distinct ways people may update the rules they use. One can search "locally", that is, consider relatively minor changes in the current rule such as adding, deleting, or replacing one or two variables, or one can search globally by considering sets of variables that have no relation whatsoever to the current set of variables. Local search may find local optima that are not global optima. Differently put, people may get "stuck" with suboptimal rules that can be improved upon only with a "paradigm shift" that considers a completely different way of looking at a problem.

Path dependence. When individuals search locally for improved rules, their reasoning is likely to exhibit path dependence. Two individuals who begin with different initial sets of variables can settle on very different rules, even after very long search times.

Regret. Our model suggests different notions of regret. In a standard model, individuals make optimal choices given the information available to them at the time they decide. In a stochastic environment, an individual may wish *ex post* that she had decided differently. However, a rational person has no reason to regret a decision she had taken since she could have done no better at the time of her decision, given the information available to her at that time. In our model there are two notions in which information can be "given", and correspondingly, two possible sources of regret. As usual, one may learn the realization of a random variable, and wish that she had decided differently. But one can also learn of a rule that one has not been aware of, even though the rule could be derived, in principle, from one's knowledge base. Should one feel regret as a result? As argued above, one could not be expected to solve NP-Complete problems, and therefore it may be argued that one could not have chosen optimally. Yet, one might expect individuals to experience a stronger sense of "I should have known" as a result of finding rules that hold in a given knowledge base, than as a result of getting new observations.

8.4.4. *Related literature*

Most of the formal literature in economic theory and in related fields is based on the Bayesian model of information processing. In this model a decision maker starts out with a prior probability, and she updates it in the face of new information by Bayes's rule. Hence, this model captures nicely changes in opinion that result from new information. But it does not deal very graciously with changes of opinion that are not driven by new information. In fact, in a Bayesian model with perfect rationality people cannot change their opinions unless new information has been received. It follows that the example we started out with cannot be explained by such models.

Relaxing the perfect rationality assumption, one may attempt to provide a pseudo-Bayesian account of the phenomena discussed here. For instance, one can use a space of states of the world to describe the subjective uncertainty that a decision maker has regarding the result of a computation, before this computation is carried out. (See Luca Anderlini and Leonardo Felli (1994) and Nabil Al-Najjar, Ramon Casadesus-Masanell, and Emre Ozdenoren (1999).) In such a model, one would be described as if one entertained a prior probability of, say p, that "democratic peace" holds. Upon hearing the rhetorical question as in our dialogue, the decision maker performs the computation of the accuracy of this rule, and is described as if the result of this computation were new information.

A related approach employs a subjective state space to provide a Bayesian account of unforeseen contingencies. (See David Kreps (1979, 1992), and Eddie Dekel, Barton Lipman, and Aldo Rustichini (1997, 1998).) Should this approach be applied to the problem of induction, each regularity that might hold in the knowledge base would be viewed as an unforeseen contingency that might arise. A decision maker's behavior will then be viewed as arising from Bayesian optimization with respect to a subjective state space that reflects her subjective uncertainty.

Our approach is compatible with these pseudo-Bayesian models. Its relative strength is that it models the process of induction more explicitly, allowing a better understanding of why and when induction is likely to be a hard problem.

Itzhak Gilboa and David Schmeidler (2001) offer a theory of case-based decision making. They argue that cases are the primitive objects of knowledge, and that rules and probabilities are derived from cases. Moreover, rules and probabilities cannot be known in the same sense, and to the same degree of certitude, that cases can. Yet, rules and probabilities may be efficient and insightful ways of succinctly summarizing many cases. The present paper suggests that summarizing knowledge bases by rules may involve loss of information, because one cannot be guaranteed to find the "optimal" rules that a given knowledge base induces.

Appendix: Proof

Proof of Theorem 1:

Let there be given $r > 0$. It is easy to see that the problem is in NP: given a suggested set $K \subset \{1,\ldots,m\}$, one may calculate R_K^2 in polynomial time in $|K|n$ (which is bounded by the size of the input, $(m + 1)n$).[16] To show that the problem is NP-Complete, we use a reduction of the following problem, which is known to be NP-Complete (see Michael Gary and David Johnson (1979), or Christos Papadimitriou (1994)):

Problem Exact Cover: Given a set S, a set of subsets of S, $\mathfrak{S}$, are there pairwise disjoint subsets in $\mathfrak{S}$ whose union equals S?

(That is, does a subset of $\mathfrak{S}$ constitutes a partition of S?)

Given a set S, a set of subsets of S, $\mathfrak{S}$, we will generate n observations of $(m + 1)$ variables, $(x_{ij})_{i \leq n, j \leq m}$ and $(y_i)_{i \leq n}$, and a natural number k, such that S has an exact cover in $\mathfrak{S}$ iff there is a subset K of $\{1,\ldots,m\}$ with $|K| \leq k$ and $R_K^2 \geq r$.

Let there be given, then, S and $\mathfrak{S}$. Assume without loss of generality that $S = \{1,\ldots,s\}$, and that $\mathfrak{S} = \{S_1,\ldots,S_l\}$ (where $s, l \geq 1$ are natural numbers). We construct $n = 2(s + l + 1)$ observations of $m = 2l$ predicting variables. It will be convenient to denote the $2l$ predicting variables by $X_1,\ldots,X_l$ and $Z_1,\ldots,Z_l$ and the predicted variable — by Y. Their corresponding values will be denoted $(x_{ij})_{i \leq n, j \leq l}$, $(z_{ij})_{i \leq n, j \leq l}$, and $(y_i)_{i \leq n}$. We will use X_j, Z_j, and Y also to denote the column vectors $(x_{ij})_{i \leq n}$, $(z_{ij})_{i \leq n}$, and $(y_i)_{i \leq n}$, respectively. Let $M \geq 0$ be a constant to be specified later. We now specify the vectors $X_1,\ldots,X_l, Z_1,\ldots,Z_l$, and Y as a function of M.

For $i \leq s$ and $j \leq l$, $x_{ij} = 1$ if $i \in S_j$ and $x_{ij} = 0$ if $i \notin S_j$;

For $i \leq s$ and $j \leq l$, $z_{ij} = 0$;

For $s < i \leq s + l$ and $j \leq l$, $x_{ij} = z_{ij} = 1$ if $i = s + j$ and $x_{ij} = z_{ij} = 0$ if $i \neq s + j$;

---

[16]Here and in the sequel we assume that reading an entry in the matrix X or in the vector Y, as well any algebraic computation require a single time unit. Our results hold also if one assumes that x_{ij} and y_i are all rational and takes into account the time it takes to read and manipulate these numbers.

For $j \leq l$, $x_{s+l+1,j} = z_{s+l+1,j} = 0$;

For $i \leq s + l$, $y_i = 1$ and $y_{s+l+1} = M$;

For $i > s + l + 1$, $y_i = -y_{i-(s+l+1)}$ and for all $j \leq l$, $x_{ij} = -x_{i-(s+l+1),j}$ and $z_{ij} = -z_{i-(s+l+1),j}$.

Observe that the bottom half of the matrix X as well as the bottom half of the vector Y are the negatives of the respective tops halves. This implies that each of the variables $X_1, \ldots, X_l$, $Z_1, \ldots, Z_l$, and Y has a mean of zero. This, in turns, implies that for any set of variables K, when we regress Y on K, we get a regression equation with a zero intercept.

Consider the matrix X and the vector Y obtained by the above construction for different values of M. Observe that the collection of sets K that maximize R_K^2 is independent of M. Hence, it is useful to define $\widehat{R}_K^2$ as the R^2 obtained from regressing Y on K, ignoring observations $s + l + 1$ and $2(s + l + 1)$. Obviously, minimizing $\widehat{R}_K^2$ is tantamount to minimizing R_K^2.

We claim that there is a subset K of $\{X_1, \ldots, X_l\} \cup \{Z_1, \ldots, Z_l\}$ with $|K| \leq k \equiv l$ for which $\widehat{R}_K^2 = 1$ iff S has an exact cover from $\mathfrak{S}$.

First assume that such a cover exists. That is, assume that there is a set $J \subset \{1, \ldots, l\}$ such that $\{S_j\}_{j \in J}$ constitutes a partition of S. This means that $\sum_{j \in J} 1_{S_j} = 1_S$ where 1_A is the indicator function of a set A. Let α be the intercept, $(\beta_j)_{j \leq l}$ be the coefficients of $(X_j)_{j \leq l}$ and $(\gamma_j)_{j \leq l}$ — of $(Z_j)_{j \leq l}$ in the regression. Set $\alpha = 0$. For $j \in J$, set $\beta_j = 1$ and $\gamma_j = 0$, and for $j \notin J$ set $\beta_j = 0$ and $\gamma_j = 1$. We claim that $\alpha \mathbf{1} + \sum_{j \leq l} \beta_j X_j + \sum_{j \leq l} \gamma_j Z_j = Y$ where $\mathbf{1}$ is a vector of 1's. For $i \leq s$ the equality

$$\alpha + \sum_{j \leq l} \beta_j x_{ij} + \sum_{j \leq l} \gamma_j z_{ij} = \sum_{j \leq l} \beta_j x_{ij} = y_i = 1$$

follows from $\sum_{j \in J} 1_{S_j} = 1_S$. For $s < i \leq s + l$, the equality

$$\alpha + \sum_{j \leq l} \beta_j x_{ij} + \sum_{j \leq l} \gamma_j z_{ij} = \beta_j + \gamma_j = y_i = 1$$

follows from our construction (assigning precisely one of $\{\beta_j, \gamma_j\}$ to 1 and the other — to 0). Obviously, $\alpha + \sum_{j \leq l} \beta_j x_{nj} + \sum_{j \leq l} \gamma_j z_{nj} = 0 = y_i = 0$. The number of variables used in this regression is l. Specifically, choose $K = \{X_j \mid j \in J\} \cup \{Z_j \mid j \notin J\}$, with $|K| = l$, and observe that $\widehat{R}_K^2 = 1$.

We now turn to the converse direction. Assume, then, that there is a subset K of $\{X_1, \ldots, X_l\} \cup \{Z_1, \ldots, Z_l\}$ with $|K| \leq l$ for which $\widehat{R}_K^2 = 1$. Since all variables have zero means, this regression has an intercept of zero ($\alpha = 0$ in the notation above). Let $J \subset \{1, \ldots, l\}$ be the set of indices of the X variables in K, i.e., $\{X_j\}_{j \in J} = K \cap \{X_1, \ldots, X_l\}$. We will show that $\{S_j\}_{j \in J}$ constitutes a partition of S. Set $L \subset \{1, \ldots, l\}$ be the set of indices of the Z variables in K, i.e., $\{Z_j\}_{j \in L} = K \cap \{Z_1, \ldots, Z_l\}$. Consider the coefficients of the variables in K used in the regression obtaining $\widehat{R}_K^2 = 1$. Denote them by $(\beta_j)_{j \in J}$ and $(\gamma_j)_{j \in L}$. Define $\beta_j = 0$ if $j \notin J$ and $\gamma_j = 0$ if $j \notin L$. Thus, we have

$$\sum_{j \leq l} \beta_j X_j + \sum_{j \leq l} \gamma_j Z_j = \Upsilon.$$

We argue that $\beta_j = 1$ for every $j \in J$ and $\gamma_j = 1$ for every $j \in L$. To see this, observe first that for every $j \leq l$, the $s + j$ observation implies that $\beta_j + \gamma_j = 1$. This means that for every $j \leq l$, $\beta_j \neq 0$ or $\gamma_j \neq 0$ (this also implies that either $j \in J$ or $j \in L$). If for some j both $\beta_j \neq 0$ and $\gamma_j \neq 0$, we will have $|K| > l$, a contradiction. Hence for every $j \leq l$ either $\beta_j \neq 0$ or $\gamma_j \neq 0$, but not both. (In other words, $J = L^c$.) This also implies that the non-zero coefficient out of $\{\beta_j, \gamma_j\}$ has to be 1.

Thus the cardinality of K is precisely l, and the coefficients $\{\beta_j, \gamma_j\}$ define a subset of $\{S_1, \ldots S_l\}$: if $\beta_j = 1$ and $\gamma_j = 0$, i.e., $j \in J$, S_j is included in the subset, and if $\beta_j = 0$ and $\gamma_j = 1$, i.e., $j \notin J$, S_j is not included in the subset. That this subset $\{S_j\}_{j \in J}$ constitutes a partition of S follows from the first s observations as above.

We now turn to define M. We wish to do so in such a way that, for every set of explanatory variables K, $R_K^2 \geq r$ iff $\widehat{R}_K^2 = 1$. Fix a set K. Denote by $\widehat{SSR}$ and $\widehat{SST}$ the explained variance and the total variance, respectively, of the regression of Υ on K without observations $s + l + 1$ and $2(s + l + 1)$, where SSR and SST denote the variances of the regression with all observations. Thus, $R_K^2 = SSR/SST$ and $\widehat{R}_K^2 = \widehat{SSR}/\widehat{SST}$. Observe that $\widehat{SST} = 2(s + l)$ and $SST = 2(s + l) + 2M^2$. Also, $SSR = \widehat{SSR}$ is independent of M.

Note that if K is such that $\widehat{R}_K^2 = 1$, then $(SSR =) \widehat{SSR} = \widehat{SST} = 2(s+l)$. In this case, $R_K^2 = \frac{2(s+l)}{2(s+l)+2M^2}$. If, however, K is such that $\widehat{R}_K^2 < 1$, then we argue that $(SSR =) \widehat{SSR} \le \widehat{SST} - \frac{1}{9}$. Assume not. That is, assume that K is such that $\widehat{SSR} > \widehat{SST} - \frac{1}{9}$. This implies that on each of the observations $1, \ldots, s+l, s+l+2, \ldots, 2(s+l)+1$, the fit produced by K is at most $\frac{1}{3}$ away from y_i. Then for every $j \le l$, $|\beta_j + \gamma_j - 1| < \frac{1}{3}$. Hence for every $j \le l$ either $\beta_j \ne 0$ or $\gamma_j \ne 0$, but not both, and the non-zero coefficient out of $\{\beta_j, \gamma_j\}$ has to be in $(\frac{2}{3}, \frac{4}{3})$. But then, considering the first s observations, we find that K is an exact cover. It follows that, if $\widehat{R}_K^2 < 1$, then $R_K^2 \le \frac{2(s+l)-\frac{1}{9}}{2(s+l)+2M^2}$.

Choose a rational M in the interval $\left(\sqrt{\frac{(1-r)(s+l)-\frac{1}{18}}{r}}, \sqrt{\frac{(1-r)(s+l)}{r}} \right)$ so that $\frac{2(s+l)-\frac{1}{9}}{2(s+l)+2M^2} < r < \frac{2(s+l)}{2(s+l)+2M^2}$, and observe that for this M, there exists a K such that $R_K^2 \ge r$ iff there exists a K for which $\widehat{R}_K^2 = 1$, that is, iff K is an exact cover.

To conclude the proof, it remains to observe that the construction of the variables $(X_j)_{j \le l}, (Z_j)_{j \le l}$, and Y can be done in polynomial time in the size of the input. $\qquad \square$

Proof of Theorem 2:

Let there be given $r > 0$. The proof follows that of Theorem 1 with the following modification. For an integer $t \ge 1$, to be specified later, we add t observations for which all the variables $((X_j)_{j \le l}, (Z_j)_{j \le l},$ and $Y)$ assume the value 0. These observations do not change the R^2 obtained by any set of regressors, as both SST and SSR remain the same. Assuming that t has been fixed (and that it polynomial in the data), let r' be the R^2 corresponding to an adjusted R^2 of r, with l regressors. That is, $(1 - r') = (1 - r)\frac{t+2s+2l+1}{t+2s+l+1}$. Define M as in the proof of Theorem 1 for r'.

We claim that there exists a set of regressors that obtains an adjusted R^2 of r iff there exists a set of l regressors that obtains an R^2 of r' (hence, iff there exists an exact cover in the original problem). The "if" part is obvious from our construction. Consider the "only if" part. Assume, then that a set of regressors obtains an adjusted R^2 of r. If it has l regressors, the same calculation shows that it obtains the desired R^2. We now argue that if no set of l regressors obtains an adjusted R^2 of r, then no set of regressors (of any cardinality) obtains an adjusted R^2 of r.

Consider first a set K_0 with $|K_0| = k_0 > l$ regressors. Observe that, by the choice of M, r' is the upper bound on all R_K^2 for all K with $|K| = l$, as r' was computed assuming that an exact cover exists, and that, therefore, there are l variables that perfectly match all the observations but $s + l + 1$ and $2(s + l + 1)$. Due to the structure of the problem, r' is also an upper bound on R_K^2 for all K with $|K| \geq l$. This is so because the only observations that are not perfectly matched (in the hypothesized l-regressor set) correspond to zero values of the regressors. It follows that the adjusted R^2 for K_0 is lower than r.

Next consider a set K_0 with $|K_0| = k_0 < l$ regressors. For such a set there exists a $j \leq l$ such that neither X_j nor Z_j are in K_0. Hence, observations $s+j$ and $2s+l+j+1$ cannot be matched by the regression on K_0. The lowest possible SSE in this problem, corresponding to the hypothesized set of l regressors, is $2M^2$. This means that the SSE of K_0 is at least $2M^2 + 2$. That is, the SSE of the set K_0 is at least $\frac{M^2+1}{M^2}$ larger than the SSE used for the calculation of r. On the other hand, K_0 uses less variables. But if $\frac{t+2s+l+1}{t+2s+k+1} < \frac{M^2+1}{M^2}$, the reduction in the number of variables cannot pay off, and K_0 has an adjusted R^2 lower than r. It remains to choose t large enough so that the above inequality holds, and to observe that this t is bounded by the polynomial of the input size. $\qquad\square$

References

Al-Najjar, N., R. Casadesus-Masanell and E. Ozdenoren (2003), "Probabilistic Representation of Complexity", *Journal of Economic Theory*, **111**: 49–87.

Anderlini, L. and L. Felli (1994), "Incomplete Written Contracts: Indescribable States of Nature", *Quarterly Journal of Economics*, **109**: 1085–1124.

Bray, M. and N. Savin (1986), "Rational Expectations Equilibria, Learning, and Model Specification", *Econometrica*, **54**(5): 1129–1160.

Dekel, E., B. Lipman and A. Rustichini (1997), "A Unique Subjective State Space for Unforeseen Contingencies", mimeo, Northwestern University.

Dekel, E., B. Lipman and A. Rustichini (1988), "Recent Developments in Modeling Unforeseen Contingencies", *European Economic Review*, **42**: 523–542.

Gary, M. and D. Johnson (1979), *Computers and Intractability: A Guide to the Theory of NP-Completeness*, San-Francisco, CA: W. Freeman and Co.

Gilboa, I. and D. Schmeidler (2001), *A Theory of Case-Based Decisions*, Cambridge: Cambridge University Press.

Goodman, N. (1965), *Fact, Fiction and Forecast*. Indianapolis: Bobbs-Merrill.

Hastie, T., R. Tibshirani and J. Friedman (2001), *The Elements of Statistical Learning*, New York: NY: Springer.

Kreps, D. (1979), "A Representation Theorem for 'Preference for Flexibility'", *Econometrica*, **47**: 565–576.

Kreps, D. (1992), "Static Choice and Unforeseen Contingencies" in *Economic Analysis of Markets and Games: Essays in Honor of Frank Hahn*, P. Dasgupta, D. Gale, O. Hart, and E. Maskin (eds.), MIT Press: Cambridge, MA, 259–281.

La Porta, R., F. Lopez-de-Silanes, A. Shleifer and R. Vishny (1998), "The Quality of Government", mimeo, Harvard University.

Lund, C. and M. Yannakakis (1994), "On the hardness of approximating minimization problems", *Journal of the ACM*, **41**: 960–981.

Mallows, C. (1973), "Some comments on Cp." *Technometrics*, **15**: 661–675.

Maoz, Z. and B. Russett (1993), "Normative and Structural Causes of Democratic Peace, 1946–1986", *American Political Science Review*, **87**(September): 640–654.

Papadimitriou, C. (1994), *Computational Complexity*, Addison-Wesley.

Raz, R. and S. Safra (1997), "A sub-constant error-probability low-degree test, and sub-constant error-probability PCP characterization of NP", *Proceedings of the 29th Annual ACM Symposium on Theory of Computation, ACM*, 475–484.

Simon, H. (1955), "A Behavioral Model of Rational Choice", *Quarterly Review of Economics*, **69**: 99–118.

Tibshirani, R. (1996), "Regression Shrinkage and Selection via the Lasso", *Journal of the Royal Statistical Society*, **58**(1): 267–288.

Chapter 9

Empirical Similarity*

Itzhak Gilboa[†], Offer Lieberman[‡] and David Schmeidler[§]

Reprinted with Permission from *Review of Economics and Statistics*,
88 (2006): 433–444.

An agent is asked to assess a real-valued variable Y_p based on certain characteristics $X_p = (X_p^1, \ldots, X_p^m)$, and on a database consisting of $(X_i^1, \ldots, X_i^m, Y_i)$ for $i = 1, \ldots, n$. A possible approach to combine past observations of X and Y with the current values of X to generate an assessment of Y is *similarity-weighted averaging*. It suggests that the predicted value of Y, $\bar{Y}_p^s$, be the weighted average of all previously observed values Y_i, where the weight of Y_i, for every $i = 1, \ldots, n$, is the similarity between the vector $X_p^1, \ldots, X_p^m$, associated with Y_p, and the previously observed vector, $X_i^1, \ldots, X_i^m$. We axiomatize this rule. We assume that, given every database, a predictor has a ranking over possible values, and we show that certain reasonable conditions on these rankings imply that they are determined by the proximity to a similarity-weighted average for a certain similarity function. The axiomatization does not suggest a particular similarity function, or even a particular form of this function. We therefore

---

*We wish to thank John Geanakoplos and Don Brown for conversations that greatly influenced this work. We are also grateful to Daron Acemoglu, Yoav Binyamini, Raul Drachman, Gabi Gayer, Mark Machina, Enno Mammen, Yishay Mansour, and two anonymous referees for comments and references. Gilboa and Lieberman gratefully acknowledge support from the Pinhas Sapir Center for Development and the Chaim Katzman Gazit-Globe Real Estate Institute at Tel-Aviv University. Gilboa and Schmeidler gratefully acknowledge support from the Polarization and Conflict Project CIT-2-CT-2004-506084 funded by the European Commission-DG Research Sixth Framework Programme and the Israel Science Foundation Grant No and 975/03.

[†]Tel-Aviv University and Yale University. igilboa@tau.ac.il
[‡]The Technion. offerl@ie.technion.ac.il
[§]Tel-Aviv University and The Ohio State University. schmeid@tau.ac.il

proceed to suggest that the similarity function be estimated from past observations. We develop tools of statistical inference for parametric estimation of the similarity function, for the case of a continuous as well as a discrete variable. Finally, we discuss the relationship of the proposed method to other methods of estimation and prediction.

JEL Codes: C1, C8, D8

Keywords: Similarity, Estimation

9.1. Introduction

9.1.1. *Motivation*

Economic agents as well as various professionals are often required to assess the value of a certain numerical variable. In many situations, available data are relevant for the assessment problem, but they do not suggest a value that is indisputably the only reasonable assessment to make. Consider the following examples.

1. A home owner considers selling her house, and she wonders how much she could get for it. Naturally, she should be basing her assessment on the prices at which other houses were sold. Yet, every house has its idiosyncratic characteristics. Hence the "market value" of her house is a variable that needs to be assessed based on observations of other transactions, but cannot be uniquely determined by these transactions in the same way that the price of a ton of wheat can.

2. An art dealer wants to sell a painting by a reasonably famous painter. Evidently, the market price of the painting is related to the prices at which other, similar paintings were sold. Yet, the painting is unique, and its price may differ from the prices of all other paintings, as well as from their average.

3. An analyst is asked to predict the rate of inflation for the coming year. Using past empirical frequencies of various inflation rates is hardly an option in this case, since every year differs from past years in several ways. Yet, it is obvious that past inflation rates are informative and should somehow be used for the prediction.[1]

---

[1] This application was suggested by Raul Drachman.

4. The same analyst is now asked to assess the probability of a stock market crash within the next six months. Again, she is expected to generate an assessment that is based on past observations. However, every two situations would typically differ in the values of certain important economic variables.

5. A physician is asked to assess the probability of success of an operation to be performed on a certain patient. Past experience with other patients is clearly relevant and should inform the assessment process. Yet, every human body is unique, and simple relative frequencies of success do not summarize all the relevant information.

6. A lawyer is asked by her client what are the chances of winning a case. Clearly, every case is idiosyncratic. Yet, the rulings in similar cases and under like-minded judges are relevant for the assessment.

In all of these problems one attempts to assess the value of a variable Y_p based on the values of relevant variables, $X_p = (X_p^1, \ldots, X_p^m)$, and on a database consisting of the variables $(X_i^1, \ldots, X_i^m, Y_i)$ for $i = 1, \ldots, n$. The question is, how do and how should people combine past observations of X and Y with the current values of X to generate an assessment of Y?

This problem is extensively studied in statistics, machine learning, and related fields. Among the numerous methodologies that have been suggested and used to solve such problems one may mention parametric and non-parametric regression, neural nets, linear and non-linear classifiers, k-nearest neighbor approaches (Fix and Hodges (1951, 1952), Cover and Hart (1967), Devroye, Gyorfi, and Lugosi (1996)), kernel-based estimation (Akaike (1954), Rosenblatt (1956), Parzen (1962), Silverman (1986), Scott (1992)), and others. Each of these methodologies has considerable success in a variety of applications. Moreover, each methodology can also be viewed as a tentative model of human reasoning. How should we choose among these approaches for descriptive and for normative applications?

Our approach to this problem is axiomatic and empirical. We start with a system of axioms that characterizes a class of assessment rules. We do not expect the axiomatic approach, or other theoretical considerations to fully specify the parameters of the assessment rule. Rather, we suggest that these parameters be estimated from data.

This estimation is done in the context of a probability model that allows statistical inference. We now turn to describe this approach in more detail.

9.1.2. *Axiomatization of similarity-weighted averaging*

For the axiomatic model we assume that, given a database $B = (X_i, Y_i)_{i \leq n}$, where $n \in \mathbb{N}$, $X_i \in \mathbb{R}^m$, and $Y_i \in \mathbb{R}$, and a new data point $X_p \in \mathbb{R}^m$, the agent has a ranking $\succsim_{B,X_p}$ over the possible values of Y_p. The interpretation of $\xi \succsim_{B,X_p} \zeta$ is that, given the database B and the new data point X_p, ξ is more likely to be observed than is ζ. We study the rankings $\succsim_{B,X_p}$ that the agent would generate given various possible databases, holding m fixed. We formulate axioms on such rankings, and show that the rankings satisfy these axioms if and only if they can be represented by similarity-weighted averaging. Specifically, the axioms are equivalent to the existence of a function $s : \mathbb{R}^m \times \mathbb{R}^m \to \mathbb{R}_{++} = (0, \infty)$ such that, given a database $B = (X_i, Y_i)_{i \leq n}$ and a new data point $X_p = (X_p^1, \ldots, X_p^m) \in \mathbb{R}^m$, two possible estimates of Y_p are ranked according to their proximity to the similarity-weighted average of all observations in the database, namely,

$$\overline{Y}_p^s = \frac{\sum_{i \leq n} s(X_i, X_p) Y_i}{\sum_{i \leq n} s(X_i, X_p)} \tag{1}$$

This rule for generating predictions is reminiscent of kernel estimation. (See Akaike (1954), Rosenblatt (1956), and Parzen (1962). See details in subsection 9.1.2 below.). We prefer the term "similarity" since it suggests a cognitive interpretation of the function, as opposed to the more technical "kernel". This is obviously only a matter of interpretation.[2]

The axioms we propose are not universal and they need not be satisfied by all types of human reasoning. Specifically, when people use the data to develop theories, and then use these theories to generate

---

[2]Our axiomatization relies on that of Gilboa and Schmeidler (2001, 2003). Yet, the former is not a special case of the latter. Moreover, the analysis conducted here employs the fact that the variable Y is real-valued.

predictions, they are unlikely to satisfy our axioms, or to follow (1). (We elaborate on this point after the presentation of the axioms in Section 9.3.) Our axioms attempt to describe the assessment of an agent who aggregates data, but who does not engage in theorizing. When agents do reason by general rules, or theories, a model such as regression analysis may be a better model than the similarity-weighted averaging we discuss here.

We also axiomatize the relation "more likely than" that corresponds to a set of agents, constituting a "market", and we show that, under our axioms, one may replace all agents with their subjective similarity functions by a "representative" agent with an appropriately defined similarity function.

9.1.3. *The empirical similarity*

The axiomatization we propose does not specify a particular similarity function, or even a particular functional form thereof.[3] Where do the similarity numbers come from? In this paper we do not attempt to provide a theoretical answer to this question. Rather, we suggest an empirical approach: given a database $B = (X_i, Y_i)_{i \leq n}$, we assume that past values Y_i were also generated in accordance with equation (1), adapted for $p = i$ and $n = i - 1$, that is,

$$\overline{Y}_i^s = \frac{\sum_{k < i} s(X_k, X_i) Y_k}{\sum_{k < i} s(X_k, X_i)} \qquad (2)$$

relative to the similarity function s of the representative agent. We then ask which similarity function $s : \mathbb{R}^m \times \mathbb{R}^m \to \mathbb{R}_{++}$ can best fit the data B under this assumption. This function, dubbed *the empirical similarity*, can then be used to generate assessments of $\overline{Y}_p^s$. These assessment will be more objective than similar assessments based on a subjective similarity function.

---

[3] Billot, Gilboa, and Schmeidler (2004) offer an axiomatization of a particular functional form of a similarity function. Assuming that an agent employs a similarity-weighted averaging as suggested here, they impose additional axioms on the agent's assessments given various databases and various new data points, which are equivalent to the existence of a norm on $\mathbb{R}^m$ such that the similarity function is a negative exponent of this norm.

In this paper we address a parametric version of the question of estimation of the similarity function. We suggest a functional form of s, and estimate its parameters by maximum likelihood estimator in a statistical model that we define shortly. However, an "empirical similarity function" may be any function that is estimated from the data, or that is chosen to fit the data according to equation (2).

Further discussion of our estimation methodology and the assumptions underlying it is deferred to Section 9.6. We now proceed to describe a statistical model within which this estimation can be analyzed.

9.1.4. *Statistical analysis*

The empirical similarity we obtain can be viewed as a point estimate of a similarity function, if we embed equation (1) in a statistical model. Specifically, we are interested in similarity functions that depend on a weighted Euclidean distance,

$$d_w(x, x') = \sqrt{\sum_{j \leq m} w_j (x^j - x'^j)^2} \qquad (3)$$

where $x = (x^1, \ldots, x^m)$ and $x' = (x'^1, \ldots, x'^m)$. The similarity function may be expected to decrease in the distance d_w, to obtain the value 1 for $d_w = 0$ and to converge to 0 as $d_w \to \infty$. Natural candidates for such a function include $s_w = e^{-d_w}$ or $s_w = \frac{1}{1+d_w}$. Billot, Gilboa, and Schmeidler (2004) assume that an agent generates assessments according to (1), and take an axiomatic approach to the problem of selecting the functional form of the similarity function. Specifically, they show that certain conditions on the assessments Υ_p generated given various databases are equivalent to the existence of a norm $\| \cdot \|$ on $\mathbb{R}^m$, such that $s_w(x, x') = e^{-\|x-x'\|}$. Since d_w is a norm on $\mathbb{R}^m$ when $w_j > 0$ for all $j \leq m$, $s_w = e^{-d_w}$ may be viewed as a special case of the similarity function axiomatized in Billot *et al.* (2004).

Observe that the weights $(w_j)_j$ are not restricted to sum to 1. This allows some flexibility in the relative weight of closer versus more remote observations. For instance, multiplying all weights $(w_j)_j$ by a constant $\lambda^2 > 0$ is tantamount to multiplying d_w by $\lambda > 0$. If $\lambda > 1$, this transformation reduces the relative impact of remote points.

For $t = 2, \ldots, n$, we assume that

$$Y_t = Y_t^s = \frac{\sum_{i<t} s(X_i, X_t) Y_i}{\sum_{i<t} s(X_i, X_t)} + \varepsilon_t \tag{4}$$

where $\varepsilon_t \overset{i.i.d.}{\sim} N(0, \sigma^2)$, and Y_1 is an arbitrary random variable.

In such a model it makes sense to ask whether the point estimates of the unknown parameters are significantly different from a pre-specified value, and in particular, from zero. In this paper we focus on maximum likelihood estimation of the parameters $(w_j)_j$, and we develop tests for such hypotheses.

For some applications, including examples 5 and 6 above, the observed values of Y_t are categorical. In this case one cannot assume a model such as (4), and the latter should be replaced with a model of the form

$$P(Y_t = 1 \mid X_1, Y_1, \ldots, X_{t-1}, Y_{t-1}) = F(Y_t^s)$$

where F is a cumulative distribution function, X_i is an m-vector, and $Y_i \in \{0, 1\}$, with $Y_i = 1$ denoting success and $Y_i = 0$ denoting failure in examples 5 and 6. This model differs from discrete choice models in a way that parallels the difference between our model for a continuous Y_t^s and linear regression. Specifically, the probability that Y_t assumes the value 1 depends on the weighted relative frequency of 1 among past values $\{Y_i\}_{i<t}$, where the weight of the value Y_i depends on the similarity between the vector X_i observed in the past and the current observation X_t. We provide a statistical model for this case, and develop tests for hypotheses about the values of the parameters $(w_j)_j$ in this model as well.

The rest of this paper is organized as follows. In Section 9.2 we discuss the relationship between our methodology and existing statistical methodologies. Section 9.3 provides the axiomatization of similarity-weighted averaging, for a single agent and for a set of agents. In Section 9.4 we develop the statistical theory for the continuous case, whereas Section 9.5 deals with the discrete case. Finally, Section 9.6 concludes.

9.2. Related Techniques

Our main focus is on human reasoning. We are interested in data that are generated by people, and we take the similarity-weighted average as a possible model of how people generate assessments. That is, we interpret our model as describing a causal relationship.

Our methodology can be applied also to databases in which the variable Y is not a result of human reasoning. In this case our model should not be interpreted causally, but one may still find a similarity function that best fits the data. Moreover, one may even conduct hypotheses tests for the parameters of the similarity function, to the extent that one believes that the data generating process may be in agreement with one of the models specified above. In other words, the empirical approach suggested here, coupled with the statistical inference that accompanies it, may be viewed as a general-purpose statistical technique dealing with the prediction of a variable Y based on variables $X_1, \ldots, X_m$ and past observations of all these variables in conjunction.

Viewed from this perspective, one might wonder how our prediction technique compares with established ones, such as regression analysis. An obvious weakness of our approach is that it does not attempt to identify trends. For instance, assume that there exists a single variable X which denotes time, and that the data lie on a line $Y = X$. This obvious trend will not be recognized by our technique, which will continue to expect the next value of Y to be a weighted average of past values of Y. The prediction technique we suggest makes sense especially when one might believe that past observations were obtained under similar circumstances.

9.2.1. *Non-linear regression*

Our approach differs from non-linear regression in that we do not assume that the data generating process follows a basic functional relationship of the form $Y = f(X^1, \ldots, X^m)$. Rather, we assume that Y is distributed around a weighted average of its past values, where the X's determine these weights.

If, however, one does assume that there exists an underlying functional relationship $Y = f(X^1, \ldots, X^m)$, our technique may still be used for prediction of Y. As long as f is sufficiently smooth, one may hope that, with a large number of observations that are evenly scattered in terms of their X values, the similarity-weighted averaging will result in reasonable predictions. Indeed, the similarity-weighted average is reminiscent of Nadaraya-Watson's estimator of a non-parametric functional relationship. Observe that, as opposed to Nadaraya-Watson's technique and related literature, we do not attempt to find an optimal kernel function based on theoretical considerations, but find the kernel/similarity function that best fits the data.

Observe that our estimation of the similarity function s is parametric. This does not imply that we restrict the function f to a parametrized family of functions, should a relationship $Y = f(X^1, \ldots, X^m)$ actually exist. Any function s may be used to generate predictions in a non-parametric problem. To simplify our estimation problem, we restrict attention to functions s within a parametrized family of similarity functions. Thus, we try to parametrically estimate how to best perform non-parametric estimation.

9.2.2. *Kernel estimation and case-based reasoning*

If we think of similarity-weighted averaging as a model of human reasoning, we find that a case-based reasoner, as modeled by this formula, can be viewed as someone who believes in a general rule of the form $Y = f(X^1, \ldots, X^m)$ but does not know the functional form of f and therefore attempts to estimate it by non-parametric techniques.

The notion that people reason by analogies dates back to Hume (1748) at the latest. In artificial intelligence, this idea was reincarnated as case-based reasoning by Schank (1986) and Schank and Riesbeck (1989). Inspired by this work, Gilboa and Schmeidler (1995, 2001, 2003) developed a formal, axiomatically-based theory of decision and prediction by analogies. In this literature it has been mentioned that case-based reasoning is a natural and flexible mode of thinking and decision making. Our statistical approach strengthens this intuition by pointing out that case-based reasoning may be a way to estimate a functional rule.

Taking an evolutionary viewpoint, assume that nature programs the mind of an organism who needs to operate in an unknown environment. The organism will need to learn certain functional rules of the form $Y = f(X^1, \ldots, X^m)$, but it is not yet known what form the function f might take. The statistical viewpoint suggest case-based assessment by similarity-weighted averaging as a procedure to predict Y, which may perform well in a variety of possible environments f. Moreover, it turns out that the similarity-weighted averaging does not explicitly resort to general rules and theories, and thus does not require abstract thinking. Case-based reasoning therefore appears to be a flexible methodology of learning rules, which can be implemented on simple machines. Admittedly, this methodology is limited and human reasoning requires also abstract thinking and the development of explicit general theories. Yet, the evolutionary viewpoint seems to support case-based reasoning as a simple but powerful technique.

9.2.3. *Interpolation*

Our prediction methodology can also be viewed as a type of interpolation. Consider first the case $m = 1$, that is, a single variable X. Every past case is a point $(x_i, y_i) \in \mathbb{R}^2$, and we are asked to assess the value of Y for a new point $x_p \in \mathbb{R}$. Assume for simplicity that x_p is in the interval $[\min_i x_i, \max_i x_i]$. Linear interpolation would generate a prediction by the line segment connecting (x_i, y_i) and (x_k, y_k) for the two values x_i and x_k that are closest to x_p in either direction. This approach may be a bit extreme since it uses only the y values for the closest x's. In this respect, it is similar to a (single) nearest neighbor technique. Other types of interpolation, such as polynomial interpolation, would take into account also other points (x_l, y_l) for x_l that is not necessarily the closest to x_p on either side.

These interpolation techniques implicitly assume that the values observed are the actual, precise values of an unknown function. If, however, we recognize that there is some inherent randomness in the process, that we may not measure certain hidden variables, or that there are measurement errors, we might opt for a technique that is less sensitive to each particular value of Y. Following this line of thought, our approach can be viewed as performing *statistical interpolation*: every observation is used in the interpolation process, where closer points have a higher impact on the predicted value. As opposed to interpolation by high-order polynomials, when many points have been observed, no particular point would have a large impact on the predicted value.

When we consider the case $m > 1$, generalizing this interpolation technique requires a multi-dimensional distance function. Our methodology might therefore be conceptualized as a multi-dimensional statistical interpolation technique, where the distance function is empirically learnt.

9.2.4. *Bayesian updating*

A special case of the formula (1) is when s is constant (say, $s \equiv 1$), and the formula boils down to the simple average (of Y) over the entire database. This could be viewed as an estimator of the unconditional expectation of Y, having not observed any X's.

By contrast, one may consider an extreme similarity function given by $s(X_i, X_p) = 1_{\{X_i = X_p\}}$, where 1 denotes the indicator function. That is, two data points are considered to be perfectly similar if they have exactly the same X values, and absolutely dissimilar otherwise.[4] In this case, the formula (1) yields the average of Y over the sub-database defined by the values X_p, and it can be viewed as an estimate of the *conditional* expectation of Y, *given* X_p.

Thus, the formula (1) provides a continuum between conditional and unconditional expectations. When $s(X_i, X_p) = 1_{\{X_i = X_p\}}$, the

---

[4]In our model the similarity function is positive everywhere. This simplifies the formula and the axiomatization alike. But one can extend the model to include similarity functions that may vanish, or consider zero similarity values as a limit case.

reasoner only considers identical cases as relevant, and all of them are then deemed equally relevant. By contrast, if $s \equiv 1$, the reasoner considers all cases as identically relevant. In between, (1) allows for various cases to have a varying degree of relevance. Given the new datapoint X_p, past points X_i are judged for their relevance, but not in a dichotomous way. In other words, Bayesian updating may be viewed as a special case of (1), where similarity is evaluated in a binary way: two observations are similar if and only if they are identical in every possible known aspect.

As compared to Bayesian updating, a reasoners who employs (1) might be viewed as a less extreme assessor of similarity. She does not use only the observations with identical X values, but also other, less relevant ones. Why would she do that? Why should she contaminate her assessment of Y for X_p with Y that were observed for other X's?

The answer is, presumably, the scarcity of data. If we are faced with a database in which the very same X_p values appear a very large number of times, it would seem reasonable to assess the conditional expectation of Y given X_p based solely on the observations that share the exact values of X_p. But one may find that these exact values were encountered very few times, if at all. Indeed, the X's might include certain variables, such as time and location, that uniquely identify the observation. In this case, no two observations ever share the exact X values, and conditioning on X_p leaves one with an empty sub-database. Even in less extreme examples, the resulting sub-database may be too meager for generating predictions. In those cases, the formula (1) offers an alternative, in which the similarity of the observations is traded off for the size of the database.

Viewed thus, the formula (1) may deserve the title "kernel updating". As in other kernel-based techniques, the relevance of an observation (X_i, Y_i) is not restricted to identical datapoints $X_p = X_i$, but is extended to other datapoints X_p, to an extent determined by the kernel values $s(X_i, X_p)$. The use of a kernel function in this case is justified by the paucity of the data, that is, by the fact that observations with precisely the same X_p are scarce. This parallels the motivation for the use of kernel functions in kernel estimation of a density function and in kernel classification.

Finally, we observe that the use of observations (X_i, Y_i) where $X_i \neq X_p$ for the prediction of Y_p may also follow from Bayesian updating if one assumes that the X variables are observed with noise.[5]

9.2.5. *Auto-regression models*

From a mathematical viewpoint, the similarity-weighted average can be regarded as a type of an auto-regression model. In auto-regression models, as well as in our case, Y_t is distributed around a linear function of past values of Y.[6] Yet, the similarity-weighted average formula differs from auto-regression models in several important ways. Mathematically, the weights that past values $\{Y_i\}_{i<t}$ have in the equation of Y_t do not depend on the time difference $(t - i)$, but on the similarity of the corresponding X values, that is on $s(X_i, X_t)$. In particular, observe that the weights of $\{Y_i\}_{i<t}$ in the determination of the expectation of Y_t are not known before time t, because these weights depend on X_t. Observe also that in our case each Y_t depends on *all* past observations. Thus, our model is an auto-regression model whose order is not bounded a-priori. Another important difference is that in our case the index t has no cardinal significance. We use it only to order the data, but our procedure does not rely on the fact that the time difference between observations $t - 1$ and t is the same as the time difference between observations $t - 2$ and $t - 1$.[7]

Conceptually, our model assumes that similar situations in the past might have a significant impact on current values of Y, even if they occurred a long time ago. When one discusses natural phenomena, such as population growth, one expects the weight of past observations to be increasing as a function of their recency. But when we deal with human reasoning, as in the case of inflationary expectations, less recent, but more similar situations in the past may have a greater impact on the

---

[5] This comment is due to Mark Machina.

[6] As pointed out to us by an anonymous referee, when the similarity function is allowed to vanish, the i.i.d. process is a special case of our process when $s(X_1, X_t) = 1$ and $s(X_j, X_t) = 0$ for $1 < j < t$, and $Y_1 = 0$.

[7] In fact, our procedure can be easily adapted to the case in which observations are only partially ordered. As we briefly mention below, a different variant of our model can deal with situations in which the observations are not ordered at all.

future than would more recent but less similar situations. In a sense, human memory may serve as a channel through which past periods can affect future periods without the mediation of the periods in between.

The above need not imply that our model ignores time completely. One may introduce time as one of the variables X^j. This would allow more recent periods to have greater impact on the prediction than less recent ones, simply because the time difference is translated, via the variable X^j, to a distance in the X space, and thus to a lower degree of similarity.

9.2.6. *How to analyze time series*

We conclude that the relationship between our model and auto-regression models is superficial. Yet, our model can be adapted to deal with time series in a way that resembles auto-regression in a more profound way. Auto-regression can be viewed, in bold strokes, as explaining a variable by its own past values, with statistical techniques such as linear regression. The natural counterpart in our case would be to predict the variable Y by equation (1) where the variables $(X^j)_j$ include lagged values of Y itself. For example, assume that Y_t is a quarterly growth rate. Introducing $Y_{t-1}, \ldots, Y_{t-k}$ as $X_t^1, \ldots, X_t^k$ would suggest that the predicted rate of growth at period t be a (weighted) average of the rates of growth in similar periods in the past, *where similarity is defined by the pattern of growth rates in the most recent k periods*. Our technique would find weights $w_1, \ldots, w_k$ that best fit the data when one uses the equation

$$\overline{Y}_t^s = \frac{\sum_{i<t} s_w((Y_{i-k}, \ldots, Y_{i-1}), (Y_{t-k}, \ldots, Y_{t-1})) Y_i}{\sum_{i<t} s_w((Y_{i-k}, \ldots, Y_{i-1}), (Y_{t-k}, \ldots, Y_{t-1}))} \tag{5}$$

where

$$s_w((Y_{i-k}, \ldots, Y_{i-1}), (Y_{t-k}, \ldots, Y_{t-1})) = e^{-\sqrt{\sum_{j\leq k} w_j(Y_{i-j} - Y_{t-j})^2}}$$

This estimation technique could be interpreted as follows. We first ask, what determines the similarity of patterns of growth? That is, is a "pattern" defined by the most recent period, or by several most recent periods, how many of these, and what are the relative weights?

The estimation of the weights w_j attempts to answer this question. While the resulting weights need not be monotonically decreasing in j (the time difference), one would expect that these weights would become small for large values of j. In fact, in determining the number of periods that define a "pattern", k, one implicitly assumes that periods more distant than k are not part of the "pattern". The selection of this k may be compared to the selection of the order p in auto-regression models of order p ($AR(p)$).

Once the weights w_j have been determined, we search the entire database for periods i such that the pattern preceding i, $(Y_{i-k}, \ldots, Y_{i-1})$, resembles the current pattern, $(Y_{t-k}, \ldots, Y_{t-1})$. For such periods, the value Y_i would have a higher weight in the prediction of Y_t than would the value corresponding to periods for which $(Y_{l-k}, \ldots, Y_{l-1})$ resembles $(Y_{t-k}, \ldots, Y_{t-1})$ to a lesser degree. Again, one may also add time as an additional variable X^{k+1} to make sure that the prediction discounts the past.

9.3. Axiomatization

9.3.1. *Single agent*

The axiomatization does not require that past data points range over all of $\mathbb{R}^m$. We assume that they belong to a non-empty subset $\Gamma \subset \mathbb{R}^m$. However, we do assume that every possible data point in Γ may have been observed together with every value $y \in \mathbb{R}$ any finite number of times. We therefore model the database as a vector of counters, denoted I, rather than the set of observations B used in the introduction.

Specifically, let $C = \Gamma \times \mathbb{R}$ denote *case types*. A case type $(x, \xi) \in C$ is interpreted as an observation of a *data point* $x \in \Gamma$ coupled with the value $\xi \in \mathbb{R}$. *Memory* is a non-zero function $I : C \to \mathbb{Z}_+$ (where $\mathbb{Z}_+$ denotes the non-negative integers) such that $\sum_{c \in C} I(c) < \infty$, specifying for every case type c how many cases of that type have appeared. Let $\mathcal{I}$ be the set of all memories.

We are currently presented with a new data point $x_p \in \Gamma$. The task is to estimate the value $\eta \in \mathbb{R}$ that corresponds to x_p. We assume that the predictor does not only choose one such η, but has a likelihood ranking over all possible predictions. Formally, for $I \in \mathcal{I}$, let $\succsim_I \subset \mathbb{R} \times \mathbb{R}$ be a binary relation over the reals. As usual, $\succ_I$ denotes the asymmetric part of $\succsim_I$. For $\xi, \eta \in \mathbb{R}$, $\xi \succsim_I \eta$ is interpreted as "Given memory I, ξ is a more likely value for the variable Y at the new data point x_p than is η". Observe that in the formal notation we suppress x_p. This new data point is fixed throughout this section.

We now state axioms on $\{\succsim_I\}_{I \in \mathcal{I}}$. The first three are identical to those appearing in Gilboa-Schmeidler (2001, 2003).

A1 Order: For every $I \in \mathcal{I}$, $\succsim_I$ is complete and transitive on $\mathbb{R}$.

A2 Combination: For every $I, J \in \mathcal{I}$ and every $\xi, \eta \in \mathbb{R}$, if $\xi \succsim_I \eta$ ($\xi \succ_I \eta$) and $\xi \succsim_J \eta$, then $\xi \succsim_{I+J} \eta$ ($\xi \succ_{I+J} \eta$).

A3 Archimedean Axiom: For every $I, J \in \mathcal{I}$ and every $\xi, \eta \in \mathbb{R}$, if $\xi \succ_I \eta$, then there exists $l \in \mathbb{N}$ such that $\xi \succ_{lI+J} \eta$.

Observe that in the presence of Axiom 2, Axiom 3 also implies that for every $I, J \in \mathcal{I}$ and every $\xi, \eta \in \mathbb{R}$, if $\xi \succ_I \eta$, then there exists $l \in \mathbb{N}$ such that for all $k \geq l$, $\xi \succ_{kI+J} \eta$.

Axiom A1 is rather standard. It requires that, given any memory, the "more likely than" relation be a weak order.

Axiom A2 is the main axiom of Gilboa-Schmeidler (1997, 2001, and 2003). Roughly, it states that, if ξ is more likely than η given each of two memories, then ξ should also be more likely than η given their union. This axiom is satisfied by a variety of statistical techniques, such as kernel estimation, kernel classification, and maximum likelihood rankings. Yet, it is by no means universal. To illustrate its limitations, consider the following example. Suppose that there is only one predictor ($m = 1$) and that the database consists of $\{(1,1), (2,2), \ldots, (5,5)\}$, and the new datapoint is $X_6 = 6$. Given each observation (i,i) for $i = 1, \ldots, 5$, the value 6 might seem less likely than the value 5. But given the entire database, where 5 points lie exactly in the line $Y = X$, the value 6 seems a much more reasonable prediction for $X_6 = 6$. Indeed, the similarity-weighted average formula that we axiomatize is doomed to predict some weighted average of the values $\{1, \ldots, 5\}$, and will not be able to predict a value higher than 5.

This example shows a major limitation of the similarity-weighted formula, for which axiom A2 carries most of the blame: this formula is incapable of identifying trends and generating predictions based on them. Axiom A2 suggests that a conclusion that holds in two databases has to hold in their union. But if there is a trend, or a pattern in the data, it may be identified only when data is amassed. A2 rules out the possibility that the union of two memories would generate new insights. Similarly, if the similarity function is being learnt by the predictor while she produces predictions, or if the estimator uses both inductive and deductive reasoning, then the combination axiom should not be expected to hold. Moreover, if the predictor knows that the data are generated by a particular model, such as a linear regression model, or a specific Bayesian model, she will generate predictions based on that model. In this case she is likely to satisfy the combination axiom when estimating the parameters of the model (and, in particular, maximum likelihood estimation will satisfy the axiom), but not at the level of specific predictions generated by the model. However, A2 appears reasonable as a requirement on simple aggregation of evidence, in the absence of a theory on the way the data are generated.

A3 states that, if memory I contains evidence that ξ is more likely than η, then, for each other memory J there exists a large enough number, l, such that l repetitions of I would be sufficient to overwhelm the evidence provided by J, and suggest that ξ is more likely than η also given the union of J and l times I. Thus A3 precludes the possibility that one piece of evidence is infinitely more weighty than another.

Gilboa-Schmeidler (1997, 2001, and 2003) also use a diversity axiom, which we do not use here. Instead, we impose a new axiom that is specific to our set-up. It states that, if memory I consists solely of cases that relate to the same data point x, then the ranking $\succsim_I$ is consistent with simple averaging. Observe that for such databases there is nothing to be learnt from the values of x since they do not change at all. In this case, it makes sense that the most likely value of y be the average of its observed values, and that possible values be ranked according to their proximity to this average.[8]

---

[8] As pointed out to us by an anonymous referee, one may obtain axiomatizations of similarity-weighted versions of other statistics, such as the median. Any statistic that, in the absence of predictors, minimizes a convex cost function (summed over the given observations) may be viewed as the most likely value according to a relation $\succsim_I$ that satisfies A1-A3, and can then be generalized to a statistic that minimizes the similarity-weighted sum of that cost function.

For $x \in \Gamma$, define $\mathcal{I}_x$ to be the set of memories in which only data point x has been observed. Formally, $\mathcal{I}_x = \{ I \in \mathcal{I} \mid I((x',y)) = 0$ for $x' \neq x \}$. For $I \in \mathcal{I}_x$, define the average $y_I \in \mathbb{R}$ by

$$y_I = \frac{\sum_{(x,y) \in C} I((x,y))y}{\sum_{(x,y) \in C} I((x,y))}.$$

The last axiom we employ is:

A4 Averaging: For every $x \in \Gamma$, every $I \in \mathcal{I}_x$, and every $\xi, \eta \in \mathbb{R}$, $\xi \succsim_I \eta$ iff $|\xi - y_I| \leq |\eta - y_I|$.

Our result can now be stated:

Theorem 1: *Let there be given Γ, and $\{\succsim_I\}_{I \in \mathcal{I}}$. Then the following two statements are equivalent:*

(i) *$\{\succsim_I\}_{I \in \mathcal{I}}$ satisfy A1-A4;*
(ii) *There is a function $s : \Gamma \to \mathbb{R}_{++}$ such that:*

$$\begin{cases} \text{for every } I \in \mathcal{I} \text{ and every } \xi, \eta \in \mathbb{R}, \\ \xi \succsim_I \eta \quad \text{iff } |\xi - y_{s,I}| \leq |\eta - y_{s,I}|, \end{cases} \quad (*)$$

where $y_{s,I} = \dfrac{\sum_{(x,y) \in C} s(x) I((x,y))y}{\sum_{(x,y) \in C} s(x) I((x,y))}$

Furthermore, in this case the function s is unique up to multiplication by a positive number.

9.3.2. *Discussion*

The theorem states that, if we rank possible predictions of Y by their proximity to the average of past values of Y whenever the values of $X^1, \ldots, X^m$ are fixed, and we wish to extend it to general databases in a way that satisfies our axioms (notably, the combination axiom), we are bound to do it by proximity to weighted averages.

The axiomatization we provide can be interpreted descriptively or normatively. From a descriptive point of view, the theorem suggests that, if an agent's rankings of possible values of a variable y given various databases satisfy our axioms, she can be ascribed a similarity function s such that her rankings are determined by proximity to a similarity-weighted average of past values of y, calculated by the similarity function s. From a normative viewpoint, the axiomatization might be used to convince an agent that similarity-weighted averaging is a reasonable way to assess the variable y given a database of past observations. Finally, the axiomatization also suggests a definition of an agent's similarity function, and method of elicitation for it.

A weighted averaging formula is also axiomatized in Billot, Gilboa, Samet, and Schmeidler (2003). In their model a reasoner is asked to name a probability vector based on a memory I. Billot *et al.* impose an appropriate version of the combination axiom to conclude that the probability vector given a memory I is the weighted average of the vectors induced by each case separately. Unfortunately, the result of Billot *et al.* only applies if there are at least 3 states of the world, that is, if the probability vector has at least two degrees of freedom. For the special case of a single-dimension probability simplex, their theorem does not hold. In this sense, the present paper complements Billot *et al.* (2003).

9.3.3. *Representative agent*

The theorem above shows under what conditions an agent's "more likely than" relation will follow the similarity-weighted average formula for an appropriately chosen similarity function. It relates the theoretical concept of "similarity" to the relation "more likely than", which is assumed to be observable.

In practice, however, one can often observe only aggregate data. For instance, one may observe market prices of houses or paintings, but not the assessments of these prices by agents. What properties should such assessments satisfy? How are individual assessments aggregated over agents? Can such aggregates also be described as similarity-weighted averages?

To answer these questions, we extend the model presented above to incorporate more than one agent. Specifically, let $P = \{1, \ldots, p\}$ be a set of agents, and re-define the case types to be $C = P \times \Gamma \times \mathbb{R}$. Case of type (i, x, y) is interpreted as "agent $i \in P$ has observed a data point $x \in \Gamma$ and a corresponding value of $y \in \mathbb{R}$". Thus, every observation in this model specifies the observer, and not only the observed.

We continue as before to define *memory* as a non-zero vector I : $C \to \mathbb{Z}_+$ such that $\sum_{c \in C} I(c) < \infty$. Let $\mathcal{I}$ be the set of all memories. We now think of memory I is a matrix of counters, specifying how many times each agent has observed any possible $(x, y) \in \Gamma \times \mathbb{R}$ combination.

The relation $\succsim_I$ is interpreted as follows. For $\xi, \eta \in \mathbb{R}$, $\xi \succsim_I \eta$ means that, if I specifies how many times each agent has seen each pair (x, y), then ξ is more likely than η to be the assessment of the set of agents. This assessment is supposed to reflect some collective opinion, and it does not reflect economic power or strategic considerations. If, for instance, we discuss the value of a painting by van Gogh, every agent is expected to have some assessment of the value of the painting, regardless of their ability or willingness to pay for it.

The axioms we use are the same axioms verbatim. The logic behind the axioms mirrors that of the single agent case, though, naturally, in the multi-agent case the axioms are more demanding.

We first state the theorem as applied to this case:

Corollary 2:   *Let there be given P, Γ, and $\{\succsim_I\}_{I \in \mathcal{I}}$. Then the following two statements are equivalent:*

(i) *$\{\succsim_I\}_{I \in \mathcal{I}}$ satisfy A1-A4;*
(ii) *There exist functions $\{s_i : \Gamma \to \mathbb{R}_{++}\}_{i \in P}$ such that:*

$$\begin{cases} \text{for every } I \in \mathcal{I} \text{ and every } \xi, \eta \in \mathbb{R}, \\ \xi \succsim_I \eta \quad \text{iff } |\xi - y_{s,I}| \leq |\eta - y_{s,I}|, \end{cases} \qquad (**)$$

where $y_{s,I} = \dfrac{\sum_{(i,x,y) \in C} s_i(x) I((i,x,y)) y}{\sum_{(i,x,y) \in C} s_i(x) I((i,x,y))}$

Furthermore, in this case the functions $\{s_i\}_{i \in P}$ are unique up to joint multiplication by a positive number.

We wish to show that, if we assume that all information is shared, then a set of agents P, characterized by functions $\{s_i\}_{i \in P}$, is indistinguishable from a representative agent whose similarity function is the average of $\{s_i\}_{i \in P}$. To this end, define $\mathcal{I}_{sh}$ as the set of memories in which all agents have the same information, that is: $\mathcal{I}_{sh} = \{ I \in \mathcal{I} \mid I((i,x,y)) = I((i',x,y)) \text{ for all } i, i' \in P \}$. We now have

Corollary 3: *Let there be given* P, Γ, *and* $\{\succsim_I\}_{I \in \mathcal{I}}$. *Assume that* $\{\succsim_I\}_{I \in \mathcal{I}}$ *satisfy A1-A4. Then there exists a function* $s : \Gamma \to \mathbb{R}_{++}$ *such that:*

$$\begin{cases} \text{for every } I \in \mathcal{I}_{sh} \text{ and every } \xi, \eta \in \mathbb{R}, \\ \xi \succsim_I \eta \quad \text{iff} \quad |\xi - y_{s,I}| \le |\eta - y_{s,I}|, \end{cases} \qquad (***)$$

where $y_{s,I} = \dfrac{\sum_{(i,x,y) \in C} s(x) I((i,x,y)) y}{\sum_{(i,x,y) \in C} s(x) I((i,x,y))}$

Furthermore, in this case the function s *is unique up to multiplication by a positive number.*

Observe that the identification of individual similarity functions s_i requires that memories $I \notin \mathcal{I}_{sh}$ be considered, that is, memories in which different agents may have observed different cases. Measuring $\succsim_I$ for $I \notin \mathcal{I}_{sh}$ and testing our axioms may be done in controlled experiments in a laboratory. It is more challenging to observe $\succsim_I$ for $I \notin \mathcal{I}_{sh}$ in empirical data. Yet, one may imagine that such relations exist, and satisfy our axioms.

As long as we restrict attention to shared information, namely, to memories in $\mathcal{I}_{sh}$, we only observe the average similarity function. Attributing this average similarity to a representative agent, we conclude that the assessment made by a set of agents will be equivalent to that made by the representative agent.

The observability of $\succsim_I$ mirrors the observability of a utility function in economics: in principle, one may measure each agent's utility function. In reality, often only aggregate data are available. Under certain conditions, one may assume that the decisions of a set of agents can be described by the decision of a single, representative agent. Similarly, in our case one may, in principle, measure each agent's similarity function. In practice, we often observe only aggregate assessments. However, under the conditions specified above, we may replace the set of agents by a single, representative agent, and obtain the same assessment for shared information. It is this similarity function, of a representative agent, that we attempt to estimate.

9.4. Statistical Inference for a Continuous Model

9.4.1. *The model and the likelihood function*

If we assume the initial condition to be $Y_1 = \varepsilon_1$, then equation (4) can be written in matrix form as

$$Sy = \varepsilon,$$

where $S = S(w) = I - B_w A_w$, I is the identity matrix of order n,

$$A_w = \begin{pmatrix} 0 & & & \cdots & \\ s_{w,2,1} & 0 & & & \\ \cdots & & \cdots & & \\ s_{w,n,1} & & s_{w,n,n-1} & 0 & \end{pmatrix},$$

$$B_w = \begin{pmatrix} 0 & 0 & \cdots & & 0 \\ 0 & (e'_2 A_w 1)^{-1} & & \cdots & \\ \cdots & & \cdots & & 0 \\ 0 & & \cdots & 0 & (e'_n A_w 1)^{-1} \end{pmatrix},$$

$s_{w,i,j} = s_w(X_j, X_i) = e^{-d_w(x_j, x_i)}$, d is based on (3), 1 is an $n \times 1$ vector of $1's$, e_j is the canonical vector of $0's$, apart from the j-th position where it is set to unity, $y = (Y_1, \ldots, Y_n)'$ and ε is an $n \times 1$ vector of i.i.d. Gaussian variables with zero mean and variance σ^2. Note that S is a lower triangular matrix that does not depend on the variables Y_i.

We set $\theta = (\theta_1, \ldots, \theta_{m+1}) = (\sigma^2, w_1, \ldots, w_m)$ and observe that $\theta \in \Theta \subset \mathbb{R}_+^{m+1}$. The maximum likelihood estimator (MLE) of θ, $\hat{\theta}_n$, maximizes

$$l(\theta) = -\frac{n}{2} \log(2\pi\theta_1) - \frac{1}{2} y' H(\theta) y,$$

where $H = S'S/\sigma^2$.

Note the difference between nonparametric regression and our approach. In the former, the postulated relationships are of the form $y = g(x) + \varepsilon$ and the Nadaraya–Watson estimator of the unknown $g(x)$ has precisely the same form as the term $\sum_{i<t} s_w(X_i, X_t) Y_i / \sum_{i<t} s_w(X_i, X_t)$ appearing in (4). In our set-up this term is part of

the data generating process. In addition, in nonparametric regression the bandwidth is selected so as to minimize some criterion, such as mean integrated square error, whereas we use maximum likelihood to estimate the weights w_j.

9.4.2. *Hypotheses tests*

Rejecting the null hypothesis $H_0 : w_j = 0$ implies that the variable X_j contributes to the determination of Υ, in the sense that the distance function, according to which Υ_t is determined in (4), does not ignore the j-th variable.

Under general conditions on the similarity function which are satisfied for exponential similarity, Lieberman (2005) proved that the MLE is weakly consistent, is locally asymptotically mixed normal and

$$\sqrt{n}F^{-1/2}(\theta_0)(-P_n(\theta_0))(\hat{\theta}_n - \theta_0) \to_d N(0, I_{m+1}), \qquad (6)$$

where

$$F(\theta_0) = \lim_{n\to\infty} E_{\theta_0}\left(\frac{1}{n}\frac{\partial l_n(\theta_0)}{\partial \theta}\frac{\partial l_n(\theta_0)}{\partial \theta'}\right),$$

$$P_n(\theta) = \frac{1}{n}\frac{\partial^2 l_n(\theta)}{\partial\theta\partial\theta'}$$

and θ_0 is the true value of θ. For simple hypotheses tests for which the parameter is at the interior of the parameter space under the null, (6) can be used to apply any of the conventional likelihood based tests (likelihood ratio, Lagrange Multiplier and Wald) in a straightforward manner. For an hypothesis of the form $H_0 : w_r = 0$ vs. $w_r > 0$ $(r = 1,\ldots,m)$, the parameter is on the boundary under H_0. For this case, Chant (1974, equation (8)) showed that the distribution of the normalized MLE is half-normal. Hence, for a one-sided t-test of the form above we reject H_0 as we do in the usual case when t is large (e.g., when it exceeds 1.645, if a 5% significance level is desired).

9.5. Statistical Inference for the Discrete Case

9.5.1. *The model and the likelihood function*

We now deal with the case in which each Y_t is categorical. In particular, consider examples 5 and 6 above, in which an expert is asked to estimate the probability of a certain event, and the observed values of Y can only be $\{0, 1\}$. A probability estimated by our formula with the empirical similarity function may be viewed as "objective" in that it does not rely on subjective similarity judgments, while still allowing different datapoints to have differing relevance to the estimation problem at hand.

When assessing probabilities, the assessed values can be anywhere in the interval $[0, 1]$. Indeed, the formula (2) may generate any value in $[0, 1]$. But in this case one cannot assume that previously observed values of Y were generated by a Normal distribution centered around a similarity-weighted average such as in model (4).[9]

We therefore assume the following model[10]

$$P(Y_t = 1 | \mathcal{F}_{t-1}) = F_t(z_t(w)), \quad t = 1, \ldots, n, \qquad (7)$$

where F_t is a continuous conditional distribution function with density f_t, $\mathcal{F}_{t-1} = \sigma(Y_{t-1}, \ldots, Y_1; X_t, \ldots, X_1)$ and

$$z_t = \frac{\sum_{i<t} s_w(X_i, X_t) Y_i}{\sum_{i<t} s_w(X_i, X_t)}. \qquad (8)$$

In this setting the X's are taken to be fixed. Letting F_t be the standard normal cumulative distribution function (cdf) leads to a probit type model whereas letting F_t be the logistic distribution leads to a logit type model. Since $z_t \in [0, 1]$, it might be sensible to let F_t be a beta distribution, or quite simply, the uniform distribution. Note that in the classical case, corresponding to the rule based model, it is postulated

---

[9]Other reasons for which model (4) is inappropriate in this case are that the R^2 of regression would typically be low and that, because of the non-Gaussian nature of the observations, OLS would be inefficient.

[10]The categorical variables we discuss here may only assume the values 0 or 1. However, the analysis that follows can be extended to the case of a categorical variable assuming more than two categories.

that $P(\Upsilon_t = 1|X) = F(X\beta)$. Unlike our case, no Υ_j's appear on the right hand side and the model (7) is nonlinear through both F_t and z_t.

In view of (7) and (8)

$$\frac{\partial P(\Upsilon_t = 1|\mathcal{F}_{t-1})}{\partial s_w(X_j, X_t)} = f_t(z_t)\frac{\sum_{i<t} s_w(X_i, X_t)(\Upsilon_j - \Upsilon_i)}{(\sum_{i<t} s_w(X_i, X_t))^2},$$

which is non-negative if $\Upsilon_j = 1$ and non-positive when $\Upsilon_j = 0$. In other words, when the similarity between Υ_t and Υ_j increases, the conditional probability that $\Upsilon_t = 1$ will not fall when $\Upsilon_j = 1$ and will not rise when $\Upsilon_j = 0$. The model thus makes sense at least in this respect.

Given our setup, the log-likelihood is given by

$$l = \ln(L) = \sum_{t=1}^{n} (\Upsilon_t \ln(F_t(z_t)) + (1 - \Upsilon_t)\ln(1 - F_t(z_t)))$$

and the MLE's are the solutions of

$$\frac{\partial l}{\partial w_j} = \sum_{t=1}^{n} \frac{\Upsilon_t - F_t}{F_t(1 - F_t)} f_t v_{t,j} = 0, \quad j = 1, \dots, m$$

where

$$v_{t,j} = \partial z_t / \partial w_j$$

$$= \frac{(\sum_{i<t} \dot{s}_{w,j}(X_i, X_t)\Upsilon_i)(\sum_{i<t} s_w(X_i, X_t))}{(\sum_{i<t} s_w(X_i, X_t))^2}$$

$$- \frac{(\sum_{i<t} \dot{s}_{w,j}(X_i, X_t))(\sum_{i<t} s_w(X_i, X_t)\Upsilon_i)}{(\sum_{i<t} s_w(X_i, X_t))^2}$$

$$\dot{s}_{w,j}(X_i, X_t) = \frac{\partial s_{w,t,i}}{\partial w_j} = -\frac{s_{w,t,i}(X_{ji} - X_{jt})^2}{2d(X_i, X_j)}$$

and d is given in (3). As in the previous section, any likelihood based procedure can be employed for hypothesis tests of the form $H_0 : w = w_0$.

9.6. Concluding Remarks

In the statistical analysis we assumed that each observation Y_t is distributed around a weighted average of past Y_i, or that $P(Y_t = 1)$ depends on such a weighted average. Such an ordering is necessary for a causal interpretation of our models. But if we consider a non-causal relationship, one may assume a model in which the distribution of each Y_i conditional on the other variables $\{Y_k\}_{k \neq i}$ is, say, normal around the weighted average of $\{Y_k\}_{k \neq i}$. Indeed, such a model may be more natural for applications in which the data are not naturally ordered. For this case, one should adapt the statistical model and the estimation of the similarity function accordingly.

The assumptions underlying our estimation process call for elaboration. The axiomatic model aims to describe how an *assessment* of Y_p, $\overline{Y}_p^s$, is generated based on *actually observed* values of the variable in question, namely, past values $(Y_i)_{i \leq n}$, such as selling prices of houses or of paintings. Applied to each past observation Y_i, it suggests that the *assessment* of Y_i, $\overline{Y}_i^s$, is generated by (2) for *actually observed* past values $(Y_k)_{k < i}$. That is, when we explain Y_i by past observations $(Y_k)_{k < i}$, we treat Y_i as if it were an assessment. When we explain Y_l for $l > i$, we treat Y_i as if it were an actual value. What justifies this confusion between the actual value of a variable and an assessment thereof?

For many applications of interest the answer lies in the notion of equilibrium. If all economic agents agree in their assessment of the price of a house or a painting, this joint assessment will indeed be its market price. Similarly, the price of a financial asset would equal its own assessment, if all agents agree on the latter. In these cases, one may assume that, as a feature of equilibrium, actually observed data coincide with their assessments.[11]

There may be applications in which one has direct access to, or indirect measurement of both actual values (Y_i) and to their assessments, say (Z_i). In these cases one may find the similarity function

---

[11]To a lesser degree, the rate of inflation and the probability of a stock market crash are also influenced by what economic agents assess them to be.

s that best fits the data according to

$$\overline{Z}_i^s = \frac{\sum_{k<i} s(X_k, X_i) Y_k}{\sum_{k<i} s(X_k, X_i)} \qquad (9)$$

namely, a function s that provides values $(\overline{Z}_i^s)_i$ that are close to $(Z_i)_i$, and then use this function to generate an estimate of Z_p, $\overline{Z}_p^s$, using actual values Y_k by equation (9) applied to $i = p$.

Yet another class of applications involves only the assessments (Z_i). Assume, for instance, that one only observes asking prices, (Z_i), and not actual selling prices, (Y_i). (This is the case in Gayer, Gilboa, and Lieberman (2004).) If everyone has access only to the asking prices (Z_i), one may apply our axiomatization to these variables, and conclude that the asking price of a new observation Z_p will be a similarity-weighted average of past asking prices $(Z_i)_{i\leq n}$. Moreover, it makes sense to assume that the same similarity function governed the generation of past values Z_i as a function of their past, $(Z_k)_{k<i}$. Hence one may estimate the similarity function in equation (2) with Z_i instead of Y_i, and use the estimated similarity for the prediction of Z_p.

Finally, there are situations in which one does not have access to the assessments (Z_i), and in which there is no theoretical reason to assume that $Z_i = Y_i$. In these cases our empirical approach could still be applied. That is, one may still ask, which function $s : \mathbb{R}^m \times \mathbb{R}^m \to \mathbb{R}_{++}$ can best fit the data under the assumption that there were generated by equation (2), and the function can then be used for prediction of Y_p by equation (1). In this type of application, (Y_i) can be viewed as proxies for (Z_i). Observe that it is only in the estimation of s that we replace (Z_i) by (Y_i). In the generation of the prediction $\overline{Z}_p^s$ using the estimated s, we use the actual values (Y_i) as indeed we should.

This paper is devoted to the theory of similarity-weighted averaging. This technique is used in Gayer, Gilboa, and Lieberman (2004) for the assessment of real estate prices, as in example 1. Their paper compared this method, representing case-based reasoning, to linear regression, representing rule-based reasoning.

Appendix: Proofs

Proof of Theorem 1:

We begin by proving sufficiency of the axioms (that is, that (i) implies (ii)), and the uniqueness of the function s. Consider a pair $\xi, \eta \in \mathbb{R}$. Restricting $\{\succsim_I\}_{I \in \mathcal{I}}$ to $\{\xi, \eta\}$, one notices that they satisfy the conditions of the representation theorem in Gilboa and Schmeidler (2001, Theorem 3.1, p. 67, or 2003, Theorem 2, p. 16). Indeed, the first three axioms follow directly from A1-A3, whereas the diversity axiom for two alternatives follows from the averaging axiom, A4. To apply this theorem we have also to define the trivial relation for the memory $I = 0$: $\succsim_0 = \mathbb{R} \times \mathbb{R}$. Hence there exists a function $v^{\xi\eta} : \Gamma \times \mathbb{R} \rightarrow \mathbb{R}$, unique up to multiplication by a positive number, such that, for every $I \in \mathcal{I}$,

$$\xi \succsim_I \eta \quad \text{iff} \quad \sum_{(x,y) \in C} I((x,y)) v^{\xi\eta}(x,y) \geq 0.$$

Next consider a triple $\{\xi, \eta, \varsigma\} \subset \mathbb{R}$. Restricting $\{\succsim_I\}_{I \in \mathcal{I}}$ to $\{\xi, \eta, \varsigma\}$ will no longer satisfy the diversity axiom in Gilboa and Schmeidler (2001). This axiom would state that for every permutation of the triple $\{\xi, \eta, \varsigma\}$ there exists $I \in \mathcal{I}$ such that $\succ_I$ agrees with that permutation. This condition does not follow from our A4. Indeed, the diversity axiom is too strict for our purposes. If $\xi > \eta > \varsigma$, then no $\succsim_I$ represented by ($*$) will satisfy $\xi \succ_I \varsigma \succ_I \eta$.

However, the proof of Gilboa and Schmeidler's theorem does not require the full strength of their diversity axiom. All that is required for three alternatives $\{\xi, \eta, \varsigma\}$ is that $v^{\xi\eta}$ not be a multiple of $v^{\eta\varsigma}$. To this end, it suffices that there be three different permutations in $\{\succ_I\}_{I \in \mathcal{I}}$ (restricted to $\{\xi, \eta, \varsigma\}$). This latter condition is guaranteed by our A4. Specifically, by the averaging axiom A4, for every distinct (ξ, η, ς), there exists $I \in \mathcal{I}$ such that $\xi \succ_I \eta, \varsigma$. Hence there are at least three permutations in $\{\succ_I\}_{I \in \mathcal{I}}$ (restricted to $\{\xi, \eta, \varsigma\}$), and the representation theorem for triples holds. Observe also that this argument does not employ all the relations $\{\succsim_I\}_{I \in \mathcal{I}}$, and it can also be used for a restricted domain $\{\succsim_I\}_{I \in \mathcal{I}_x}$ for any $x \in \Gamma$.

It follows that, for every triple $\{\xi, \eta, \varsigma\}$, one can find $v^\xi, v^\eta, v^\varsigma$: $\Gamma \times \mathbb{R} \to \mathbb{R}$ such that, for every $a, b \in \{\xi, \eta, \varsigma\}$, for every $I \in \mathcal{I}$,

$$a \succsim_I b \quad \text{iff} \quad \sum_{(x,y) \in C} I((x,y)) v^a(x,y) \geq \sum_{(x,y) \in C} I((x,y)) v^b(x,y)$$

$$\Leftrightarrow \sum_{(x,y) \in C} I((x,y))[v^a(x,y) - v^b(x,y)] \geq 0. \tag{B1}$$

In this case, the matrix $(v^\xi, v^\eta, v^\varsigma)$ is unique up to multiplication by a positive constant and addition of a constant to each row. Fix one such matrix $(v^\xi, v^\eta, v^\varsigma)$.

Fix $x \in \Gamma$ and consider $\mathcal{I}_x$. Restrict attention to $\{\succsim_I\}_{I \in \mathcal{I}_x}$. Since (B1) applies to all $I \in \mathcal{I}$, it definitely holds for all $I \in \mathcal{I}_x \subset \mathcal{I}$. However, we claim that, even on this restricted domain, the matrix $(v^\xi, v^\eta, v^\varsigma)$ is unique as above. To see this, recall that our derivation of (B1), coupled with the uniqueness result, holds true for $\{\succsim_I\}_{I \in \mathcal{I}_x}$ for any $x \in \Gamma$.

Observe that the relations $\{\succsim_I\}_{I \in \mathcal{I}_x}$ are completely specified by A4. Specifically, for every $a, b \in \mathbb{R}$, for every $I \in \mathcal{I}_x$,

$$a \succsim_I b \quad \text{iff} \quad |a - y_I| \leq |b - y_I| \tag{B2}$$

where

$$y_I = \frac{\sum_{(x,y) \in C} I((x,y))y}{\sum_{(x,y) \in C} I((x,y))}.$$

That is, $a \succsim_I b$ iff a is closer to the average y_I than is b. Consider

$$f_I(\alpha) = \sum_{(x,y) \in C} I((x,y))(\alpha - y)^2.$$

The function $f_I(\alpha)$ is quadratic (in α), and it has a minimum at $\alpha = y_I$. It follows that for every a, b, for every $I \in \mathcal{I}_x$,

$$f_I(a) \leq f_I(b) \quad \text{iff} \quad |a - y_I| \leq |b - y_I|.$$

Combining this fact with the definition of f_I and with (B2), we conclude that, for every $a, b \in \{\xi, \eta, \varsigma\}$ and for every $I \in \mathcal{I}_x$,

$$a \succsim_I b \quad \text{iff} \quad \sum_{(x,y) \in C} I((x,y))(a-y)^2 \leq \sum_{(x,y) \in C} I((x,y))(b-y)^2$$

$$\Leftrightarrow \sum_{(x,y) \in C} I((x,y))[(a-y)^2 - (b-y)^2] \leq 0. \qquad (B3)$$

The uniqueness of the representation in (B1) and (B3) imply that there exists a constant $s(x) > 0$ such that

$$v^a(x, y) - v^b(x, y) = -s(x)[(a-y)^2 - (b-y)^2] \qquad (B4)$$

for every $a, b \in \{\xi, \eta, \varsigma\}$ and for every $y \in \mathbb{R}$. Obviously, once $v^a(x, y), v^b(x, y)$ are fixed, $s(x)$ is uniquely determined by (B4).

We now turn to discuss various x's, while still focusing on the triple $\{\xi, \eta, \varsigma\}$. Consider I in $\mathcal{I}$ (but not necessarily in $\mathcal{I}_x$ for any x). Combine (B1) and (B4) to conclude that, for $a, b \in \{\xi, \eta, \varsigma\}$ and for all $I \in \mathcal{I}$,

$$a \succsim_I b \quad \text{iff} \quad \sum_{(x,y) \in C} I((x,y))s(x)(a-y)^2$$

$$\leq \sum_{(x,y) \in C} I((x,y))s(x)(b-y)^2$$

$$\Leftrightarrow \sum_{(x,y) \in C} I((x,y))s(x)[(a-y)^2 - (b-y)^2] \leq 0. \qquad (B5)$$

Define

$$g_I(\alpha) = \sum_{(x,y) \in C} I((x,y))s(x)(\alpha - y)^2.$$

As the function f_I above, the function $g_I(\alpha)$ is also quadratic (in α), and it has a minimum at

$$\alpha = y_{s,I} = \frac{\sum_{(x,y) \in C} s(x)I((x,y))y}{\sum_{(x,y) \in C} s(x)I((x,y))}.$$

It follows that for every a, b, for every $I \in \mathcal{I}$,

$$g_I(a) \leq g_I(b) \quad \text{iff} \quad |a - y_{s,I}| \leq |b - y_{s,I}|. \qquad (B6)$$

Combining (B5) with (B6) we obtain

$$a \succsim_I b \quad \text{iff} \quad |a - y_{s,I}| \leq |b - y_{s,I}|,$$

that is, $\{s(x)\}_x$ satisfies ($*$) for the triple $\{\xi, \eta, \varsigma\}$.

Observe that $\{s(x)\}_x$ are unique up to multiplication by a positive number. In fact, we argue that if s and s' both satisfy ($*$) for particular $a, b \in \mathbb{R}$, $a \neq b$, then there exists $\lambda > 0$ such that $s'(x) = \lambda s(x)$ for all $x \in \Gamma$. Indeed, assume that s and s' both satisfy ($*$) for particular a, b. This would imply that they both satisfy (B5) for these a, b, and then the uniqueness of $v^a(x, y) - v^b(x, y)$ in (B1), combined with (B4), implies that there exists $\lambda > 0$ such that $s'(x) = \lambda s(x)$ for all $x \in \Gamma$.

It remains to show that the function $s(x)$ does not depend on the choice of the triple $\{\xi, \eta, \varsigma\} \subset \mathbb{R}$. Consider the triple $\{\xi, \eta, \tau\}$ where $\tau \neq \varsigma$. Since (B6) applied to ξ and η holds both for the function s of the triple $\{\xi, \eta, \varsigma\}$ and that of the triple $\{\xi, \eta, \tau\}$, these two functions have to be positive multiples of each other. Using this argument inductively implies that all functions s derived from different triples differ only by a constant. Since s can always be multiplied by a positive constant and still satisfy ($*$), one may choose an s of one triple $\{\xi, \eta, \varsigma\}$ arbitrarily and use it for all other triples as well.

We need to prove the necessity of the axioms, that is, that (ii) implies (i). The necessity of A1, A2, and A3 is proved as in Gilboa and Schmeidler (2001, 2003), whereas the necessity of A4 follows directly from ($*$). □

Proof of Corollary 2
This result is a re-writing of Theorem 1 for the special case in which C is the product of two sets. □

Proof of Corollary 3
Use Corollary 2 and define $s(x) = \frac{1}{p} \sum_{i \in P} s_i(x)$. For $I \in \mathcal{I}_{sh}$,

$$\frac{\sum_{(i,x,y) \in C} s_i(x) I((i, x, y)) y}{\sum_{(i,x,y) \in C} s_i(x) I((i, x, y))} = \frac{\sum_{(i,x,y) \in C} s(x) I((i, x, y)) y}{\sum_{(i,x,y) \in C} s(x) I((i, x, y))} = y_{s,I}$$

which concludes the proof. □

References

Akaike, H. (1954), "An Approximation to the Density Function", *Annals of the Institute of Statistical Mathematics*, **6**: 127–132.

Basawa, I. V., P. D. Feigin and C. C. Heyde (1976), "Asymptotic Properties of Maximum Likelihood Estimators for Stochastic Processes", *Sankhya* A, **38**: 259–270.

Billot, A., I. Gilboa, D. Samet and D. Schmeidler (2005), "Probabilities as Similarity-Weighted Frequencies", *Econometrica*, **73**: 1125–1136.

Billot, A., I. Gilboa and D. Schmeidler (2004), "An Axiomatization of an Exponential Similarity Function", mimeo.

Chant, D. (1974), "On Asymptotic Tests of Composite Hypotheses in Nonstandard Conditions", *Biometrika*, **61**: 291–298.

Cover, T. and P. Hart (1967), "Nearest Neighbor Pattern Classification", IEEE *Transactions on Information Theory*, **13**: 21–27.

Devroye, L., L. Gyorfi and G. Lugosi (1996), *A Probabilistic Theory of Pattern Recognition*, New York: Springer-Verlag.

Fix, E. and J. Hodges (1951), "Discriminatory Analysis. Nonparametric Discrimination: Consistency Properties". Technical Report 4, Project Number 21-49-004, USAF School of Aviation Medicine, Randolph Field, TX.

— (1952), "Discriminatory Analysis: Small Sample Performance". Technical Report 21-49-004, USAF School of Aviation Medicine, Randolph Field, TX.

Gayer, G., I. Gilboa and O. Lieberman (2004) "Rule-Based and Case-Based Reasoning in Housing Prices", mimeo.

Gilboa, I. and D. Schmeidler (1995), "Case-Based Decision Theory", *Quarterly Journal of Economics*, **110**: 605–639.

— (1997), "Act Similarity in Case-Based Decision Theory", *Economic Theory*, **9**: 47–61.

— (2001), *A Theory of Case-Based Decisions*, Cambridge: Cambridge University Press.

— (2003) "Inductive Inference: An Axiomatic Approach", *Econometrica*, **71**: 1–26. Reprinted as Chapter 4 of this volume.

Hume, D. (1748), *Enquiry into the Human Understanding*, Oxford: Clarendon Press.

Lieberman, O. (2005), "Asymptotic Theory for Empirical Similarity Models", mimeo. Published in *Econometric Theory*, **26** (2009): 1032–1059.

Parzen, E. (1962), "On the Estimation of a Probability Density Function and the Mode", *Annals of Mathematical Statistics*, **33**: 1065–1076.

Riesbeck, C. K. and R. C. Schank (1989), *Inside Case-Based Reasoning*, Hillsdale, NJ: Lawrence Erlbaum Associates, Inc.

Rosenblatt, M. (1956), "Remarks on Some Nonparametric Estimates of a Density Function", *Annals of Mathematical Statistics*, 27: 832–837.

Schank, R. C. (1986), *Explanation Patterns: Understanding Mechanically and Creatively*, Hillsdale, NJ: Lawrence Erlbaum Associates.

Scott, D. W. (1992), *Multivariate Density Estimation: Theory, Practice, and Visualization*, New York: John Wiley and Sons.

Silverman, B. W. (1986), *Density Estimation for Statistics and Data Analysis*, London and New York: Chapman and Hall.

Chapter 10

Axiomatization of an Exponential Similarity Function*

Antoine Billot[†], Itzhak Gilboa[‡] and David Schmeidler[§]

Reprinted from *Mathematical Social Sciences*, 55 (2008): 107–115.

An individual is asked to assess a real-valued variable y based on certain characteristics $x = (x^1, \ldots, x^m)$, and on a database consisting of n observations of $(x^1, \ldots, x^m, y)$. A possible approach to combine past observations of x and y with the current values of x to generate an assessment of y is *similarity-weighted averaging*. It suggests that the predicted value of y, y^s_{n+1}, be the weighted average of all previously observed values y_i, where the weight of y_i is the similarity between the vector $x^1_{n+1}, \ldots, x^m_{n+1}$, associated with y_{n+1}, and the previously observed vector, $x^1_i, \ldots, x^m_i$. This paper axiomatizes, in terms of the prediction y_{n+1}, a similarity function that is a (decreasing) exponential in a norm of the difference between the two vectors compared.

10.1. Introduction

In many prediction and learning problems, an individual attempts to assess the value of a real variable y based on the values of relevant

---

*We wish to thank Jerome Busemeyer for comments and references. Gilboa and Schmeidler gratefully acknowledge support from the Polarization and Conflict Project CIT-2-CT-2004-506084 funded by the European Commission-DG Research Sixth Framework Programme. Gilboa and Schmeidler gratefully acknowledge support from the Israel Science Foundation (Grant Nos. 790/00 and 975/03).
[†]University de Paris II, IUF, and CERAS-ENPC. billot@u-paris2.fr
[‡]Tel-Aviv University and Yale University. igilboa@post.tau.ac.il
[§]Tel-Aviv University and The Ohio State University. schmeid@tau.ac.il

variables, $x = (x^1, \ldots, x^m)$, and on a database, B, consisting of past observations of the variables $(x_i, y_i) = (x_i^1, \ldots, x_i^m, y_i)$, $i = 1, \ldots, n$. Some examples for the variable y include the weather, the behavior of other people, and the price of an asset. The relevant variables x may represent meteorological conditions, psycho-social cues, or the attributes of the asset, respectively.

There are many well-known approaches for the prediction of y given x and the database B. For instance, regression analysis is such a method. k-nearest neighbor techniques (Fix and Hodges, 1951, 1952) would be another method, predicting the value of y at a point x by the values that y has assumed for points close to x. In fact, the literature in statistics and in machine learning offers variety of methods for this problem, which encompasses a wide spectrum of problems that people encounter in their daily lives as well as in professional endeavors.

One approach to deal with the classical learning/prediction problem is to use a similarity-weighted average: fix a similarity function $s : \mathbb{R}^m \times \mathbb{R}^m \to \mathbb{R}_{++}$ and, given the database B and the new data point $x \in \mathbb{R}^m$, generate the prediction

$$y^s = \frac{\sum_{i \le n} s(x_i, x) y_i}{\sum_{i \le n} s(x_i, x)}$$

This formula was suggested and axiomatized in Gilboa, Lieberman, and Schmeidler (2004).[1,2] They assume that, for every $n \in \mathbb{N}_{++}$, any database B (consisting of $n \ge 1$ observations in $\mathbb{R}^{m+1}$), and every new point $x \in \mathbb{R}^m$, a predictor has an ordering over $\mathbb{R}$, $\gtrsim_{B,x}$, interpreted as "more likely than". They show that these orderings satisfy certain axioms if and only if there exists a similarity function such that the ordering ranks possible predictions y according to their proximity to y^s.

---

[1] It is reminiscent of derivations in Gilboa and Schmeidler (2003) and in Billot, Gilboa, Samet, and Schmeidler (2005). It also bears resemblance to kernel-based methods of estimations, as in Akaike (1954), Rosenblatt (1956), Parzen (1962) and others. See Silverman (1986) and Scott (1992) for surveys.
[2] The term similarity at this point does not impose any restriction on the function. It just indicates that this function is used in a formula like the one above.

In this paper we investigate the explicit form of the similarity function s, in the context of the similarity-weighted formula. That is, we assume that Υ is assessed according to

$$\Upsilon(B, x) = \frac{\sum_{i \leq n} s(x_i, x) y_i}{\sum_{i \leq n} s(x_i, x)} \tag{1}$$

where the function $\Upsilon(\cdot, \cdot)$ is defined on the all databases, $\mathbb{B} = \cup_{n \geq 1}(\mathbb{R}^{m+1})^n$, and for all $x \in \mathbb{R}^m$. The derivation of formula (1) by Gilboa, Lieberman, and Schmeidler (2004) is done for each x separately, considering the rankings of possible values of $\Upsilon(B, x)$ for various databases B, but for a fixed $x \in \mathbb{R}^m$. Hence, they obtain a separate function $s(\cdot, x)$ for each x. This function is strictly positive and it is unique up to multiplication by a positive number. For concreteness, we here normalize this function such that $s(x, x) = 1$ for every x. With this convention, s is unique.

We consider the behavior of $\Upsilon(\cdot, \cdot)$ when one varies its arguments. We suggest certain consistency conditions on Υ, referred to as "axioms", which characterize an exponential functional form, namely, a similarity function s that satisfies, for every $x, z \in \mathbb{R}^m$,

$$s(z, x) = \exp[-\nu(x - z)] \tag{2}$$

for some norm ν on $\mathbb{R}^m$. Assuming that the assessment Υ are observable, our result may be interpreted as showing what observable implications are there to the assumption of exponential similarity (2) in the context of the similarity-weighted average formula (1).

The notion of a similarity function which is decaying exponentially as a function of distance is rather natural, and appears in other contexts as well. For instance, Shepard (1987) derives an exponential similarity function which measures the probability of generalizing a response from one stimulus to another. An exponential decay function is used to model the probability of recall (see, for instance, Bolhuis, Bijlsma, and Ansmink, 1986), which may be interpreted as a measure of the similarity between two points of time. The present paper shows that exponential decay, relative to some norm, has, and is characterized by rather appealing properties also when similarity is used for the computation of similarity-weighted average as in (1).

The axioms and the main result are stated in the next section. They are followed by comments on several special cases of the norm v, the special case of a single-dimensional space, and a general discussion. Proofs are to be found in an appendix.

10.2. Main Result

Suppose that there are given functions $\Upsilon : \mathbb{B} \times \mathbb{R}^m \to \mathbb{R}$ and $s : \mathbb{R}^m \times \mathbb{R}^m \to \mathbb{R}_{++}$ as in formula 1. (The positive integer m is fixed throughout the paper.) We impose the following axioms on Υ:

A1 Shift Invariance: For every $B = (x_i, y_i)_{i \leq n} \in \mathbb{B}$, and every x, $w \in \mathbb{R}^m$,

$$\Upsilon((x_i + w, y_i)_{i \leq n}, x + w) = \Upsilon((x_i, y_i)_{i \leq n}, x).$$

A1 states that the prediction does not depend on the absolute location of the points (x_i), x in $\mathbb{R}^m$, but only on their relative location. More precisely, it demands that a shift in all independent variables in the database, accompanied by the same shift in the new independent variable for which prediction is required, will not affect the predicted value Υ.

The next axiom requires that evidence that was obtained for further points has lower impact. It is restricted to a rather uncontroversial definition of "being further away": it is only required to hold along rays emanating from zero, when prediction is required for the point $x = 0$. (To avoid confusion we will denote the origin in $\mathbb{R}^m$ by a bold $0, 0$.)

A2 Ray Monotonicity: For every $x, z \in \mathbb{R}^m$, $\Upsilon(((\lambda x, 1), (z, -1)), 0)$ is strictly decreasing in $\lambda \geq 0$.

A2 considers databases consisting of two points, one, λx, at which the value 1 was observed, and another, z, at which the value -1 was observed. Obviously, equation (1) would generate a value Υ in $(-1, 1)$ for such a database. When we vary λ, the value of Υ will be higher, the more similar is λx considered to be to 0. A2 states that, if we move λx further away from 0 (along the ray through x), it will be considered less similar to 0, and the prediction Υ will decrease. (i.e., it will move away from 1 toward -1.)

A3 Symmetry: For every $x \in \mathbb{R}^m$,

$$\Upsilon(((x, 1), (0, 0)), 0) = \Upsilon(((0, 1), (x, 0)), x).$$

A3 considers two situations. In the first, one has observed the value 1 for $x \in \mathbb{R}^m$, and the value 0 for $0 \in \mathbb{R}^m$, and one is asked to make a prediction for $0 \in \mathbb{R}^m$. In the second situation, the roles are reversed: the value 1 was observed at $0 \in \mathbb{R}^m$, the value 0 at $x \in \mathbb{R}^m$, and the prediction is requested for x. A3 then requires that the prediction be the same in these two situations. Intuitively, it demands that the impact an observation at x has on observation at 0 is the same as the impact of the same observation at 0 has on observation at x.

Axiom 4 is reminiscent of A1, but the antecedent is more restrictive and the conclusion stronger. It applies to a database where all the independent variables are on a ray through the origin. A shift along this ray leaves the prediction unchanged although the independent variable for which the predictions are made is the origin before and after the shift. Formally,

A4 Ray Shift Invariance: Let there be given $B = (\alpha_i v, y_i)_{i \leq n} \in \mathbb{B}$, for some $v \in \mathbb{R}^m$ and $\alpha_i \geq 0$ ($i \leq n$). Then, for every $\theta > 0$, $\Upsilon((\alpha_i v + \theta v, y_i)_{i \leq n}, 0) = \Upsilon((\alpha_i v, y_i)_{i \leq n}, 0)$.

Our last axiom is,

A5 Self-Relevance: For every $x, z \in \mathbb{R}^m$,

$$\Upsilon(((0, 1), (x, 0)), z) \leq \Upsilon(((0, 1), (x, 0)), 0).$$

A5 considers a simple database B consisting of two points: the value 1 was observed for the point 0, while the value 0 was observed for the point x. Given such a database, any prediction generated by equation (1) is necessarily in $[0, 1]$. Intuitively, the prediction generated given this database, for every z, is higher the higher is the similarity of z to 0 relative to its similarity to x. Self-Relevance requires that this relative similarity be maximized at $z = 0$. That is, no other point $z \neq 0$ can be more similar to 0 than to x, as compared to 0 itself.

Recall that a *norm* on $\mathbb{R}^m$ is a function $v: \mathbb{R}^m \to \mathbb{R}_+$ satisfying:

(i) $v(\xi) = 0$ iff $\xi = 0$;
(ii) $v(\lambda\xi) = |\lambda|v(\xi)$ for all $\xi \in \mathbb{R}^m$ and $\lambda \in \mathbb{R}$;
(iii) $v(\xi + \zeta) \leq v(\xi) + v(\zeta)$ for all $\xi, \zeta \in \mathbb{R}^m$.

We can now state our main result:

Theorem 1:   *Let there be given a function Υ as in formula 1, where s is normalized by, $s(x,x) = 1$ for all $x \in \mathbb{R}^m$. The following are equivalent:*

(i) *Υ satisfies A1–A5;*
(ii) *There exists a norm $v : \mathbb{R}^m \to \mathbb{R}_+$ such that*

$$(*) \quad s(x,z) = \exp[-v(x-z)] \quad \text{for every } x, z \in \mathbb{R}^m$$

We observe that, given s, the norm v is uniquely defined by $(*)$, and vice versa.

The shift axioms (A1) enables us to state the rest of the axioms for $\Upsilon(\cdot, 0)$ rather than for $\Upsilon(\cdot, w)$ for every $w \in \mathbb{R}^m$. As will be clear from the proof of the theorem, one may drop A1, strengthen the other axioms so that they hold for every $w \in \mathbb{R}^m$, and obtain a similar representation that depends on a more general distance function (that is not necessarily based on a norm).

It will also be clear from the proof that our result can be generalized at no cost to the case that the data points x_i belong to any linear space (rather than $\mathbb{R}^m$). This is true also of the axiomatization in Gilboa, Lieberman, and Schmeidler (2004). Taken together, the two results may be viewed as axiomatically deriving a norm on a linear space, based on predictions Υ.

The similarity function obtained in Gilboa, Lieberman, and Schmeidler has no structure whatsoever. The only property that follows from their axiomatization is the positivity of s. An important feature of our result is that observable conditions on predictions Υ imply that v is a norm, and this, in turn, imposes restrictions on the similarity function. First, since for a norm v, $v(\xi) = v(-\xi)$, we conclude that $s(x,z) = s(z,x)$, that is, that s is symmetric.

Another important feature of norms is that they satisfy the triangle inequality. This would imply that s satisfies a certain notion of transitivity. Specifically, it is not hard to see that, given the representation ($*$), the triangle inequality for v implies that for every $x, z, w \in \mathbb{R}^m$,

$$s(x, w) \geq s(x, z)s(z, w)$$

Thus, if both x and w are similar to z to some degree, x and w have to be similar to each other to a certain degree. Specifically, if both $s(x, z)$ and $s(w, z)$ are at least ε, then $s(x, w)$ is bounded below by ε^2.

10.3. Special Cases

One may impose additional conditions on Υ that would restrict the norm that one obtains in the theorem. For instance, consider the following axiom:

A6 Rotation: Let P be an $m \times m$ orthonormal matrix. Then, for every $B = (x_i, y_i)_{i \leq n}$, $\Upsilon((x_i, y_i)_{i \leq n}, 0) = \Upsilon((x_i P, y_i)_{i \leq n}, 0)$.

A6 asserts that rotating the database around the origin would not change the prediction at the origin. It is easy to see that in this case the norm v coincides with the standard norm on $\mathbb{R}^m$.

For certain applications, one may prefer a norm that is defined by a weighted Euclidean distance, rather than by the standard one. We obtain a derivation of such a norm, we need an additional definition.

For two points $z, z' \in \mathbb{R}^m$, we write $x \sim x'$ if the following holds: for every $B \in \mathbb{B}$, and $y \in \mathbb{R}$, $\Upsilon((B, (x, y)), 0) = \Upsilon((B, (x', y)), 0)$, where $(B, (x, y))$ denotes the database obtained by concatenation of B with (x, y). In light of equation (1), it is easy to see that two vectors x and x' are considered $\sim$-equivalent if and only if $s(x, 0) = s(x', 0)$. Using this fact, or using the definition directly, one may verify that $\sim$ is indeed an equivalence relation.

In the presence of axiom A1, two vectors x and x' are considered $\sim$-equivalent if observing y at a point that is x-removed from the new point has the same impact on the prediction as observing y at a point that is x'-removed from the new point.

For $j \leq m$, let $e_j \in \mathbb{R}^m$ be the j-th unit vector in $\mathbb{R}^m$ (that is, $e_j^k = 1$ for $k = j$ and $e_j^k = 0$ for $k \neq j$). we can now state

A7 Elliptic Rotation: Assume that, for $j, k \leq m$ and $\beta > 0$, $e_j \sim \beta e_k$. Let $\theta, \mu > 0$ be such that $\beta\theta^2 + \mu^2 = \beta$. Then for every $x = (x^1, \ldots, x^m)$, $x + e_j \sim x + \theta e_j + \mu e_k$.

A7 requires that $\sim$-equivalence classes would be elliptic. Specifically, it compares a unit vector on the j-th axis to a multiple of the unit vector on the k-axis. It assumes that β is the appropriate multiple of e_k that would make it equivalent to e_j. It then considers the ellipse connecting these points, and demands that this ellipse would lie on an equivalence curve of $\sim$. It can be verified that A7 will imply that v is defined by a weighted Euclidean distance.

More generally, one may use the equivalence relation above to state axioms that correspond to various specific norms. In particular, any L_p norm can be derived from an axiom that parallels A7.

10.4. A Single Dimension

An interesting special case is where there is only one predictor, i.e., when $m = 1$. A prominent example would be when the data are indexed by time. In this case, the point for which a prediction is required is larger, that is, further into the future, than any point in the database. In this case, not all the axioms are needed for our main result. Moreover, in this case the exponential similarity function can also be justified on different grounds. We begin by stating the appropriate versions of the axioms that are needed in the case $m = 1$.

Let $\mathbb{B}' = \{((x_i, y_i)_{i \leq n})|(x_i, y_i) \in \mathbb{R}^2, x_i \geq x_j \text{ for } i > j\}$. Denote by $\mathbb{B}'_0$ the union of $\mathbb{B}'$ and the set containing the empty database (corresponding to $n = 0$). Assume that Υ is defined on

$$\mathbb{D} = \{((x_i, y_i)_{i \leq n}, x)|(x_i, y_i)_{i \leq n} \in \mathbb{B}', x \in \mathbb{R}, x \geq x_n\}.$$

Re-write the axioms as follows.

A1' Shift Invariance: For every $((x_i, y_i)_{i \leq n}, x) \in \mathbb{D}$, and every $w \in \mathbb{R}$, $\Upsilon((x_i + w, y_i)_{i \leq n}, x + w) = \Upsilon((x_i, y_i)_{i \leq n}, x)$.

A2′ Monotonicity: $\Upsilon(((-1,1),(\lambda,-1)),1)$ is strictly decreasing in $\lambda \in [-1,1]$.

A4′ Ray Shift Invariance: For every $((x_i, y_i)_{i \le n}, x) \in \mathbb{D}$, and every $w \ge 0$, $\Upsilon((x_i, y_i)_{i \le n}, x + w) = \Upsilon((x_i, y_i)_{i \le n}, x)$.

The Shift Invariance axiom states that shifting the entire database, as well as the new point, does not affect the prediction. The monotonicity axiom states that the closer is a datapoint (λ) to the new prediction (1), the higher is its impact, that is, the -1 associated with λ, has a greater weight in the prediction for $x = 1$ as compared to another datapoint (1 observed at -1). Finally, the Ray Shift Invariance states that if a prediction is required for a later point ($x + w$ rather than x), but no new datapoint have been observed, the prediction does not change.

Interpreting the single predictor as time, the axioms have quite intuitive justifications: Shift Invariance states that the point at which we start measuring time is immaterial. Monotonicity simply requires that a more recent experience have a greater impact on current predictions. Finally, Ray Shift Invariance can be viewed as stating that the predictor does not change her prediction simply because time has passed. If no new datapoints were added, no change in prediction would result.

In a single dimension, the exponential similarity function allows one to summarize a database by a single case, such that, for all future observations and all future prediction problems, the summary case would serve just as well as the entire database. Specifically, we formulate a new condition:

Summary: For every $((x_i, y_i)_{i \le n}) \in \mathbb{B}$, there exists $(\bar{x}, \bar{y}) \in \mathbb{R}^2$, such that for every $((x_i', y_i')_{i \le m}) \in \mathbb{B}_0$ with $x_1' \ge x_n$ (if $m > 0$), and every $x \ge x_m'$, $\Upsilon(((x_i, y_i)_{i \le n}, (x_i', y_i')_{i \le m}), x) = \Upsilon(((\bar{x}, \bar{y}), (x_i', y_i')_{i \le m}), x)$.

We can now state

Proposition 2: *Let there be given a function Υ as in formula 1, where s is normalized by, $s(x, x) = 1$ for all $x \in \mathbb{R}^m$. The following are equivalent:*

(i) *Υ satisfies A1′, A2′, and A4′;*
(ii) *Υ satisfies A1′, A2′, and Summary;*
(iii) *There exists $\theta \in \mathbb{R}_+$ such that $s(x, z) = \exp[-\theta(z - x)]$ for every $z \ge x$.*

Appendix: Proof

Proof of Theorem 1:

It is convenient to prove that (i) is equivalent to (ii) by imposing one axiom at a time. This will also clarify the implication of A1, A1 and A2, etc.[1]

It is easy to see that A1 is equivalent to the existence of a function $f : \mathbb{R}^m \to \mathbb{R}_{++}$, with $f(0) = 1$, such that $s(x, z) = f(x - z)$ for every $x, z \in \mathbb{R}^m$. Indeed, if such an f exists, A1 will hold. Conversely, if A1 holds, one may define $f(x) = s(x, 0)$ and use the shift axiom to verify that $s(x, z) = f(x - z)$ holds for every $x, z \in \mathbb{R}^m$.

Next consider A2. Since $f(x) = s(x, 0)$, it is easy to see that A2 holds if and only if f is strictly decreasing along any ray emanating from the origin. Explicitly, A1 and A2 hold if and only if $s(x, z) = f(x - z)$ for every $x, z \in \mathbb{R}^m$ and $f(\lambda x)$ is strictly decreasing in $\lambda \geq 0$ for every $x \in \mathbb{R}^m$, and $f(0) = 1$.

It is easily seen that symmetry (A3) is equivalent to the fact that $f(x) = f(-x)$ for every $x \in \mathbb{R}^m$.

We now turn to A4. Consider a ray originating from the origin, $\{\lambda x | \lambda \geq 0\}$, for a given $x \in \mathbb{R}^m$ ($x \neq 0$). We observe that for Ray Invariance to hold, in the presence of Monotonicity, $s(\lambda x, 0)$ has to be exponential in λ. To see this, observe that Ray Invariance implies that the ratio $s(k\lambda x, 0)/s((k + 1)\lambda x, 0)$ is independent of k for every λ. This guarantees that $s(\lambda x, 0)$ is exponential on the rational values of λ. Given monotonicity (A2) we conclude that for every $x \in \mathbb{R}^m$ there exists a number v_x such that $s(\lambda x, 0) = \exp[-\lambda v_x]$. Obviously, $v_{\lambda x} = \lambda v_x$ for $\lambda \geq 0$. A2 also implies that $v_x > 0$ for $x \neq 0$.

Combining these observations with the previous ones, we conclude that A1–A4 are equivalent to the existence of a function $f : \mathbb{R}^m \to \mathbb{R}_{++}$, such that $s(x, z) = f(x - z)$ before every $x, z \in \mathbb{R}^m$, where $f(0) = 1$, $f(x) = f(-x)$ for every $x \in \mathbb{R}^m$, and, for every $x \in \mathbb{R}^m$ there exists a non-negative number v_x such that $f(x) = \exp[-\lambda v_x]$ and $v_{\lambda x} = \lambda v_x$ for $\lambda \geq 0$. Further, $v_x = 0$ only for $x = 0$. Defining

---

[1] We will follow the order A1–A4. The exact implication of each subset of axioms separately can be similarly analyzed.

$v(x) = v_x$ we obtain the representation $(*)$ for a function v that satisfies all the condition of a norm, apart from the triangle inequality.

To conclude the proof, we need to show that v satisfies $v(x + z) \leq v(x) + v(z)$ if and only if A5 holds. Consider arbitrary $x, z \in \mathbb{R}^m$. A5 states that

$$\Upsilon(((0,1),(x,0)),z) \leq \Upsilon(((0,1),(x,0)),0)$$

which implies that

$$\frac{s(0,z)}{s(0,z) + s(x,z)} \leq \frac{s(0,0)}{s(0,0) + s(x,0)}$$

or

$$\frac{s(0,z)}{s(0,z) + s(x,z)} \leq \frac{1}{1 + s(x,0)}$$

Equivalently, we have

$$\frac{s(0,z) + s(x,z)}{s(0,z)} \geq 1 + s(x,0)$$

which is equivalent, in turn to

$$\frac{s(x,z)}{s(0,z)} \geq s(x,0)$$

and to

$$s(x,z) \geq s(x,0)s(0,z)$$

Observe that A5 is equivalent to this form of multiplicative transitivity independently of the other axiom. While we obtain the multiplicative transitivity condition only at 0, an obvious strengthening of A5 will imply that $s(x,z) \geq s(x,w)s(w,z)$ for every $x, z, w \in \mathbb{R}^m$.

Using the representation of s, we conclude that A5 is equivalent to the claim that, for every $x, z \in \mathbb{R}^m$,

$$\exp[-v(x - z)] \geq \exp[-v(x) - v(-z)]$$

or

$$v(x - z) \leq v(x) + v(-z)$$

Setting $\xi = x$ and $\zeta = -z$, we conclude that A5 holds if and only if v satisfies the triangle inequality.

This completes the proof of the theorem. $\square$

Proof of Proposition 2:

The equivalence of (i) and (iii) is proved as in the general case (see the proof of Theorem 1 above). We wish to show that Summary may replace A4'. First, observe that Summary is a stronger condition than is A4'. This follows from restricting Summary to the case $m = 0$, and observing that $\Upsilon((\bar{x}, \bar{y}), x) = \bar{y}$ for all x. Conversely, it is easy to verify that (iii) implies Summary. $\square$

References

Akaike, H. (1954), "An Approximation to the Density Function", *Annals of the Institute of Statistical Mathematics*, **6**: 127–132.

Bolhuis, J. J., S. Bijlsma and P. Ansmink (1986), "Exponential decay of spatial memory of rats in a radial maze", *Behavioral Neural Biology*, **46**: 115–22.

Billot, A., I. Gilboa, D. Samet and D. Schmeidler (2005), "Probabilities as Similarity-Weighted Frequencies", *Econometrica*, **73**: 1125–1136.

Fix, E. and J. Hodges (1951), "Discriminatory Analysis. Nonparametric Discrimination: Consistency Properties". Technical Report 4, Project Number 21-49-004, USAF School of Aviation Medicine, Randolph Field, TX.

— (1952), "Discriminatory Analysis: Small Sample Performance". Technical Report 21-49-004, USAF School of Aviation Medicine, Randolph Field, TX.

Gilboa, I. and D. Schmeidler (2003) "Inductive Inference: An Axiomatic Approach", *Econometrica*, **71**: 1–26. Reprinted as Chapter 4 of this volume.

Gilboa, I., O. Lieberman and D. Schmeidler (2004) "Empirical Similarity", *Review of Economics and Statistics*, **88** (2006): 433–444. Reprinted as Chapter 9 of this volume.

Parzen, E. (1962), "On the Estimation of a Probability Density Function and the Mode", *Annals of Mathematical Statistics*, **33**: 1065–1076.

Rosenblatt, M. (1956), "Remarks on Some Nonparametric Estimates of a Density Function", *Annals of Mathematical Statistics*, **27**: 832–837.

Scott, D. W. (1992), *Multivariate Density Estimation: Theory, Practice, and Visualization*, New York: John Wiley and Sons.

Shepard, R. N. (1987), "Toward a Universal Law of Generalization for Psychological Science", *Science*, **237**: 1317–1323.

Silverman, B. W. (1986), *Density Estimation for Statistics and Data Analysis*, London and New York: Chapman and Hall.

Chapter 11

On the Definition of Objective Probabilities by Empirical Similarity*

Itzhak Gilboa[†], Offer Lieberman[‡] and David Schmeidler[§]

Reprinted from *Synthese*, *172* (2009): 79–95.

We suggest to define objective probabilities by similarity-weighted empirical frequencies, where more similar cases get a higher weight in the computation of frequencies. This formula is justified intuitively and axiomatically, but raises the question, which similarity function should be used? We propose to estimate the similarity function from the data, and thus obtain objective probabilities. We compare this definition to others, and attempt to delineate the scope of situations in which objective probabilities can be used.

11.1. Definitions of Probability

How should we assign probabilities to events? What is the meaning of a statement of the form, "Event A will occur with probability p"? Or, to be more cautious, under which conditions can we assign probabilities to events, and under which conditions do possible meanings of the term apply?

---

*We wish to thank Gabi Gayer, Jacob Leshno, Arik Roginsky, and Idan Shimony for comments and references. This project was supported by the Pinhas Sapir Center for Development and Israel Science Foundation Grants Nos. 975/03 and 355/06.

[†]Tel-Aviv University, HEC, and Cowles Foundation, Yale University. igilboa@tau.ac.il

[‡]University of Haifa. offerl@econ.haifa.ac.il

[§]Tel-Aviv University and The Ohio State University. schmeid@tau.ac.il

To address these questions we consider a few examples. For concreteness, we embed these examples in decision problems, and one may further suppose that the decision makers in question are interested in assigning probabilities to events in order to subsequently maximize expected utility. Yet, our focus is on the concept of "probability" as such. In discussing these problems, observe also that our approach is epistemological rather than ontological. We do not purport to discuss the "true" nature of probability, but only the notion of probability insomuch as it can be measured and quantified. We are interested in the type of circumstances in which a statement "Event A will occur with probability p" can be made, and the meaning that should then be attached to such a statement.

Example 1 A coin is about to be tossed. Sarah is offered a bet on the outcome of the toss. She wonders what is the probability of the event "The coin lands on Head."

Example 2 John normally parks his car on the street. He is offered an insurance policy that will cover theft. To decide whether he should buy the policy, he wonders what is the probability that his car will be stolen during the coming year.

Example 3 Mary has to decide whether to undergo a medical operation that is supposed to improve her quality of life, but that may also involve serious risks. Trying to make a rational decision, Mary asks her physician what is the probability of various events, such as death.

Example 4 George considers an investment opportunity, and he figures out that the investment will not be very successful if there is another war in the Middle East over the next year. He then attempts to assess the probability of such a war in this time frame.

The term "probability" has various meanings and definitions. Among these, at least three seem to be widely accepted:

The "Classical" approach suggests that all possible outcomes have the same probability. This approach has been used in early writings on games of chance, and it has been explicitly formulated by Laplace as a principle, later dubbed the "Principle of Insufficient Reason" or the "Principle of Indifference."

The **"Frequentist" approach** offers the empirical frequency of an event in past observations as a definition of its probability. Bernoulli's (1713) law of large numbers guarantees that independent and identical repetitions of an experiment will result, with probability 1, in a relative frequency of occurrence of an event that converges to the event's probability. In fact, this limit relative frequency is often used as an intuitive definition of probability. The relative frequency in a finite sample can thus be a good estimate, or even a definition of the probability of the event.

The **"Subjective" approach** views probability as a numerical measure of degree of belief that is constrained to satisfy certain conditions (or "axioms"). Subjective probability has been discussed from the very first days of probability theory, and it is used already by Pascal in his famous "wager." Ramsey (1931), de Finetti (1937), and Savage (1954) have promoted it, and suggested axioms on observed behavior, that would necessitate the existence and uniqueness of a subjective probability measures. Specifically, Savage (1954) provided a set of axioms on choices between alternative courses of actions, which imply that the decision maker behaves as if she wished to maximize the expectation of a certain function with respect to a certain probability measure. Interpreting the function as "utility" and the measure as "subjective probability," his theorem provides a behavioral definition of subjective probability, coupled with the principle of expected utility maximization.

We now turn to examine how each of these three approaches deals with the four examples described above.

11.1.1. *The classical approach*

The classical approach applies in Example 1. When Sarah is considering the bet on the outcome of the coin toss, she might, in the absence of any other information, assign a probability of 50% to each possible outcome. However, the same approach does not seem to be tenable in the other examples. If John were to say, "either my car is stolen, or it isn't, hence it has probability of 50% of being stolen," he would hardly be rational. Nor can Mary or George assign 50% to surviving the operation, or to a war in the Middle East, respectively. Clearly, there is too much information in these examples to apply the principle of insufficient reason. In fact, even in the absence of information, this principle has come under attack on various grounds. For instance, it is very sensitive to the representation of the state space, especially when the latter in infinite.

Despite these attacks, in Example 1 the principle of insufficient reason is acceptable, whereas in Examples 2–4 it is completely inappropriate. One may wonder what are precisely the features of Example 1 that distinguish it from the others in this respect. That is, one may wish to delineate the scope of applicability of the principle of insufficient reason. Here we merely conclude that this approach for the assignment of probabilities is not very useful in most real life decision problems.

11.1.2. *The frequentist approach*

The frequentist approach appears to be more promising. Like the classical approach, it can deal with the coin problem in Example 1: if one has many observations of tosses of the same coin in the past, conducted under similar conditions, one may take the observed relative frequency as a definition of the probability of the coin landing on Head. Indeed, it is quite possible that this approach will coincide with the principle of insufficient reason, as will be the case if the coin is fair. But the relative frequency approach would apply equally well also if the coin is not fair. This approach does not assume any symmetries in the problem, and it is therefore robust to alternative representations of the state space. The frequentist approach deals with Example 2 in basically the same way as with Example 1: if there are many observations of cars parked overnight, and if these observations were taken under practically the same conditions, that is, in the same neighborhood, for the same type of car, and so forth, then it makes sense to take the average rate of theft as the probability of a particular car being stolen.

How would the frequentist approach be applied to Example 3? In principle, it should follow the same pattern: Mary should ask her physician how often the operation has succeeded in the past, and use the ratio of successes to trials as the "probability" of success. But Mary may find her physician uneasy about a straightforward quote of relative frequencies. After all, the physician might say, the data were collected over a variety of individuals, who differ from Mary in many relevant ways, including age, gender, weight, blood pressure, and so forth. They were operated on in different hospitals and by different surgeons. In fact, the physician might say, since no two cases are identical, you can choose which dataset to look at, and thereby affect the "probability" you obtain. Thus, the objectivity of the empirical frequency approach is compromised by the subjectivity of the choice of the sample.

The applicability of the frequentist approach to Example 4 is even more dubious. This approach would call for the listing of past cases in which war has or has not erupted, and taking the number of wars divided by the overall number of cases to be the probability of war. One difficulty that becomes obvious in this example is the precise delineation of a "case" in time. Should we take each year to be a separate case? If so, how would we deal with a war that lasted more than one year, or with a year in which more than one war has occurred? Should we perhaps lump periods together in larger chunks? Or should we define cases as starting and ending by historically meaningful events? Clearly, splitting and merging cases will affect the relative frequencies of wars, and thereby our probability assessments.

A second difficulty that George would encounter in Example 4 is that encountered by Mary in Example 3: the choice of the dataset, or the relevant "sample" is not obvious. What should count as a case, relevant for the relative frequency of the occurrence of wars? Should we go back to wars in the Middle East in biblical times? These might be relevant when certain geographical or strategic considerations are concerned, but their relevance seems limited, as well as our degree of confidence in their veracity. Should we perhaps restrict attention to modern times? But if so — how can one define "modern times" objectively? Does it make more sense to restrict one's attention to post-WWII period, or to rely on a larger dataset that predates WWII? Similarly, one may further wonder which other features of past cases should matter. Should one consider only cases in which the involved parties had similar military might, similar regime, or similar economic conditions? Clearly, as in Mary's medical example, George also faces a situation that is unique. History repeats itself, but never in precisely the same form, and the current case has enough features that distinguish it from all past cases. If George were to take all these considerations into account, he will end up with an empty dataset. If he ignores them completely, his dataset is large but very uneven in terms of relevance. Thus, the choice of the dataset becomes a subjective one, which ends up affecting the assessed probability.

There is yet another difficulty that is unique to George's problem: in Example 4, past cases cannot be assumed to be causally independent.[1] Thus, the relative frequency approach may be ignoring important mechanisms that are at work. Observe that, in Example 3, Mary could ignore possible causal dependencies. She could argue that the success of the operation on other patients does not directly affect its success in her case. This is clearly an assumption about the world. Moreover, it may not be true, if, for instance, Mary is going to be operated on by a surgeon who failed in his previous operation and may have even been sued for malpractice. Still, the independence assumption seems like a reasonable one for Mary, and it allows her to view relative frequencies as proxies for "probability." This is not the case in George's problem. Recent wars are intricately related in various causal relationships to the possible next war. There are political and military lessons that are being learned, there are goals that have and have not been obtained, and so forth. Hence, simply considering relative frequencies may be completely misleading.

11.1.3. *The subjective approach*

The subjective approach appears to be immune to all the difficulties mentioned above. According to this approach, probabilities are subjective, or "personal," and therefore they need not derive from past data

---

[1]We use the term "causation" in an intuitive sense. Of the various definitions of this term, some resort to probability as a primitive (cf. Pearl, 2000). Such definitions cannot be used in our context, as we are attempting to define the term "probability." However, we do not use causation as part of our suggested definition. It will only be used in the informal meta-discussion, attempting to characterize the scope of applicability of our definition.

or from perceived symmetries. Rather, they reflect intuition, and model it in a precise way. One may have a degree of belief in the eruption of war just as one has a degree of belief in a coin landing on Head, and the formal probability model can help sharpen this intuition and put it to use. Moreover, making decisions in accordance with certain sets of axioms implies that one makes decisions as if one were to use a probability measure.[2]

The subjective approach is conceptually very neat. Rather than coping with the essence of objectivity, with the meaning of factual knowledge, and with the possibility of processing data in an objective way, this approach steps back, gives up any claim to objectivity, and rearranges its defense around universality: probability is only subjective, but, as such, it may apply to any source of uncertainty, irrespective of the amount of relevant data gathered.

However, this approach has been attacked on several grounds. Ellsberg's experiments (Ellsberg, 1961) have shown that people often behave as if they do not have a subjective probability measure that may summarize their beliefs. Several authors have also attacked the subjective approach on normative grounds. (See Shafer, 1986, and Gilboa, Postlewaite, and Schmeidler, 2006.) In particular, it has been argued that in the absence of information, it may not be rational to choose a single probability measure, a choice that is bound to be arbitrary. Moreover, the behavioral derivations of probabilistic beliefs have also been criticized on the normative appeal of their underlying axioms.

In this paper we do not take issue with the subjective approach. Rather, our focus is on the possibility of defining objective probabilities. We therefore consider Examples 3 or 4, and ask whether some intuitive notion of objective probabilities can be defined in these examples.

---

[2] Probability may be used in various decision rules, the most famous of which is expected utility maximization. But a decision maker may be following a well-defined subjective probability measure also when using other rules. Machina and Schmeidler (1992) defined and axiomatized "probabilistic sophistication," which may be defined as "having a subjective probability measure and making decisions based solely on the distributions that this probability induces." Rostek (2006) suggested axioms that imply that the decision maker has a subjective probability measure, and that she makes decisions so as to maximize the median utility with respect to that probability.

11.1.4. *Extending frequentism*

Let us consider Example 3 again. The main difficulty that Mary was facing in applying the frequentist approach was that past cases differed in many ways, and that each case was basically unique. It is useful to observe that, in principle, the same objection may apply to the application of the frequentist approach in Examples 1 and 2 as well. In Example 2, for instance, one might argue that no two cars are identical, just as no two patients are in Example 3. Even in Example 1 we may have to admit that no two tosses of a coin are precisely identical. Various factors distinguish one toss from another, such as meteorites that might affect Earth's gravitational field, the mood of the person tossing the coin, and so forth.

More generally, every case is unique, if only because it can be defined by its exact time and location. Thus, the differences between Examples 1, 2, and 3, as far as the frequentist approach is concerned, are differences of degree, not of kind. All cases are inherently unique, but in examples such as 1 and 2 one may make the simplifying assumption that a certain "experiment" was repeated many times. In other words, it is a judgment of similarity that allows the frequentist approach to be used.

This observation paves the way to a natural generalization of the frequentist approach: if cases are not identical, or if there aren't sufficiently many cases that may be assumed identical, one may bring forth the similarity between cases and use it in the definition of probability. This would be in line with Hume's (1748) focus on similarity as key to prediction. Specifically, the probability of an event can be defined by its weighted relative frequency in past cases, where each case is weighed by its similarity to the present case. Thus, a success in an operation of another patient in the past makes a success in Mary's case more likely, but the degree to which the past case matters depends on the similarity between Mary and the patient in the past observation. We devote Section 11.2 to a more formal description of this approach, as well as to further discussion of the similarity-weighted frequency formula and its axiomatic derivations.

11.1.5. *Is it objective?*

The similarity-weighted frequency approach may thus overcome some of the difficulties encountered by the frequentist approach. But will we not give up objectivity in this process? A given dataset will result in a large range of possible "probabilities," depending on the similarity function that we choose to employ. If the similarity function is a matter of subjective judgment, so is the resulting probability. It would therefore appear that the similarity-weighted frequency approach has, at best, translated the question, "Which probability should we use?" to "Which similarity should we use?"

However, we maintain that this translation is a step forward. In fact, we argue that the choice of the similarity function need not be arbitrary or subjective: we propose to estimate the similarity function from the data. The basic idea is to try explaining past data by a similarity-weighted frequency formula, and, in this context, to ask which similarity function best explains the data we have observed. Section 11.3 describes this estimation procedure in more detail.

We thus suggest similarity-weighted frequencies, employing the empirical similarity function derived from the data, as a definition of objective probabilities. We hold that this is a reasonable definition in certain domains of application, such as described in Example 3. In Section 11.4, we compare our definition to alternative definitions that are based on statistical techniques. We argue that our approach is more appropriate, mostly because it is a natural extension of the frequentist approach, and because it is axiomatically based.

Yet, we do not view our definition as universally applicable. In fact, the axiomatizations of our formula are also helpful in identifying classes of situations in which it might be inappropriate. Example 4 is such a situation. We are not aware of any method for the assignment of objective probabilities that would be intuitive in situations such as Example 4. Our definition certainly isn't. We devote Section 11.5 to limitations of our approach. Finally, Section 11.6 discusses possible directions for extending our definition to a wider class of cases.

11.2. Similarity-Weighted Relative Frequencies

11.2.1. *The formula*

For concreteness, we stick to Example 3 in the exposition. Let the variable of interest be $Y \in \{0, 1\}$, indicating success of a medical procedure. The characteristics of patients are $X = (X^1, \ldots, X^m)$. These variables are real-valued, but some (or all) of them may be discrete. We are given a database consisting of past observations of the variables $(X, Y) = (X^1, \ldots, X^m, Y)$, denoted $(X_i, Y_i)_{i \leq n}$. A new case is introduced, with characteristics $X_{n+1} = (X_{n+1}^1, \ldots, X_{n+1}^m)$, and we are asked to assess the probability that $Y_{n+1} = 1$. •

Assume that we are also equipped with a "similarity" function s such that, for two vectors of characteristics, $X_i = (X_i^1, \ldots, X_i^m)$ and $X_j = (X_j^1, \ldots, X_j^m)$, $s(X_i, X_j) > 0$ measures the similarity between a patient with characteristics X_i and another patient with characteristics X_j. The similarity function s will later be estimated from the data. We propose to define the probability that $Y_{n+1} = 1$, given the function s, by[3]

$$\hat{Y}_{n+1}^s = \frac{\sum_{i \leq n} s(X_i, X_{n+1}) Y_i}{\sum_{i \leq n} s(X_i, X_{n+1})}. \tag{11.1}$$

That is, the probability that Y_{n+1} be 1, i.e., that the procedure will succeed in the case of patient X_{n+1}, is taken to be the s-weight of all past successes, divided by the total s-weight of all past cases, successes and failures alike.

11.2.2. *Intuition*

Formula (11.1) is obviously a generalization of the notion of empirical frequency. Indeed, should the function s be constant, so that all observations are deemed equally relevant, (11.1) boils down to the

---

[3]For simplicity, we assume that s is strictly positive, so that the denominator of (11.1) never vanishes. But in certain situations, such as conditional frequencies, one may wish to allow zero similarity values. Leshno (2007), who extends the axiomatization to this case, and allows a sequence of similarity functions that are used lexicographically as in (11.1).

relative frequency of $\Upsilon_i = 1$ in the database. If, however, one defines $s(X_i, X_j)$ to be the indicator function of $X_i = X_j$ (allowing for the value 0 in case the vectors differ from each other, and setting it to be 1 in case they are equal), then formula (11.1) becomes the conditional relative frequency of $\Upsilon_i = 1$, that is, its relative frequency in the sub-database defined by X_{n+1}. It follows that (11.1) suggests a continuous spectrum between the two extremes: as opposed to conditional relative frequencies, it allows us to use the entire database. This is particularly useful in the medical example, where the database defined by $X_i = X_{n+1}$ may be very small or even empty. At the same time, it does not ignore the variables X, as does simple relative frequency over the entire database. Thus, formula (11.1) uses the entire database, but it still allows a differentiation among the cases depending on their relevance.

11.2.3. *Axiomatic derivations*

Formula (11.1) has been axiomatized in Gilboa, Lieberman, and Schmeidler (GLS, 2006) for the case discussed here, namely, the estimation of the probability of a single event, or, equivalently, of the distribution of a random variable with two possible values. Billot, Gilboa, Samet, and Schmeidler (2005) provide an axiomatization of the same formula in the case that the random variable under discussion may assume at least three distinct values.[4] Gilboa, Lieberman, and Schmeidler (2007) extend this axiomatization to the assessment of a density function of a continuous variable. While these axiomatic derivations differ in the framework, as well as in the assumptions regarding which data are observable, they all use a "combination" axiom, which states, roughly, that if a certain conclusion should be arrived at given two disjoint databases, then this conclusion should also be the result of the union of these databases.[5]

The basic logic of the axiom is as follows. Assume that Mary asks her physician whether the operation is more likely to succeed

---

[4]The axiomatization in Billot *et al.* (2005) relies on the fact that space of probability vectors has at least two dimensions, and it therefore cannot be adapted to the case or a binary variable (i.e., the one-dimensional case).

[5]Such an axiom was also used in Gilboa and Schmeidler (2001, 2003).

than not. Suppose that the physician says that "chances are" it will, meaning that success is more likely than failure. Mary decides to seek a second opinion. She consults another doctor, who has been working in a different hospital for many years. Let us assume that both doctors have the same inference algorithm, and that they only differ in the databases they have been exposed to. Suppose that the second doctor also thinks that success is more likely than failure. Should Mary ask the two to get together and exchange databases?

If Mary does not feel that the two doctors should exchange data, she implicitly believes that a conclusion, which has been warranted given each of the two databases, will also be warranted given their union. Casual observation suggests that people are generally reassured when they find that the advice of different experts converge. Hence we find the basic logic of combination axiom rather natural.

Gilboa and Schmeidler (2003) show that several well-known statistical techniques satisfy the combination axiom. These include likelihood ranking by empirical frequencies, kernel estimation of a density function, kernel classification, and maximum likelihood ranking of distributions. The fact that all these techniques obey the same principle, namely the combination axiom, may be taken as an indirect piece of evidence that the axiom is a good starting point for a theory of belief formation. Having said that, there are several important classes of applications where the combination axiom is unreasonable. We discuss these in Section 11.5.

11.3. Empirical Similarity

11.3.1. *Best fit*

As mentioned above, the probability obtained from similarity-weighted frequencies depends on the similarity function one employs. Which function should we use? In an attempt to avoid arbitrary choices, and in the hope of retaining objectivity, we define the *empirical similarity* to be the similarity function that best explains the database, assuming that we use it as in (11.1). To simplify the estimation problem, we choose a particular functional form, and thus render the problem parametric.

Specifically, we specify a vector of positive weights $w = (w_1, \ldots, w_m)$, consider the weighted Euclidean distance corresponding to it,

$$d_w(\bar{x}, \bar{x}') = \sqrt{\sum_{j \leq m} w_j (x_j - x'_j)^2}$$

and use as a similarity function the negative exponential of the weighted distance[6]:

$$s_w = e^{-d_w}.$$

Given the database, for each vector w, one may calculate, for each $i \leq n$, the value

$$\hat{Y}_i^{s_w} = \frac{\sum_{j \neq i} s_w(X_i, X_{n+1}) Y_j}{\sum_{j \neq i} s_w(X_i, X_{n+1})}. \tag{11.2}$$

The goodness of fit can be measured by

$$SSE(w) = \sum_{i \leq n} (\hat{Y}_i^{s_w} - Y_i)^2.$$

It then makes sense to ask, which vector w minimizes the sum of squared errors, $SSE(w)$. The minimizer of this function is then used in (11.1) to define the probability that $Y_{n+1} = 1$. When we use (11.1) in conjunction with the SSE-minimizing vector w, we obtain probability estimates that are "objective" in the same sense that classical statistics generally is: one may resort to statistical considerations for the choice of the general procedure, but no specific knowledge relating to the application is needed to implement the procedure. In Example 3, Mary needs to consult a statistician for the choice of the functional form of the similarity function, as well as for the measure of goodness of fit. But she does not need to consult a physician. Indeed, the resulting probability assessments are independent of the physician's subjective judgment or

---

[6]The exponential function was characterized in Billot, Gilboa, and Schmeidler (2005). They provide simple conditions on assessments, presumably generated by similarity-weighted averages, and show that these conditions are equivalent to the existence of a norm on $\mathbb{R}^m$ such that the similarity function between two vectors is the negative exponent of the norm of the difference between these vectors. The choice of the weighted Euclidean distances out of all possible norms is made for simplicity.

expertise. These assessments follow directly from the database, and they may serve an inexperienced doctor just as an experienced one.

11.3.2. *Comparison with the notion of IID random variables*

The textbook examples of classical statistics have to do with i.i.d. random variables, that is, random variables that are identically and independently distributed. These properties guarantee the laws of large numbers, the central limit theorem, and all the results that derive from these. Importantly, the laws of large numbers offer a natural definition of probability by relative frequency. Both the identical distribution and the statistical independence assumptions may be relaxed to a certain extent without undermining the laws of large numbers. But these assumptions cannot be dropped completely. If there is neither statistical independence, nor a certain weakening thereof, the size of the sample does not guarantee that the relative frequency would converge to anything at all, let alone to a number that can be interpreted as probability. Worse still, if the distributions of the random variable are not identical, or close to identical, it is not at all clear what is the "probability" that the relative frequency should converge to.

The mapping between the assumptions of i.i.d. observations in statistics and our model is not straightforward. Both notions, "identical" and "independent," are defined in probabilistic terms, whereas in our model no probability is assumed. Indeed, the model attempts to invest this concept with meaning, and therefore cannot assume it as primitive. Yet, our model suggests intuitive counterparts to these assumptions. Identicality of distribution is somewhat akin to identicality of the circumstances, i.e., of all observed variables x. One might argue that we cannot directly observe the distribution of the variable of interest, y, and if all observed variables x assume the same values, this is the closest that we can get to "identical distribution" in an empirical study. Stochastic independence of random variables is a rather strong condition, which implies the absence of causal relationships between the variables in question. This causal independence is implicit in our combination axiom.

Viewed from this perspective, our approach suggests that the independence assumption is, in a very vague sense, more fundamental than the identical distribution assumption. Our model drops the assumption that all observations are taken under identical conditions. In doing so, it foregoes the notion of a probability number that exists in some abstract or platonic sense, independent of our sample, and to which relative frequencies might converge. In our model, the probability of the event $y_t = 1$ occurring at observation t is a number that differs with t, depending of the x_t values. Our approach is therefore not an attempt to measure a quantity whose existence is external to the sample. Rather, our approach defines certain rules, by which the term "probability" can be used in an objective and well-defined way. And we argue that for this notion of "objective probability" one need not assume any notion of "identical repetition."

By contrast, our model heavily relies on the combination axiom, which may be viewed as retaining some notion of independence. Indeed, in the presence of causal dependencies neither our axioms nor our formula are very plausible. Thus, our approach suggests that in order to discuss objective probabilities, one need not resort to any notion of identical repetition, but one does need some notion of independence.

11.3.3. *Statistical theory*

Finding the parameters that minimize the sum of squared errors is an accepted way of selecting the "best" model. But in order to employ statistical inference techniques such as hypotheses tests and confidence intervals, one needs to couch the similarity-weighted frequency formula in a statistical model. In GLS (2006) we analyze the following statistical model.

For $t = 2, \ldots, n$, we assume that

$$Y_t^{s_w} = \frac{\sum_{i<t} s_w(X_i, X_t) Y_i}{\sum_{i<t} s_w(X_i, X_t)} + \varepsilon_t \qquad (11.3)$$

where $\varepsilon_t \sim N(0, \sigma^2)$, independently of the other variables.

In such a model it makes sense to ask whether the point estimates of the unknown parameters are significantly different from a pre-specified value, and in particular, from zero. In GLS (2006) we focus on maximum likelihood estimation of the parameters $(w_j)_j$, and we develop tests for such hypotheses. Observe, however, that the statistical model (11.3) differs from (11.2) in that, in the former, each observation is assumed to depend only on observations that precede it in the database. This assumes a certain order of the datapoints. When no such order is naturally given, such an order may be chosen at random. In this case, the statistical analysis should be refined to reflect this additional source of randomness.

11.4. Related Definitions

The problem of predicting a variable Y based on observable variables $X^1, \ldots, X^m$ is extensively studied in statistics, machine learning, and related fields. Among the numerous methods that have been suggested and used to solve such problems one may mention linear and non-linear regression, neural nets, linear and non-linear classifiers, k-nearest neighbor approaches (Fix and Hodges, 1951, 1952), kernel-based estimation (Akaike, 1954; Silverman, 1986; Scott, 1992), and others. Indeed, kernel-based methods are very similar to similarity-weighted frequencies.

When the variable of interest, Y, is binary (0 or 1), the approach that is probably the most popular in medical research is logistic regression. This method, introduced by McFadden (1974), uses the measurable variables in a linear formula, which is transformed in a monotonic way to a number between 0 and 1. This number can be computed for each given set of coefficients of the variables, and it is taken to be the predicted probability that the event in question will materialize in the next observation. Logistic regression finds the coefficients that result in the "best fit" that can be obtained between the predicted probabilities and the actual observations. This process may be viewed as a possible definition of the term "probability" in Example 3.

The probability numbers generated by logistic regression depend on the predicting variables in these observations. At the same time, these probabilities can be thought of as "objective," because they do not resort to a physician's subjective assessments. Rather, they rely solely on observed data. Admittedly, statistics' claim to objectivity is always qualified. Different choices that a statistician makes in the estimation process will result in different outcomes. Yet, logistic regression, as well as the empirical similarity method we propose are objective in the sense that they do not require the statistician to consult with a medical expert in order to generate predictions or estimate probabilities.

It is generally expected that each method for the assessment of probabilities will be more successful in certain applications and less in others. Finding how well each method performs in a particular type of application is an empirical question that is beyond the scope of this paper. At the theoretical level, we hold that the method offered here has several advantages over the alternatives mentioned above. First, the similarity-weighted frequencies are an intuitive extension of simple frequencies, and can thus be offered as a definition of the notion "probability." By contrast, some of the alternative methods might be useful predicting tools, but they are not intuitively interpretable as "probability." Second, our basic formula (11.1) appears to be the only one which is axiomatically derived. Third, not all alternative method have the statistical theory required for statistical inference (in particular, hypotheses testing). Finally, in contrast to, say, logistic regression, our approach does not assume any functional relationship between the observed variables and the predicted variable. In this sense, the empirical similarity approach is more "epistemically humble," in that it allows the data to determine not only the actual probability assessment, but also the way that this assessment is computed (via determination of the similarity function).

11.5. Limitations

Gilboa and Schmeidler (2003) contain an extensive discussion of the combination axiom, including an attempt to characterize several classes of counter-examples, that is, of situations in which the axiom appears unreasonable. We will not replicate this discussion here, but we will mention a few classes of problems in which one should not expect the axiom to hold, and, consequently, one should not use the methods that are restricted to satisfy it.

The first class of counter-examples involves theorizing, that is, inductive inferences from cases to underlying theories, and then deductive inference back from these theories to future cases. For example, when one uses past observations to learn the parameter of a coin, p, and then uses probability theory to make predictions regarding sequences of tosses of that coin using the best estimate of p, the combination axiom is unlikely to hold.

Similarly, if one believes that the observations are generated by a linear function, uses the data to estimate a linear regression formula, and then uses the estimated formula to make predictions, one is unlikely to satisfy the combination axiom. Indeed, it is evident from the basic similarity-weighted average formula (11.1) that it does not make any attempt to identify trends. To consider an extreme example, assume that $m = 1$ and that the database contains many points with $x = 1$, for which the relative frequency of $\Upsilon = 1$ was 0.1. There are also many points with $x = 2$ and a relative frequency of 0.2 for $\Upsilon = 1$, and so on for $x = 3$ and $x = 4$. Next assume that we are asked to make a prediction for a new case in which $x_t = 5$. It seems patently plausible to suggest that the probability that $\Upsilon_t = 1$ be 0.5. It makes sense to identify a trend, by which the probability of $\Upsilon = 1$ goes up with x. In fact, logistic regression seeks precisely this kind of relationships. But the similarity-weighted average method will fail to produce a value that is outside the observed range of $[0.1, 0.4]$. It is important to recall that the similarity-weighted average does not seek trends, and, more generally, does not attempt to theorize about the data. It engages only in case-to-case induction, but not in case-to-rule induction coupled with rule-to-case deduction.

The second class of counter-examples to the combination axiom involves changes in the similarity function. In particular, if the probabilistic reasoner perform so-called second-order induction (Gilboa and Schmeidler, 2001), that is, if she learns which similarity function should be used to learn from past cases about the future, then the combination axiom is again an unlikely principle. Learning of the similarity function may involve qualitative insights, if, say, a physician, after examining a large database, says, "and it suddenly dawned on me that the common feature to all these cases was...." But such learning may also be quantitative and follow from purely statistical reasons: with the accumulation of more data, the need to use more remote observations is reduced, and one may obtain better assessments by restricting attention to close cases. More generally, learning of the similarity function, that is, any process that adapts the similarity function as a function of the data should be expected to violate the combination axiom.

Obviously, our approach faces a difficulty here. On the one hand, we justify the similarity-weighted average based on the combination axiom, which is violated when the similarity function is learnt itself. On the other hand, the notion of empirical similarity is precisely one of learning the similarity function. Thus, a probabilistic reasoner who would follow our advice to compute the empirical similarity will thereby violate our recommendation to satisfy the combination axiom, and will consequently not be sure that the similarity-weighted average method makes sense to begin with.

One possible resolution is to assume that the similarity function is updated only at certain periods, and between each two such consecutive periods the combination axiom holds. We conjecture that the axiomatizations mentioned above would have approximate counterparts with bounded databases, which would allow the use of formula (11.1) between updating periods. Admittedly, this resolution is rather awkward, and more elegant axiomatizations of similarity-weighted frequencies with a similarity function that is learned are called for.

11.6. Future Directions

The discussion above suggests several directions in which one may extend the empirical similarity approach to the definition of probabilities. First, one should have more satisfactory theories, allowing the similarity function to be learnt and refined in the process, in a way that parallels the choice of a kernel function in non-parametric estimation. (See Silverman, 1986.) Second, analogical, case-based reasoning which is incorporated in similarity-weighted frequencies should be combined with deductive, rule-based reasoning. For instance, one may extend the similarity-weighted formula so that each observation (x_i, y_i) will give support not only to the value observed y_i, but also to various functions $f(x)$ that the observation approximately satisfies, i.e., to functions f such that $y_i \approx f(x_i)$. Thus, if all points observed lie near the graph of a certain function f, this function will gain support from each of the observations, and will thus offer itself as a natural generalization of the cases to a rule. The class of functions f one allows into this analysis has to be limited to make the analysis meaningful (and to avoid "overfitting" by finding a function that matches all the data precisely, but that does poorly in prediction). It is a challenge to find natural limitations on the class of functions that will retain a claim to objectivity.

Another extension might combine Bayesian reasoning with similarity-weighted averages. One may start with a Bayesian network (see Pearl, 1986), reflecting possible dependencies among variables, and assess probabilities on each edge in the network by the empirical similarity technique. These probability numbers will then be used by the Bayesian network to generate probabilistic predictions that make full use of the power of Bayesian reasoning. In this case, again, part of the challenge is to find ways to develop Bayesian networks that will be "objective."

However, it is not at all obvious that these and other extensions, or, in fact, any other approach, can come up with a reasonable definition of objective probability in Example 4 above. The main difficulty appears to be the causal dependence between cases. When cases are causally independent, an observation of one case may teach us something about the likelihood of the occurrence of an event in another case. As long as the latter is a fixed target, one may have a hope that, with sufficiently many observations, one may learn more about this likelihood, to a degree that it can be quantified in a way that most people would agree on. But when causal dependence is present, an observation of a particular case not only reveals information about another case, it also changes its likelihood. Thus, we are after a moving target, and find it difficult to separate the process of observation from the process observed.

Idiosyncrasy of cases and causal relationships do not allow us to define objective probabilities by empirical frequencies. These two phenomena also make it difficult to assign observable meaning to counterfactual propositions. It is relatively easy to understand what it meant by the statement "If I were to drop this glass, it would break." There are many cases of practically identical glasses being held and being dropped, and since these cases are assumed to be causally independent, this counterfactual statement has a verifiable meaning. Correspondingly, one can design an experiment that will be viewed as a test of this statement. By contrast, it is much harder to judge the veracity of the counterfactual, "Had Hitler crossed the channel, he would have won the war." First, historical cases of war are never identical. Second, they are seldom causally independent. Our approach suggests that a theory of counterfactuals might more easily deal with the first problem than with the second.

At present, we are not convinced that the notion of objective probability can be meaningfully defined in situations involving intricate causal relationships between observations. However, it seems obvious that when causal independence holds, the definitions discussed here can be greatly improved upon.

References

Akaike, H. (1954), "An Approximation to the Density Function", *Annals of the Institute of Statistical Mathematics*, 6: 127–132.

Bernoulli, J. (1713), Ars Conjectandi. 1713.

Billot, A., I. Gilboa, D. Samet and D. Schmeidler (2005), "Probabilities as Similarity-Weighted Frequencies", *Econometrica*, 73: 1125–1136.

Billot, A., I. Gilboa and D. Schmeidler (2005), "Exponential Similarity", mimeo. Published in *Mathematical Social Sciences*, 55 (2008): 107–115. Reprinted as Chapter 10 of this volume.

de Finetti, B. (1937), "La Prevision: Ses Lois Logiques, Ses Sources Subjectives", *Annales de l'Institute Henri Poincare*, 7: 1–68.

Ellsberg, D. (1961), "Risk, Ambiguity and the Savage Axioms", *Quarterly Journal of Economics*, 75: 643–669.

Fix, E. and J. Hodges (1951), "Discriminatory Analysis. Nonparametric Discrimination: Consistency Properties". Technical Report 4, Project Number 21-49-004, USAF School of Aviation Medicine, Randolph Field, TX.

— (1952), "Discriminatory Analysis: Small Sample Performance". Technical Report 21-49-004, USAF School of Aviation Medicine, Randolph Field, TX.

Gilboa, I., O. Lieberman and D. Schmeidler (2006), "Empirical Similarity", *Review of Economics and Statistics*, forthcoming. Published in vol. **88** (2000): 433–444. Reprinted as Chapter 9 of this volume.

— (2007), "Similarity Based Prediction", mimeo. Published as "A Similarity-Based Approach to Prediction", *The Journal of Econometrics*, **162** (2011): 124–131.

Gilboa, I., A. Postlewaite and D. Schmeidler (2006), "Rationality of Belief", mimeo. *Synthese*, forthcoming.

Gilboa, I. and D. Schmeidler (1995), "Case-Based Decision Theory", *Quarterly Journal of Economics*, **110**: 605–639. Reprinted as Chapter 1 of this volume.

— (2001), *A Theory of Case-Based Decisions*, Cambridge: Cambridge University Press.

— (2003), "Inductive Inference: An Axiomatic Approach", *Econometrica*, **71**: 1–26. Reprinted as Chapter 4 of this volume.

Hume, D. (1748), *Enquiry into the Human Understanding*, Oxford: Clarendon Press.

Leshno, J. (2007), "Similarity-Weighted Frequencies with Zero Values", mimeo.

Machina, M. and D. Schmeidler (1992), "A More Robust Definition of Subjective Probability", *Econometrica*, **60**: 745–780.

McFadden, D. (1974), "Conditional Logit Analysis of Qualitative Choice Behavior". In *Frontiers in Econometrics*, ed. P. Zarembka, 105–142, New York: Academic Press.

Pearl, J. (1986), "Fusion, Propagation, and Structuring in Belief Networks", *Artificial Intelligence*, **29**: 241–288.

Pearl, J. (2000), *Causality*, Cambridge University Press.

Ramsey, F. P. (1931), "Truth and Probability", in *The Foundation of Mathematics and Other Logical Essays*, New York: Harcourt, Brace and Co.

Rostek, M. (2006), "Quantile Maximization in Decision Theory", mimeo.

Savage, L. J. (1954), *The Foundations of Statistics*, New York: John Wiley and Sons. (Second addition in 1972, Dover)

Scott, D. W. (1992), *Multivariate Density Estimation: Theory, Practice, and Visualization*, New York: John Wiley and Sons.

Shafer, G. (1986), "Savage Revisited", *Statistical Science*, **1**: 463–486.

Silverman, B. W. (1986), *Density Estimation for Statistics and Data Analysis*, London and New York: Chapman and Hall.

Chapter 12

Simplicity and Likelihood: An Axiomatic Approach*

Itzhak Gilboa[†] and David Schmeidler[‡]

Reprinted from *Journal of Economic Theory*, 145 (2010): 1757–1775.

We suggest a model in which theories are ranked given various databases. Certain axioms on such rankings imply a numerical representation that is the sum of the log-likelihood of the theory and a fixed number for each theory, which may be interpreted as a measure of its complexity. This additive combination of log-likelihood and a measure of complexity generalizes both the Akaike Information Criterion and the Minimum Description Length criterion, which are well known in statistics and in machine learning, respectively. The axiomatic approach is suggested as a way to analyze such theory-selection criteria and judge their reasonability based on finite databases.

JEL Classification Numbers: C1, D8

Keywords: Maximum likelihood; Simplicity; Model selection; Akaike information criterion; Minimum description length; Axioms.

---

*We thank Yoav Binyamini, Offer Lieberman, two anonymous referees and the associate editor for comments and suggestions. This project was supported by the Pinhas Sapir Center for Development and Israel Science Foundation Grants Nos. 975/03 and 355/06.
[†]HEC, Paris, Tel-Aviv University, and Cowles Foundation, Yale University. tzachigilboa@gmail.com
[‡]Tel-Aviv University and The Ohio State University. schmeid@tau.ac.il

12.1. Introduction

The selection of a theory based on observations is a fundamental problem that cuts across several disciplines. Finding the "right" way to select theories given evidence is at the heart of philosophy of science, statistics, and machine learning. It is also highly relevant to rational models of learning, trying to capture the way that rational agents can make sense of the data available to them.

Two fundamental criteria for the selection of theories are simplicity and goodness of fit. The preference for simple theories is well known, and is often attributed to William of Occam (see Russell, 1946). While the notion of simplicity is partly subjective and depends on language,[1] it is often surprising how much agreement one finds between the simplicity judgment of different people. For example, most people tend to agree that, other things being equal, a theory with fewer parameters is simpler than a theory with more parameters, or that a theory with a shorter description is simpler than a theory with a longer one. Whereas such claims depend on an agreement about a language, or a set of languages within which simplicity is measured, they do not seem to be vacuous statements. The suggestion that people tend to prefer simple theories to more complex ones can therefore be a meaningful empirical claim.

However, simplicity can only serve as an a priori argument for or against certain theories. How well a theory performs in explaining observed data should certainly also factor into our considerations in selecting theories. Sometimes, one may categorize theories dichotomously into theories that fit the data as opposed to theories that are refuted by the data, and then choose the simplest theory among the former. But in most problems in science, as well as in everyday life, theories are never categorically refuted. There is typically room for a measurement error, or, more generally, for probabilistic prediction. Therefore, a theory typically cannot be refuted by observations. Instead, theories can be ranked according to their goodness of fit, namely, the extent to which they match observations. In particular, the likelihood principle

---

[1]See Goodman, 1955 and Sober, 1975.

suggests to rank theories according to their likelihood function, that is, the a priori probability that the theory used to assign to the observed data before these data were indeed observed.

Viewed from a statistical point of view, the likelihood principle is a fundamental idea that neatly captures the notion of "goodness of fit" while relying on objective data alone. Choosing the theory that maximizes the conditional probability of the actually observed sample does not rely on any subjective a priori preferences, hunches, or intuitions of the reasoner. But for that very reason, the maximum likelihood principle cannot express preferences for simplicity. Due to this limitation, the applicability of this criterion is restricted to set-ups in which the set of possible theories is restricted a priori to a given class, within which complexity considerations might be ignored. When no such a priori restriction is available, the maximum likelihood principle is insufficient. More explicitly, if one considers all conceivable theories, one will always be able to find a theory that matches the observations perfectly. Such a theory will obtain the maximum conceivable likelihood value of 1, but it is likely to be "overfitting" the data. We tend not to trust a theory that matches the data perfectly if it appears very complex. Thus, maximum likelihood does not suffice to describe the totality of considerations that enter the theory selection process.[2]

We are therefore led to the conclusion that a reasonable criterion for the selection of theories based on observations has to take into account both the likelihood of a theory, or some other measure of goodness of fit, and its simplicity, or some other a priori preference for some theories versus others.[3] Indeed, combinations of likelihood and some measure of complexity are well known in the literature on statistics and

---

[2] See Gilboa and Samuelson (2009) who suggest an evolutionary argument for the preference for simplicity.

[3] Another relevant crietrion is the theory's generality. In this paper we ignore the more involved three-way trade-off between goodness of fit, simplicity, and generality. We will only mention in passing that if we "normalize" the theories under comparison so that they have the same scope of applicability, preference for generality can be derived from preference for simplicity. Specifically, if theory *a* is less general than theory *b*, one may augment *a*, for instance by using *b* when *a* was not defined. The resulting theory would be more cumbersome, and may be less preferred than *b* based on simplicity considerations.

on machine learning. Specifically, linear combinations of the logarithm of the likelihood function and a complexity measure appeared in both literatures. Akaike Information Criterion (AIC, Akaike, 1974), suggests that, when comparing different statistical models, one adopts a model a that obtains the highest value for

$$\log\left(L(a)\right) - 2k$$

where $L(a)$ is the likelihood function of a, and k is the number of parameters used in model a.[4]

The machine learning literature often adopts Kolmogorov's complexity measure (Kolmogorov, 1963, 1965, Chaitin, 1966), which suggests that the complexity of a theory be measured by the minimal length of a program (say, a description of a Turing machine) that can be used to generate the theory's predictions. Solomonoff (1964) suggested to use such a complexity measure as a basis for a theory of philosophy of science. Related concepts are the Minimal Message Length (MML, Wallace and Boulton, 1968) and the Minimum Description Length (MDL) of a theory. Recent applications often trade-off a theory's likelihood with its simplicity by considering criteria of the form

$$\log\left(L(a)\right) - MDL$$

where MDL is the minimum description length. (See Wallace and Dowe, 1999, and Wallace, 2005 for a more recent survey.)

Clearly, there could be many ways to trade-off a theory's likelihood with its complexity. Indeed, Schwartz Information Criterion (SIC, also known as BIC), suggests that the number of parameters be divided by the logarithm of the number of observations. How should we judge among such criteria? How should we trade off likelihood and complexity?

The present paper addresses this question in an axiomatic way. Our axiomatic approach does not presuppose particular measures of

---

[4]Observe that, as the sample size, n, grows to ∞, the expression above would typically tend to $-\infty$ for all models. One often divides this expression by n to obtain limits that can be meaningfully compared. Division by n obviously does not alter the ordinal ranking of models.

goodness of fit or of likelihood, let alone a particular combination thereof. Rather, we consider an abstract problem in which observations and theories are formal entities that are a priori unrelated, and are also devoid of any explicit content or mathematical structure. In particular, no statistical model is a priori assumed, and no likelihood functions are given. We only assume that a reasoner can rank theories given various databases of observations. Such rankings are modeled as weak orders (binary relations that are complete and transitive), and interpreted as "at least as plausible as" relations. We formulate certain conditions, or "axioms" on these weak orders, which can be viewed as notions of internal consistency: the axioms relate the rankings of theories given different databases of observations. The axioms do not restrict the inferences the reasoner may draw from any particular database, but they do exclude certain patterns of plausibility rankings given different databases. The main result of this paper is that the axioms imply the existence of a statistical model and a constant for each theory, such that for every database, theories are ranked according to the sum of the constant and their log-likelihood function.

Formally, theories are elements of a set $\mathbb{A}$ and observations — of a set $\mathbb{X}$. Neither set is endowed with any mathematical structure, and the two are a priori unrelated. Yet, the intended interpretation of the sets is that elements in $\mathbb{A}$ are distributions or densities on $\mathbb{X}$ with full support. As mentioned above, our formal result may well extend beyond this application. But it is useful to bear it in mind when judging the axioms.

A database I is a function $I : \mathbb{X} \to \mathbb{Z}_+$, (where $\mathbb{Z}_+$ stands for the non-negative integers) with $\sum_{x \in \mathbb{X}} I(x) < \infty$, and $I(x)$ is interpreted as the number of times an observation x has appeared in the database described by I. We assume that, for each such database I, the reasoner has a ranking over theories, $\succsim_I \subset \mathbb{A} \times \mathbb{A}$, where $a \succsim_I b$ is interpreted as "given the observations in database I, theory a is at least as plausible as theory b." When applied to the empty database, $I = 0$, $\succsim_I$ would reflect the reasoner's a priori ranking over the theories, in the absence of any data.

We impose several axioms on the collection of rankings $\{\succsim_I\}_I$, that imply the following representation: for every theory a there exists

$w(a) \in \mathbb{R}$ and, for every observation x, also a number $v(a,x) \in \mathbb{R}$, such that, for any database I, and any two theories $a, b \in \mathbb{A}$, $a \succsim_I b$ iff

$$w(a) + \sum_{x \in \mathbb{X}} I(x)v(a,x) \geq w(b) + \sum_{x \in \mathbb{X}} I(x)v(b,x). \qquad (12.1)$$

In this representation, one may interpret $v(a,x)$ as the log of $\Pr(x|a)$, and then $\sum_{x \in \mathbb{X}} I(x)v(a,x)$ is simply the log-likelihood of the theory a given the database I. If the theories are indeed given as distributions or densities over $\mathbb{X}$, it is natural to assume that the number $v(a,x)$, derived from the reasoner's rankings, will indeed coincide with the logarithm of the probability (or density) of observation x given theory a. However, in a formal model where theories have this additional structure, one would also need an additional assumption that would guarantee this equality.

The constant $w(a)$ reflects an a priori bias for the theory a, and it can be interpreted as a measure of the theory's simplicity, or some other subjective criterion for theory selection. Specifically, if there are finitely many theories, and the reasoner has an a priori subjective probability $p(a)$ that theory a actually governs the data generating process, then the a posteriori ranking of the theories given database I will follow (12.1) with $w(a) = \log(p(a))$ and $v(a,x) = \Pr(x|a)$ as above. While the Bayesian interpretation is not our preferred one (see the Discussion below), it is compatible with our axioms in many situations.

The axiomatic treatment may serve as a reason to select additive likelihood-complexity trade-offs such as AIC and MDL, and perhaps to prefer them over other criteria that do not satisfy the axioms. It also serves to clarify the commonalities among simplicity-based criteria and the Bayesian approach to model selection.

This paper may be viewed as a contribution to the axiomatic analysis of statistical techniques. In Gilboa and Schmeidler (2003) we provided an axiomatization of kernel estimation of density functions, kernel classification, as well as of maximum likelihood rankings.[5] Billot, Gilboa, Samet, and Schmeidler (2005) and Gilboa, Lieberman, and

---

[5]As explained below, the present paper heavily relies on the results in Gilboa and Schmeidler (2003).

Schmeidler (2006) axiomatize kernel estimation of probabilities. One rationale for these papers is the attempt to ground statistical and machine learning methods in axiomatic derivations. The axiomatic approach offers consistency criteria that may help one select theories based on their abstract properties. Such criteria might be of interest especially when finite samples are concerned, and asymptotic behavior may not suffice as the sole guide for the selection of theories.

The rest of this paper is organized as follows. The next section describes the model and the result. The following one is devoted to a general discussion. Proofs and related analysis are to be found in an appendix.

12.2. Model and Result

Let $\mathbb{X}$ be the set of (types of) *observations*. The set of *databases* is defined as

$$\mathbb{D} \equiv \left\{ I \mid I : \mathbb{X} \to \mathbb{Z}_+, \quad \sum_{x \in \mathbb{X}} I(x) < \infty \right\}.$$

A database $I \in \mathbb{D}$ is interpreted as a counter vector, where $I(x)$ counts how many observations of type x appear in the database represented by I.

Algebraic operations on $\mathbb{D}$ are performed pointwise. Thus, for $I, J \in \mathbb{D}$ and $k \geq 0$, $I + J \in \mathbb{D}$, and $kI \in \mathbb{D}$ are well-defined. Similarly, the inequality $I \geq J$ is read pointwise.

Let $\mathbb{A}$ be the set of *theories*. For $I \in \mathbb{D}$, $\succsim_I \subset \mathbb{A} \times \mathbb{A}$ is a binary relation on theories, where $a \succsim_I b$ is interpreted as "given the database I, theory a is at least as plausible as theory b." The asymmetric and symmetric parts of $\succsim_I$, $\succ_I$ and $\sim_I$, respectively, are defined as usual.

We now turn to describe our conditions. The first three, A1–A3, are "axioms" on the plausibility rankings. They are supposed to suggest appealing properties of the theory-selection criterion.[6] The last two are richness conditions. These conditions have no claim to suggest

---

[6]Thus, our main interpretation is normative. Alternatively, the axioms can also be interpreted descriptively.

desirable properties of such criteria. Rather, they are conditions on the set-up of the model needed for our result to hold. For simplicity of notation, we refer to the richness conditions as "A4" and "A5," despite the fact that they are not proposed as "axioms." Correspondingly, axioms A1–A3 are also necessary for the representation (12.1), while A4-A5 are not. Weakenings or alternatives to A4 and A5 that will give rise to the representation (12.1) will certainly be of interest. (See the discussion in the appendix.)

Formally, the only relation between the sets $\mathbb{A}$ and $\mathbb{X}$ is provided indirectly by the set of rankings $\{\succsim_I\}_{I\in\mathbb{D}}$. However, to fix ideas we ask the reader to bear in mind the classical statistical set-up, in which theories are simply distributions (or densities) over observations. We briefly comment on more general set-ups.

A1 Order: For every $I \in \mathbb{D}$, $\succsim_I$ is complete and transitive on $\mathbb{A}$.

A1 is a standard axiom in decision theory. Transitivity is typically considered to be a basic axiom of rationality: if theory a is at least as plausible as theory b, and the latter – at least as plausible as theory c, one would find it hard to argue that c is more plausible than a.

Completeness requires that any two theories can be compared for their plausibility, given any database. Typically, completeness is justified by necessity: once a database is given, the reasoner is asked to make some choice regarding which theory she will use for prediction. The completeness axiom requires that this choice be brought forth and explicitly modeled.

When completeness is applied to a database I consisting of one observation only (that is, $I(x) = 1$ and $I(y) = 0$ for $y \neq x$), it requires that the theories be "about" the observations. In the benchmark case, where theories are distributions over observations, a single observation x naturally induces a ranking of theories based on an a priori bias and the likelihood function. More generally, the completeness axiom still requires that, given a single observation, the reasoner will have a meaningful ranking of the theories. In particular, if the theories are about patterns of observations, rather than about single ones, completeness may not hold.

A2 Recombination: Suppose that $I, J, K, L \in \mathbb{D}$ are such that $I + J = K + L$. Then there are no $a, b \in \mathbb{A}$, for which $a \succsim_I b$ and $a \succsim_J b$, but $b \succsim_K a$ and $b \succ_L a$.

The essence of this axiom is that evidence gathered from observations is simply accumulated, and that there is no additional learning from the co-occurrence of different observations. Considering a violation of the axiom might serve to explain the type of learning that it rules out.[7] Suppose that a is a common disease, and that b is a rather rare disease. Disease a might manifest itself in symptom x or y, but it is very rare to have both symptoms present. In fact, when both x and y are observed, it is more likely to be disease b rather than a. Next assume that database I consists of two consecutive observations (for the same patient) of symptom x, i.e., $I(x) = 2$, $I(y) = 0$, and database J — of two observations of symptom y, $J(x) = 0$, $J(y) = 2$. Let $K = L$ with $K(x) = K(y) = 1$. Then $I + J = K + L$. Yet, according to our assumptions, each of I, J renders a more likely than b, whereas each of K, L suggests the opposite ranking, in violation of A2.

Thus, the recombination axiom requires that the learning from observations be done case-by-case, where no general picture is allowed to emerge from the totality of the observations. This will be satisfied if the observations are statistically independent. Our model does not assume any probabilistic model, and does not allow us to define independence in the standard statistics sense. But A2 may be viewed as a type of independence, stated in the language of rankings given databases.

The recombination axiom is a version of the "combination" axiom in Gilboa and Schmeidler (2003). The latter implied that $a \succsim_I b$ and $a \succsim_J b$ would necessitate $a \succsim_{I+J} b$. That is, a conclusion (theory a is at least as plausible as theory b) that is warranted given two disjoint databases separately should also be warranted given their union (modeled as $I + J$). This axiom is satisfied by maximum likelihood rankings. But it may be too restrictive when complexity considerations

---

[7]The following example is based on a suggestion of one of the referees.

290 Case-Based Predictions

are introduced. Specifically, a simple theory a may be considered more likely than a more complex theory b given each of the databases I and J, separately, even if b fits the data in each database better. But when the two databases are considered in conjunction, the better fit provided by b may overwhelm the complexity considerations, rendering b more plausible than a given $I + J$. The recombination axiom we impose here considers a fixed set of observations, given by $I + J = K + L$. The axiom states that the same set of observations cannot be partitioned twice into two disjoint databases, such that in one partition both databases render a at least as plausible as b, and in the other — one renders b at least as plausible as a, and the other — strictly more plausible.

A3 Archimedean Axiom: Assume that $I, J \in \mathbb{D}$ and $a, b \in \mathbb{A}$ satisfy $b \succ_J a$ and $a \succsim_{J+I} b$. Then for every $K \in \mathbb{D}$ there exists $l \in N$ such that $a \succ_{K+lI} b$.

The antecedent of the Archimedean axiom assumes that, complexity considerations aside, database I renders a more likely than b: starting from $b \succ_J a$, the addition of the observations in I reverses the plausibility ranking. Since complexity considerations and other a priori biases for one theory over another do not change when we compare the database J to the database $J + I$, the switch from b to a can only be attributed to the fact that theory a provides a better fit to the observations in I than does theory b. In this case, the axiom demands that, for every database K, the addition of sufficiently many replicas of database I should make a more plausible than b. That is, if a fits the data I better than does b, and we observe more and more databases identical to I, eventually we should prefer theory a to theory b, even if initial data (embodied in K) and complexity considerations originally gave preference to b.

If each theory is a distribution (or a density) over the observations, and the observations are i.i.d., the Archimedean axiom will be satisfied as long as the distributions have full support, that is, as long as no observation can completely refute a theory, that is, drive its likelihood function to zero. More generally, if one already assumes A2, the Archimedean axiom requires that the evidence gathered from different databases will always be comparable.

The last two axioms, or conditions, are not justified on a priori grounds. As mentioned above, they are used only because of the mathematical necessity and may well be weakened or replaced by other axioms. Having said that, we do not find them conceptually objectionable.

The first states that, for every list of four theories and any database, there is a possible continuation of the database that would rank the theories according to the order in the list.

A4 Diversity: For every list (a, b, c, d) of distinct elements of $\mathbb{A}$ and every $J \in \mathbb{D}$, there exists $I \in \mathbb{D}$, $I \geq J$ such that $a \succ_I b \succ_I c \succ_I d$. If $|\mathbb{A}| < 4$, the same applies to any permutation of the elements of $\mathbb{A}$.

A4 excludes, for instance, a situation in which one theory is always more plausible than another, regardless of the database. In particular, it excludes from the analysis theories that are tautologically true or tautologically false. More importantly, it does not allow us to include in $\mathbb{A}$ two theories a, b such that a is a generalization of b. Indeed, if each theory is a distribution (or a density) over the observations $\mathbb{X}$, one theory cannot be a generalization of another, and it cannot be always more plausible than another.

The diversity condition also imposes a certain richness condition on the observations. For instance, assume that the observations are the tosses of a coin, $\mathbb{X} = \{0, 1\}$. Suppose that the reasoner believes that the tosses are i.i.d., but does not know the parameter of the coin, so that the set of theories is $\mathbb{A} = [0, 1]$. In this case A4 will not hold, since the likelihood function over $[0, 1]$ has to be single-peaked. In this example there is a continuum of theories, but there aren't sufficiently many observations to allow us to differentiate among them in the sense of A4. More generally, this condition requires that the set of observations be sufficiently rich. If the theories are given by distribution (or density) functions, the diversity condition will be shown to require that the log-distribution (or log-density) of a theory not be weakly dominated by an affine combination of (up to) three other log-distributions (or log-density).

The reason that this condition is required to hold for every four theories but not for more is technical and will be clear in the course of the proof. It will also be clear, as explained in Gilboa and Schmeidler (2003), that this condition can be somewhat weakened at the expense of simplicity. In that paper we also show why this axiom is needed: without it, one can construct counter-examples to the representation we seek. The same counter-examples can be used in the present context.

The second richness condition, which is our last condition, requires that for every database and every three theories there is a continuation of the database that renders the three theories equally plausible.

A5 Solvability: For every $\{a, b, c\} \subset A$, and every $J \in \mathbb{D}$, there exists $I \in \mathbb{D}, I \geq J$ such that $a \sim_I b \sim_I c$.

The basic import of A5 is that for any three theories there be observations relative to which the "plausibility rankings" are in rational proportions. A counter-example in the appendix shows why this axiom is necessary, using two observations and log-likelihood functions whose ratios are irrational. We view A5 as a richness condition because, like A4, it takes as given a subset of theories, and requires that there be at least one database that induces a particular ranking over these theories.

We can finally state our main result.

Theorem 1:   *Let there be given* $\mathbb{X}$, A, *and* $\{\succsim_I\}_{I \in \mathbb{D}}$ *as above. Assume that* $\{\succsim_I\}_{I \in \mathbb{D}}$ *satisfy A1–A5. Then there is a matrix* $v : A \times \mathbb{X} \to \mathbb{R}$ *and a vector* $w : A \to \mathbb{R}$ *such that:*

$$(*) \quad \begin{cases} \text{for every } I \in \mathbb{D} \text{ and every } a, b \in A, \\[4pt] a \succsim_I b \quad \text{iff} \quad w(a) + \sum_{x \in \mathbb{X}} I(x)v(a, x) \geq w(b) \\[4pt] \qquad\qquad\qquad\qquad + \sum_{x \in \mathbb{X}} I(x)v(b, x), \end{cases}$$

Furthermore, in this case the matrix v *and the vector* w *are unique in the following sense:* (v, w) *and* (u, y) *both satisfy* $(*)$ *iff there are a scalar* $\lambda > 0$, *a matrix* $\beta : A \times \mathbb{X} \to \mathbb{R}$ *with identical rows (i.e., with constant columns) and a number* δ *such that* $u = \lambda v + \beta$ *and* $y = \lambda w + \delta$.

Observe that, in the tradition of axiomatizations in decision theory, the representation theorem above only suggests a possible representation. A reasoner whose rankings $\{\succsim_I\}_{I \in \mathbb{D}}$ satisfy our axioms can be thought of *as if* she had a likelihood function (whose logarithm is given by v) and a simplicity measure (given by w) such that she prefers theories with higher values of the sum of the log-likelihood and the simplicity measure. If it so happens that the theories involved are a priori given by a statistical model, so that a likelihood function $l(a|x)$ exists for $a \in \mathbb{A}$ and $x \in \mathbb{X}$, it does not follow that $v(a, x) = \log(l(a|x))$. Indeed, since the axioms make no reference to the likelihood function $l(a|x)$, such a conclusion would be impossible. To derive it, one has to impose additional axioms, relating the relations $\{\succsim_I\}_{I \in \mathbb{D}}$ to the supposedly given $\{l(a|x)\}_x$.

The situation is akin to Savage's derivation of a Bayesian prior: Savage's (1954) axioms imply that there exists a probability measure such that the decision maker behaves in accordance with it (via the expected utility formula). If we observe a decision maker who faces a roulette wheel with given, objective probabilities, it stands to reason that her subjective prior would coincide with the measure governing the wheel's behavior. Yet such a conclusion requires an additional assumption and does not follow from the representation theorem itself.

Much of the appeal of Savage's theorem is in that it does not assume an objectively given probability measure, but derives one from preferences. Thus he defines subjective probabilities where no objective probabilities are given. By the same token, our theorem can be said to derive a statistical model (or a likelihood function) even when such a model is not a priori given.

12.3. Discussion

12.3.1. *The recombination axiom*

The statement of the recombination axiom (A2) might bring to mind Simpson's paradox (Simpson, 1951), which appears to constitute a violation of the axiom. Consider, for example, the famous Berkeley Sex Bias Case, in which the percentage of men admitted to graduate school is higher than the percentage of women admitted, while the converse is true for each department separately. (Historically, the converse was true in *almost* all departments.[8]) For simplicity, assume that there are only

---

[8]See http://en.wikipedia.org/wiki/Simpson's_paradox#_note-3.

two departments. In this case, splitting the database by departments would yield two databases (say, I and J) in each of which women appear to be favored to men. By contrast, splitting the same database randomly would yield two other databases (say, K and L), each supporting the opposite conclusion.

However, this application of our model is inappropriate, because the single observations are not directly related to the theories discussed. In fact, in this example even the completeness axiom is problematic: given a single case, be it of a man or a woman, admitted or not, it is not at all obvious how one should rank two theories such as "women are favored at admission" vs. "men are favored at admission." These theories are about comparisons of *sets* of observations (to be precise, comparisons of percentages of admitted applicants within two sub-populations), and they do not directly say anything about a particular observation.

To deal with the Berkeley Sex Bias Case, one would have to consider "observations" that are directly relevant to the theories. For example, an observation might be a pair of candidates that are similar in all respects apart from their gender, one of whom was admitted by a certain department and the other — denied admission by the same department. Such an observation would indeed constitute a direct evidence of unequal treatment of the genders. But it is easy to see that Simpson's paradox cannot be replicated using such observations of disjoint pairs, as the paradox relies on unequal proportions of women and men applying to different departments.

Similar difficulties with the recombination axiom might arise when one considers various theories that are not about specific observations, but rather about patterns of observations. For instance, if one is to judge whether a sequence of observations is random, one may easily construct counter-examples to the recombination axiom. Again, in such examples the theories discussed do not say anything about specific observations, only about patterns thereof. This is highlighted by similar difficulties with the completeness axiom applied to databases of single observations. Having but a single observation, one cannot rationally judge whether it comes from a random sequence or not.

To conclude, our model should only be applied to theories and observations that are directly related, in the sense that every theory is relevant to every observation. Differently put, every single observation should have meaningful implications about the plausibility of the theories. When attention is restricted to such applications, the completeness axiom is not too demanding, and the recombination axiom appears reasonable.

12.3.2. *Methods of classical statistics*

It appears that maximum likelihood is a reasonable criterion only when the set of theories is a priori restricted in one way or another. For instance, one may face a regression problem and consider only linear or quadratic theories. But in this case the set of theories under discussion is subjectively chosen. That is, the model does not purport to explain why the particular set of theories — say, linear — was chosen to begin with. Assuming the model as given, likelihood maximization offers an objective ranking of theories. But the choice of the model itself remains subjective, and sometimes arbitrary.

Statistical theory offers a variety of tools to cope with the problem of overfitting data as a result of likelihood maximization. The trade-off between a good fit and the theory's complexity is familiar from model selection criteria in parametric set-ups (such as adjusted R^2, LASSO, Ridge Regression, and others) as well as in non-parametric set-ups (Akaike Information Criterion, BIC, etc.). The present paper addresses this question axiomatically, describing an inductive learning process that does not impose arbitrary restrictions on the set of theories.

12.3.3. *Bayesian analysis*

As mentioned in the Introduction, the ranking by (12.1) can be interpreted as a Bayesian ranking where $w(a)$ is taken to be the logarithm of theory a's prior probability. However, several distinctions should be borne in mind. First, the numbers $w(a)$ in our set-up are not unique. It is readily seen that they can all be multiplied by a positive constant (alongside the numbers $v(a, x)$) without changing the rankings in

(12.1). Hence, a reasoner who satisfies our axioms can be viewed as Bayesian, but her Bayesian beliefs are not uniquely determined by her rankings $\{\succsim_I\}_{I \in \mathbb{D}}$.

Among the pieces of information that are missing in $\{\succsim_I\}_{I \in \mathbb{D}}$ in order to determine the reasoner's prior probability are the rankings of subsets of theories. A Bayesian reasoner, who has a prior over the space of theories, has a prior probability for every measurable subset of theories. By contrast, our reasoner is only assumed to rank specific theories.

12.3.4. *The measurement of complexity*

The measurement of complexity is not a trivial issue. It is very appealing to use some notion of Kolmogorov's complexity, namely the length of the minimal program that implements a theory. But the minimal description length of a theory gives equal weight to bits that describe the algorithm of the program and to bits that describe arbitrary parameters. For instance, the MDL of the theory $y = 1.30972x$ is much higher than the MDL of the theory $y = 2x$. For applications to everyday human reasoning, as well as to scientific reasoning in the social sciences, a "simple" parameter such as 2 need not have any privileged status as compared to a "complicated" parameter such as 1.30972. Differently put, if the bits needed to describe 1.30972 were used to encode logical computation steps, one may have a theory that is much more complicated than the linear relationship $y = 1.30972x$. This suggests that the length of the description of a program in bits, including all numerical parameters, isn't an intuitive measure of the theory's complexity. The appropriate choice of a measure of complexity is beyond the scope of the axiomatic investigation taken in this paper.

12.4. Appendix: Proofs and Related Analysis

12.4.1. *A basic result*

We will rely on the following result, which appears in Gilboa and Schmeidler (2001, 2003). To state it, we first define a matrix v : $\mathbb{A} \times \mathbb{X} \to \mathbb{R}$ to be *diversified* if there are no elements $a, b, c, d \in A$

with $b, c, d \neq a$ and $\lambda, \mu, \theta \in \mathbb{R}$ with $\lambda + \mu + \theta = 1$ such that $v(a, \cdot) \leq \lambda v(b, \cdot) + \mu v(c, \cdot) + \theta v(d, \cdot)$. That is, v is diversified if no row in v is dominated by an affine combination of three (or fewer) other rows. The axioms used for the theorem are:

A1* Order: For every $I \in \mathbb{D}$, $\succsim_I$ is complete and transitive on $\mathbb{A}$.

A2* Combination: For every $I, J \in \mathbb{D}$ and every $a, b \in \mathbb{A}$, if $a \succsim_I b$ ($a \succ_I b$) and $a \succsim_J b$, then $a \succsim_{I+J} b$ ($a \succ_{I+J} b$).

A3* Archimedean Axiom: For every $I, J \in \mathbb{D}$ and every $a, b \in \mathbb{A}$, if $a \succ_I b$, then there exists $l \in N$ such that $a \succ_{lI+J} b$.

A4* Diversity: For every list (a, b, c, d) of distinct elements of $\mathbb{A}$ there exists $I \in \mathbb{D}$ such that $a \succ_I b \succ_I c \succ_I d$. If $|\mathbb{A}| < 4$, then for any strict ordering of the elements of $\mathbb{A}$ there exists $I \in \mathbb{D}$ such that $\succ_I$ is that ordering.

Theorem 2: *Let there be given* $\mathbb{X}$, $\mathbb{A}$, *and* $\{\succsim_I\}_{I \in \mathbb{D}}$ *as above. Then the following two statements are equivalent:*

(i) $\{\succsim_I\}_{I \in \mathbb{D}}$ *satisfy A1*–A4*;*
(ii) *There is a diversified matrix* $v : \mathbb{A} \times \mathbb{X} \to \mathbb{R}$ *such that:*

$$(**) \quad \begin{cases} \text{for every } I \in \mathbb{D} \text{ and every } a, b \in \mathbb{A}, \\ a \succsim_I b \quad \text{iff} \quad \sum_{x \in \mathbb{X}} I(x)v(a, x) \geq \sum_{x \in \mathbb{X}} I(x)v(b, x), \end{cases}$$

Furthermore, in this case the matrix v *is unique in the following sense: v and u both satisfy* ($**$) *iff there are a scalar* $\lambda > 0$, *a matrix* $\beta :$ $\mathbb{A} \times \mathbb{X} \to \mathbb{R}$ *with identical rows (i.e., with constant columns) such that $u = \lambda v + \beta$.*

12.4.2. *Proof of theorem 1*

The strategy of the proof is as follows. We define a set of auxiliary relations, $\{\succsim'_I\}_I$ on $\mathbb{A}$, interpreted as follows: $a \succsim'_I b$ suggests that the observations contained in I are at least as probably under a than under b. Thus, if we were to ignore complexity considerations or other a priori biases for one theory over the other, we would expect a to be more plausible than b given I. The relation $\succsim'_I$ will correspond to the summation of the v entries in our representation. That is, $a \succsim'_I b$ will turn out to be equivalent to

$$\sum_{x\in X} I(x)v(a, x) \geq \sum_{x\in X} I(x)v(b, x)$$

which is the numerical representation we seek if the w's are all set to zero.

The first step in the proof consists of showing that the relations $\{\succsim_I'\}_I$ satisfy the conditions of the Theorem 2. This identifies the matrix v up to the transformations allowed by Theorem 2, namely, up to addition of constants to columns and multiplication of the entire matrix by a positive number. We fix one such representing matrix v. This step does not make use of axiom A5.

The next step in the proof is to show that for every two theories a, b there exists a number α^{ab}, with $\alpha^{ba} = -\alpha^{ab}$, such that, for every I, $a \succsim_I b$ iff

$$\alpha^{ab} + \sum_{x\in X} I(x)v(a, x) \geq \sum_{x\in X} I(x)v(b, x),$$

which is the desired representation for the case of two theories. Finally, the we wish to prove that for each theory a there exists a number $w(a)$ such that, for every a, b, $\alpha^{ab} = w(a) - w(b)$.

12.4.2.1. *Step 1: The matrix v*

For $a, b \in \mathbb{A}$ and $I \in \mathbb{D}$, define $a \succ_I' b$ if there exists $J \in \mathbb{D}$ such that $b \succsim_J a$ and $a \succ_{J+I} b$. That is, $a \succ_I' b$ if the evidence contained in I is sufficient to reverse the ordering between a and b.

Lemma 1:   For $a, b \in \mathbb{A}$ and $I \in \mathbb{D}$, it is impossible that both $a \succ_I' b$ and $b \succ_I' a$.

Proof: Assume, to the contrary, that there are $J, K \in \mathbb{D}$ such that $b \succsim_J a$, $a \succ_{J+I} b$, $a \succsim_K b$, and $b \succ_{K+I} a$. Since $J + (K + I) = (J + I) + K$, this contradicts A2.   □

Lemma 2:   For $a, b \in \mathbb{A}$ and $I \in \mathbb{D}$, if there exists $J \in \mathbb{D}$ such that $b \succ_J a$ and $a \succsim_{J+I} b$, then $a \succ_{J+2I} b$.

Proof: If not, $b \succsim_{J+2I} a$, and then by defining $K = L = J + I$ and $I' = J + 2I$, we obtain $a \succsim_K b$, $a \succsim_L b$, $b \succsim_{I'} a$, $b \succ_J a$ while $K + L = I' + J = 2J + 2I$, a contradiction to A2. $\qquad\square$

Lemma 3: *For $a, b \in \mathbb{A}$ and $I \in \mathbb{D}$, $a \succ'_I b$ iff there exists $J \in \mathbb{D}$ such that $b \succ_J a$ and $a \succsim_{J+I} b$.*

Proof: Assume first that there exists $J \in \mathbb{D}$ such that $b \succ_J a$ and $a \succsim_{J+I} b$. If $a \succ_{J+I} b$, then $a \succ'_I b$ follows from the definition of $\succ'_I$. Otherwise, $a \sim_{J+I} b$. Define $J' = J + I$, and note that $b \succsim_{J'} a$. But Lemma 2 implies that $a \succ_{J'+I} b$, which yields $a \succ'_I b$.

Conversely, assume that $a \succ'_I b$. By A4 there exists L such that $b \succ_L a$. By A3, there exists k such that $a \succ_{L+kI} b$. Let k' be the minimal $k \geq 1$ such that $a \succsim_{L+kI} b$ and define $J = L + (k-1)I$. $\qquad\square$

Define, for $a, b \in \mathbb{A}$ and $I \in \mathbb{D}$, $a \sim'_I b$ if neither $a \succ'_I b$ nor $b \succ'_I a$. Clearly, $\sim'_I$ is reflexive and symmetric. We observe the following.

Lemma 4: *For $a, b \in \mathbb{A}$ and $I \in \mathbb{D}$, the following are equivalent:*

(i) $a \sim'_I b$
(ii) *for every $J \in \mathbb{D}$*

$$a \succsim_J b \Leftrightarrow a \succsim_{J+I} b$$

(iii) *for every $J \in \mathbb{D}$*

$$a \succsim_J b \Leftrightarrow a \succsim_{J+I} b$$

   and

$$b \succsim_J a \Leftrightarrow b \succsim_{J+I} a.$$

Proof: We prove that (i)$\Rightarrow$(iii)$\Rightarrow$(ii)$\Rightarrow$(i). Since (iii)$\Rightarrow$(ii) is obvious, only two steps are needed.

To prove that (i)$\Rightarrow$(iii), assume that $a \sim'_I b$. Consider $J \in \mathbb{D}$. If $a \succsim_J b$ but $a \succsim_{J+I} b$ fails to hold, then $b \succ_{J+I} a$ and $b \succ'_I a$ by definition of $\succ'_I$, contradicting $a \sim'_I b$. If $a \succsim_{J+I} b$ but $a \succsim_J b$ doesn't hold, we have $b \succ_J a$ and then Lemma 3 implies that $a \succ'_I b$, again a contradiction. Similarly, $b \succsim_J a \Leftrightarrow b \succsim_{J+I} a$.

To prove that (ii)⇒(i), assume that for every $J \in \mathbb{D}$ we have $a \succsim_J b \Leftrightarrow a \succsim_{J+I} b$. If $a \sim'_I b$ does not hold, then either $a \succ'_I b$ or $b \succ'_I a$. If $b \succ'_I a$, by definition of $\succ'_I$ there exists J with $a \succsim_J b$ but $b \succ_{J+I} a$, contradicting $a \succsim_J b \Rightarrow a \succsim_{J+I} b$. If, however, $a \succ'_I b$, by Lemma 3 there exists J such that $b \succ_J a$ and $a \succsim_{J+I} b$, a contradiction to $b \succsim_{J+I} a \Rightarrow b \succsim_J a$. $\qquad\square$

Lemma 5:   *For $a, b \in \mathbb{A}$ and $I \in \mathbb{D}$, the following are equivalent:*

(i) $a \succ'_I b$
(ii) *there exist $J \in \mathbb{D}$ and $k \geq 1$ such that $b \succsim_J a$ and $a \succ_{J+kI} b$*
(iii) *there exist $J \in \mathbb{D}$ and $k \geq 1$ such that $b \succ_J a$ and $a \succsim_{J+kI} b$*
(iv) *for every $J \in \mathbb{D}$ there exists $k \geq 0$ such that for every $l \geq 0$*

$$a \succ_{J+lI} b \Leftrightarrow l \geq k.$$

Proof: We show that (i) is equivalent to each of (ii), (iii), and (iv).

We begin with (i)⇔(ii). If (i) holds, then (ii) holds for $k = 1$. Conversely, if (ii) holds, let $l = \min\{ l \mid a \succ_{J+lI} b \}$, where $l > 0$ because $b \succsim_J a$. Denoting $J' = J + (l-1)I$ we have $b \succsim_{J'} a$ but $a \succ_{J'+I} b$, that is, $a \succ'_I b$.

The proof that (i)⇔(iii) is almost identical, defining $l = \min\{ l \mid a \succsim_{J+lI} b \}$ and invoking Lemma 3

We now show (i)⇔(iv). Assume (i) holds. Given J, consider $N = \{ l \geq 0 \mid a \succ_{J+lI} b \}$. By A3, $N \neq \varnothing$. Let k be the minimal element in N. If, for $l > k$, $b \succsim_{J+lI} a$, then, by the implication (iii)⇒(i), we obtain $b \succ'_I a$, a contradiction to Lemma 1. Hence $a \succ_{J+lI} b$ iff $l \geq k$.

Conversely, assume that (iv) holds. By A4 there exists J such that $b \succ_J a$. Let k be defined by (iv), and use the implication (ii)⇒(i). $\qquad\square$

Define $a \succsim'_I b$ if $a \succ'_I b$ or $a \sim'_I b$.

Lemma 6:   *For $a, b \in \mathbb{A}$ and $I, J \in \mathbb{D}$*

(i) $a \succ_J b$ and $a \succsim'_I b$ imply $a \succ_{J+kI} b$ for all $k \geq 1$
(ii) $a \succsim_J b$ and $a \succ'_I b$ imply $a \succ_{J+kI} b$ for all $k \geq 1$
(iii) $a \sim_J b$ and $a \sim'_I b$ imply $a \sim_{J+kI} b$ for all $k \geq 1$
(iv) $a \succsim_J b$ and $a \succsim'_I b$ imply $a \succsim_{J+kI} b$ for all $k \geq 1$
(v) $a \sim_J b$ and $a \sim_{J+kI} b$ for some $k \geq 1$ imply $a \sim'_I b$.

Proof:

(i) Assume $a \succ_J b$ and $a \succsim'_I b$. If for some $k \geq 1$, $b \succsim_{J+kI} a$, then Lemma 5 $((iii) \Rightarrow (i))$ implies that $b \succ'_I a$, a contradiction.

(ii) If $a \succ_J b$, the conclusion follows from (i). Assume, then, that $a \sim_J b$ and $a \succ'_I b$. By Lemma 5 $((ii) \Rightarrow (i))$ we know that $a \succsim_{J+kI} b$ for all $k \geq 1$. Also, Lemma 5 $((i) \Rightarrow (iv))$ implies that exists $k \geq 1$ such that for every $l \geq 0$, $a \succ_{J+lI} b \Leftrightarrow l \geq k$ and therefore $a \sim_{J+lI} b \Leftrightarrow l < k$. If $k > 1$, consider J, $I' = J + kI$, $K = J + I$, and $L = J + (k-1)I$. Observe that $J + I' = K + L = 2J + kI$. Moreover, $a \sim_J b$, $a \sim_K b$, $a \sim_L b$, but $a \succ_{I'} b$, in contradiction to A2.

(iii) Follows from Lemma 4.

(iv) Follows from (i)–(iii).

(v) Follows from (ii). □

We now show that $\{\succsim'_I\}_I$ satisfy axioms A1*–A4* of Theorem 2.

Lemma 7: *For every $I \in \mathbb{D}$, $\succsim'_I$ is a weak order.*

Proof: Completeness of $\succsim'_I$ follows from its definition. We need to prove transitivity. Assume that $a, b, c \in \mathbb{A}$ satisfy $a \succsim'_I b$ and $b \succsim'_I c$, and show $a \succsim'_I c$. We distinguish between four cases:

Case 1: $a \succ'_I b$ and $b \succ'_I c$.

By A4, there exists J such that $c \succ_J b \succ_J a$. Since $a \succ'_I b$, by Lemma 5 there exists k_1 such that $a \succ_{J+lI} b$ for $l \geq k_1$. Similarity, $b \succ'_I c$ implies that there exists k_2 such that $b \succ_{J+lI} c$ for $l \geq k_2$. Hence, there exists l (for instance, $l = \max(k_1, k_2)$) such that $a \succ_{J+lI} b \succ_{J+lI} c$, hence $a \succ_{J+lI} c$. By Lemma 5, $a \succ'_I c$.

Case 2: $a \succ'_I b$ and $b \sim'_I c$.

By A4, there exists J such that $b \succ_J c \succ_J a$. Let k be such that $a \succ_{J+kI} b$. By Lemma 4, $b \sim'_I c$ and $b \succ_J c$ imply that $b \succ_{J+kI} c$. By transitivity, $a \succ_{J+kI} c$, and $a \succ'_I c$ follows from Lemma 5.

Case 3: $a \sim'_I b$ and $b \succ'_I c$.

By A4, there exists J such that $c \succ_J a \succ_J b$. Let k be such that $b \succ_{J+kI} c$. By Lemma 4, $a \sim'_I b$ and $a \succ_J b$ imply that $a \succ_{J+kI} b$. Hence $a \succ_{J+kI} c$, and $a \succ'_I c$ follows as above.

Case 4: $a \sim'_I b$ and $b \sim'_I c$.

If $a \succ'_I c$, then applying Case 2 (with the roles of b and c reversed) implies $a \succ'_I b$, a contradiction. Similarly, $c \succ'_I a$ would imply $c \succ'_I b$. □

Lemma 8: $\{\succsim'_I\}_I$ *satisfy the Combination Axiom A2*.*

Proof: We need to show that, for every $I, J \in \mathbb{D}$ and every $a, b \in \mathbb{A}$, if $a \succsim'_I b$ $(a \succ'_I b)$ and $a \succsim'_J b$, then $a \succsim'_{I+J} b$ $(a \succ'_{I+J} b)$.

Assume first that $a \sim'_I b$ and $a \sim'_J b$. In this case, Lemma 4 implies that, for every K, $a \succsim_K b \Leftrightarrow a \succsim_{K+I} b$ and $a \succsim_K b \Leftrightarrow a \succsim_{K+J} b$. We wish to show that, for every $K \in \mathbb{D}$, $a \succsim_K b \Leftrightarrow a \succsim_{K+I+J} b$, thus establishing (by Lemma 4 again) that $a \sim'_{I+J} b$.

Let there be given such K. If $a \succsim_K b$, we have $a \succsim_{K+I} b$, and, by considering $K' = K + I$, also $a \succsim_{K+I+J} b$. Conversely, if $a \succsim_{K+I+J} b$ but $a \succsim_K b$ fails to hold, we have $b \succ_K a$. In this case $b \succ_{K+I} a$ (or else $a \succ'_I b$) and then also $b \succ_{K+I+J} a$ (otherwise $a \succ'_J b$), a contradiction. It follows that the combination axiom holds in this case.

We now turn to the case in which one of the relations $a \succsim'_I b$ and $a \succsim'_J b$ is strict. Without loss of generality, assume that $a \succ'_I b$. Hence there exists $K \in \mathbb{D}$ such that $b \succsim_K a$ but $a \succ_{K+I} b$. If $b \succsim_{K+I+J} a$, then $b \succ'_J a$ by Lemma 3. Hence, $a \succ_{K+I+J} b$. Combined with $b \succsim_K a$, this implies $a \succ'_{I+J} b$. □

Lemma 9: $\{\succsim'_I\}_I$ *satisfy the Archimedean Axiom A3*.*

Proof: We need to show that, for every $I, J \in \mathbb{D}$ and every $a, b \in \mathbb{A}$, if $a \succ'_I b$, then there exists $l \in N$ such that $a \succ'_{lI+J} b$. Consider K with $b \succ_K a$. If $a \succsim_{K+J} b$, then by Lemma 6 (ii), (for $k = 1$) we have $a \succ_{K+J+I} b$, and it follows that $a \succ'_{I+J} b$, i.e., the conclusion is obtained for $l = 1$. Otherwise, we have $b \succ_{K+J} a$. In this case, apply Lemma 5 ((i)⇒(iv) and $J' = K + J$) to conclude that there exists $l \geq 1$ such that $a \succ_{K+J+lI} b$, which, combined with $b \succ_K a$, implies that $a \succ'_{lI+J} b$. □

Lemma 10: $\{\succsim'_I\}_I$ *satisfy the Diversity Axiom A4*.*

Proof: Assume first that $|\mathbb{A}| \geq 4$. (The proof for the case $|\mathbb{A}| < 4$ is identical.) We need to show that, for every list (a, b, c, d) of distinct elements of $\mathbb{A}$ there exists $I \in \mathbb{D}$ such that $a \succ'_I b \succ'_I c \succ'_I d$. By A4 there exists J such that $d \succ_J c \succ_J b \succ_J a$. Using A4 again, this time for J, we conclude that there exists $K \in \mathbb{D}$, $K \geq J$ such that $a \succ_K b \succ_K c \succ_K d$. Since $K \geq J$, we can define $I = K - J \in \mathbb{D}$. Observe that $a \succ'_I b \succ'_I c \succ'_I d$. □

We therefore conclude that $\{\succsim'_I\}_I$ satisfy axioms A1*–A4*. By Theorem 2, there exists a diversified matrix $v : \mathbb{A} \times \mathbb{X} \to \mathbb{R}$ such that:

$$
(**) \quad \left\{
\begin{array}{c}
\text{for every } I \in \mathbb{D} \text{ and every } a, b \in \mathbb{A}, \\
a \succsim'_I b \;\; \text{iff} \;\; \sum_{x \in \mathbb{X}} I(x)v(a, x) \geq \sum_{x \in \mathbb{X}} I(x)v(b, x),
\end{array}
\right.
$$

Furthermore, the matrix v is unique in the following sense: v and u both satisfy $(*)$ iff there are a scalar $\lambda > 0$, a matrix $\beta : \mathbb{A} \times \mathbb{X} \to \mathbb{R}$ with identical rows (i.e., with constant columns) such that $u = \lambda v + \beta$. We fix a particular matrix v for the rest of the existence proof.

12.4.2.2. *Step 2: Representation for pairs of theories*

In order to uniquely identify the constants α^{ab} such that, for every I,

$$
a \succsim_I b \;\; \text{iff} \;\; \alpha^{ab} + \sum_{x \in \mathbb{X}} I(x)v(a, x) \geq \sum_{x \in \mathbb{X}} I(x)v(b, x), \quad (12.2)
$$

and to further find a vector w such that $\alpha^{ab} = w(a) - w(b)$, we need to use A5. (See the following sub-section for examples illustrating the difficulties one encounters in the absence of A5.)

Fix $a, b \in \mathbb{A}$. Given matrix v, define

$$
v_{ab}(I) = \sum_{x \in \mathbb{X}} I(x)v(a, x) - \sum_{x \in \mathbb{X}} I(x)v(b, x) \in \mathbb{R}. \quad (12.3)
$$

Evidently, $v_{ab}(I) = -v_{ba}(I)$. Observe that, by $(**)$, $v_{ab}(I) \geq (>)0$ if and only if $a \succsim'_I (\succ'_I) b$.

Using this notation, the representation we seek is

$$a \succsim_I b \quad \text{iff} \quad \alpha^{ab} + v_{ab}(I) \geq 0. \tag{12.4}$$

Choose $I \in \mathbb{D}$ with $a \sim_I b$. Define

$$\alpha^{ab} = -v_{ab}(I).$$

Define also $\alpha^{ab} = -\alpha^{ba}$. We wish to show that this α^{ab} satisfies (12.4).

Lemma 11:   *For every $J \in \mathbb{D}$,*

(i) $v_{ab}(J) + \alpha^{ab} > 0$ *implies that $a \succ_J b$*
(ii) $v_{ab}(J) + \alpha^{ab} = 0$ *implies that $a \sim_J b$*
(iii) $v_{ab}(J) + \alpha^{ab} < 0$ *implies that $b \succ_J a$.*

Proof: Let there be given $J \in \mathbb{D}$. Consider $K = I + J$. By A5, there exists $L \in \mathbb{D}_{\geq K}$ such that $a \sim_L b$. Since $K \geq I, J$ and $L \geq K$, $I' \equiv L - I, J' = L - J \in \mathbb{D}$.

Since $a \sim_I b$ and $a \sim_L b$, Lemma 6 (v) implies that $a \sim_{I'} b$. Hence $v_{ab}(I') = 0$. Also, $v_{ab}(L) = v_{ab}(I) + v_{ab}(I') = v_{ab}(I)$. We now separate the three cases.

(i) The assumption on J is that $v_{ab}(J) > v_{ab}(I)$. Since $v_{ab}(I) = v_{ab}(L) = v_{ab}(J) + v_{ab}(J')$, we obtain $v_{ab}(J') < 0$, that is, $b \succ_{J'} a$. If $b \succsim_J a$, Lemma 6 (ii) would imply $a \succ_L b$, a contradiction. Hence $a \succ_J b$ is established.

(ii) In this case, $v_{ab}(J) = v_{ab}(I)$ and it follows that $v_{ab}(J') = 0$ and $b \sim_{J'} a$. If $a \succ_J b$ $(b \succ_J a)$, $a \succ_L b$ $(b \succ_L a)$ would follow by Lemma 6 (i). Hence $a \sim_J b$.

(iii) If $v_{ab}(J) < -\alpha^{ab} = v_{ab}(I)$, $v_{ab}(J') > 0$ and $a \succ_{J'} b$. If $a \succsim_J b$, Lemma 6 (ii) would imply $a \succ_L b$, hence $b \succ_J a$.   □

Observe that we also have $b \succsim_I a$ iff $\alpha^{ba} + v_{ba}(I) \geq 0$.

Finally, we note that, given the matrix v, α^{ab} and α^{ba} are unique. Moreover, if $u = \lambda v + \beta$ also satisfies $(**)$, the constants α_u^{ab} corresponding to u is $\alpha_u^{ab} = \lambda \alpha^{ab}$.

12.4.2.3. *Step 3: Representation for all theories*

Given v satisfying $(\ast\ast)$, $(\alpha^{ab})_{a,b\in\mathbb{A}}$ are defined as above. Consider a triple $a, b, c \in \mathbb{A}$. Let I satisfy $a \sim_I b \sim_I c$. Then, by Lemma 11,

$$\alpha^{ab} + v_{ab}(I) = 0$$
$$\alpha^{bc} + v_{bc}(I) = 0$$
$$\alpha^{ca} + v_{ca}(I) = 0.$$

Summing up, and noticing that, for every a, b, c and every I,

$$v_{ab}(I) + v_{bc}(I) + v_{ca}(I) = 0$$

we obtain that

$$\alpha^{ab} + \alpha^{bc} + \alpha^{ca} = 0.$$

Fix $a \in \mathbb{A}$ and set $w(a) = 0$. For $b \neq a$ define $w(b) = w(a) - \alpha^{ab}$. Thus,

$$\alpha^{ab} = w(a) - w(b).$$

For $b, c \neq a$, observe that

$$\alpha^{bc} = -\alpha^{ab} - \alpha^{ca}$$
$$= (w(b) - w(a)) + (w(a) - w(c))$$
$$= w(b) - w(c).$$

Hence, for all $a, b \in \mathbb{A}$,

$$a \succsim_I b$$

iff

$$w(a) + \sum_{x\in X} I(x)v(a,x) \geq w(b) + \sum_{x\in X} I(x)v(b,x).$$

Clearly, the vector w is unique up to a shift by an additive constant, leaving the differences $w(a) - w(b) = \alpha^{ab}$ unchanged. This completes the proof of the theorem.

12.4.3. *Necessity and counter-examples*

The theorem does not provide an exact characterization of the collections of relations $\{\succsim_I\}_{I\in\mathbb{D}}$ that satisfy A1–A5. While axioms A1–A3 are clearly necessary for the representation ($*$), A4 and A5 are not.

As shown in Theorem 2, A4 holds only if the matrix v is diversified. Correspondingly, if $\{\succsim_I\}_{I\in\mathbb{D}}$ satisfy A1–A5, the resulting matrix v will also be diversified.

However, not every diversified v will guarantee that the relations $\{\succsim_I\}_{I\in\mathbb{D}}$ defined by v and a vector w via ($*$) will also satisfy A5. In fact, the matrix-vector pairs (v, w) that guarantee A5 as well are precisely those that satisfy the following condition:

(v, w)-**solvability**: For every $a, b, c \in \mathbb{A}$ there exists $I \in \mathbb{D}$ such that

$$w(a) + \sum_{x\in\mathbb{X}} I(x)v(a, x)$$

$$= w(b) + \sum_{x\in\mathbb{X}} I(x)v(b, x)$$

$$= w(c) + \sum_{x\in\mathbb{X}} I(x)v(c, x).$$

Adding diversity of v and (v, w)-solvability, one may obtain a version of Theorem 1 which is a precise characterization. Since the main point of the theorem from a conceptual viewpoint is the sufficiency result, and since it is also the less trivial direction, we chose to omit this condition from the statement of the theorem, leaving it with only one implication.

To see that (v, w)-solvability is not too restrictive, consider the following condition: for every $a_1, a_2, a_3 \in \mathbb{A}$ there are $x_1, x_2, x_3 \in \mathbb{X}$ such that all the numbers $\{w(a_i), v(a_i, x_j)\}_{i,j\leq 3}$ are rational (or, to be precise, generate only rational ratios).

However, dropping A5, our result may not hold. In the following, we retain the following notation from the proof: given $\{\succsim_I\}_I$, the relations $\{\succsim'_I\}_I$ derived from them as above. For given $\mathbb{A}$ and $\mathbb{X}$, v

denotes a real-valued matrix, $v : \mathbb{A} \times \mathbb{X} \to \mathbb{R}$. In the following examples, v will represent the relations $\{\succsim'_I\}_I$ by ($**$). We also retain the notation

$$v_{ab}(I) = \sum_{x \in \mathbb{X}} I(x)v(a,x) - \sum_{x \in \mathbb{X}} I(x)v(b,x) \in \mathbb{R}$$

for $I \in \mathbb{D}$, $a, b \in \mathbb{A}$.

We first show that in the absence of A5 uniqueness may fail.

Example 1: Let $\mathbb{A} = \{a, b\}$, $\mathbb{X} = \{x, y\}$ and

$$v = \begin{pmatrix} 1 & 0 \\ 0 & 1 \end{pmatrix}.$$

For every I, define $a \succ_I b$ if $v_{ab}(I) \geq 0$ (i.e., $I(x) \geq I(y)$) and $b \succ_I a$ otherwise. In this case, $\{\succsim_I\}_{I \in \mathbb{D}}$ can be represented by (v, w) via ($*$) for v above and for every w with

$$w(a) - w(b) \in (0, 1).$$

That is, the representation is not unique. Using the representation, we know that $\{\succsim_I\}_{I \in \mathbb{D}}$ satisfy A1–A3, and A4 can readily be verified. Clearly, A5 is violated in this example. □

Second, the following example shows that without A5 representation as in (12.1) may not be possible:

Example 2: Let $\mathbb{A} = \{a, b\}$, $\mathbb{X} = \{x, y\}$ and

$$v = \begin{pmatrix} 1 & 0 \\ 0 & \sqrt{2} \end{pmatrix}.$$

For every I, define $a \succ_I b$ if $v_{ab}(I) \geq 0$ and $b \succ_I a$ otherwise.

Observe that, for all $I \neq 0$, $v_{ab}(I) \neq 0$. Hence one may use the matrix v and the constants $w(a) = w(b) = 0$ to represent $\{\succsim_I\}_{I \neq 0}$ via ($*$). However, (v, w) cannot represent all of $\{\succsim_I\}_{I \in \mathbb{D}}$ because $v_{ab}(0) = 0$, hence $a \succ_0 b$, but $v_{ab}(0) = w(b) - w(a)$.

We claim that no other pair, (v', w'), may represent $\{\succsim_I\}_{I \in \mathbb{D}}$ via ($*$). To see this, assume that such a pair (v', w') is given. Normalize v' such that the minimal value in each column is 0 and the maximal value in column x is 1. Hence, $v' = v$. Observe that

$$range(v_{ab}) = \{v_{ab}(I) \mid I \in \mathbb{D}\}$$
$$= \{k - l\sqrt{2} \mid k, l \in \mathbb{Z}_+\}$$

is dense in $\mathbb{R}$. If $w'(b) - w'(a) > 0$, there exists $I \neq 0$ such that $v_{ab}(I) \in (0, w'(b) - w'(a))$ and then (v, w') cannot represent $\succsim_I$ (because (v, w) does). Similarly, $w'(b) - w'(a) < 0$ implies the existence of $I \neq 0$ with $v_{ab}(I) \in (w'(b) - w'(a), 0)$ and the same conclusion follows.

To conclude the proof we need to verify that $\{\succsim_I\}_{I \in \mathbb{D}}$ satisfy A1–A4. In the presence of only two alternatives, A1 only means completeness, which is directly verified from the definition. To see that A2 holds, assume that I, J, K, L are given, with $I + J = K + L$. Assume further that $a \succsim_I b$ and $a \succsim_J b$, but $b \succsim_K a$ and $b \succ_L a$. Observe that $a \succsim_I b$, which is only possible if $a \succ_I b$, implies that $v_{ab}(I) \geq 0$, with a strict equality unless $I = 0$. Hence $a \succsim_I b$ and $a \succsim_J b$ imply $v_{ab}(I), v_{ab}(J) \geq 0$, and $b \succsim_K a$, $b \succ_L a$ imply $v_{ab}(K), v_{ab}(L) \leq 0$. Since $v_{ab}(I) + v_{ab}(J) = v_{ab}(K) + v_{ab}(L)$, this is possible only if $v_{ab}(I) = v_{ab}(J) = v_{ab}(K) = v_{ab}(L) = 0$, and therefore $I = J = K = L = 0$. But then $b \succsim_K a$ and $b \succ_L a$ can't hold. To see that A3 holds, assume that $I, J \in \mathbb{D}$ satisfy $b \succ_J a$ and $a \succsim_{J+I} b$. In this case, $I \neq 0$ and $v_{ab}(I) > 0$ follows. Hence, for every $K \in \mathbb{D}$ there exists $l \in N$ such that $a \succ_{K+lI} b$. Finally, A4 clearly holds because no row in v dominates another.   □

References

Akaike, H. (1974), "A New Look at the Statistical Model Identification". *IEEE Transactions on Automatic Control*, **19**(6): 716–723.

Billot, A., I. Gilboa, D. Samet and D. Schmeidler (2005), "Probabilities as Similarity-Weighted Frequencies", *Econometrica*, **73**: 1125–1136.

Chaitin, G. J. (1966), "On the Length of Programs for Computing Binary Sequences", *J. Assoc. Comp. Machines*, **13**: 547–569.

de Finetti, B. (1937), "La Prevision: Ses Lois Logiques, Ses Sources Subjectives", *Annales de l'Institute Henri Poincare*, **7**: 1–68.

Goodman, N. (1955), *Fact, Fiction, and Forecast*, Cambridge, MA: Harvard University Press.

Gilboa, I., O. Lieberman and D. Schmeidler (2006), "Empirical Similarity", *Review of Economics and Statistics*, **88**: 433–444. Reprinted as Chapter 9 in this volume.

Gilboa, I. and L. Samuelson (2009), "Subjectivity in Inductive Inference", mimeo. *Theoretical Economics*, forthcoming.

Gilboa, I. and D. Schmeidler (2001), *A Theory of Case-Based Decisions*, Cambridge: Cambridge University Press.

— (2003), "Inductive Inference: An Axiomatic Approach", *Econometrica*, **71**: 1–26. Reprinted as Chapter 4 in this volume.

Hume, D. (1748), *Enquiry into the Human Understanding*, Oxford: Clarendon Press.

Kolmogorov, A. N. (1963), "On Tables of Random Numbers", *Sankhya Ser. A* 369–376.

— (1965), "Three Approaches to the Quantitative Definition of Information", *Probability and Information Transmission*, **1**: 4–7.

Ramsey, F. P. (1931), "Truth and Probability", in The Foundation of Mathematics and Other Logical Essays, New York: Harcourt, Brace and Co.

Rissanen, J. J. (1978), "Modelling by Shortest Data Description", *Automatica*, **14**: 465–471.

Russell, B. (1946), *A History of Western Philosophy*, Great Britain: Allen & Unwin.

Savage, L. J. (1954), *The Foundations of Statistics*, New York: John Wiley and Sons. (Second addition in 1972, Dover).

Silverman, B. W. (1986), *Density Estimation for Statistics and Data Analysis*, London and New York: Chapman and Hall.

Simpson, E. H. (1951), "The Interpretation of Interaction in Contingency Tables". *Journal of the Royal Statistical Society, Ser. B*, **13**: 238–241.

Sober, E. (1975), *Simplicity*, Oxford: Clarendon Press.

Solomonoff, R. (1964), "A Formal Theory of Inductive Inference I, II", *Information Control*, 7: 1–22, 224–254.

Wallace, C. S. (2005), *Statistical and Inductive Inference by Minimum Message Length*, Series: Information Science and Statistics, Springer.

Wallace, C. S. and D. M. Boulton (1968), "An Information Measure for Classification", *Comput. J.*, **13**: 185–194.

Wallace, C. S. and D. L. Dowe (1999), "Minimum Message Length and Kolmogorov Complexity", *The Computer Journal*, **42**: 270–283.